MATHEMATICS RESEARCH DEVELOPMENTS

FOUNDATIONS OF ISO-DIFFERENTIAL CALCULUS

ORDINARY ISO-DIFFERENTIAL EQUATIONS

VOLUME 3

MATHEMATICS RESEARCH DEVELOPMENTS

Additional books in this series can be found on Nova's website
under the Series tab.

Additional e-books in this series can be found on Nova's website
under the e-book tab.

MATHEMATICS RESEARCH DEVELOPMENTS

FOUNDATIONS OF ISO-DIFFERENTIAL CALCULUS

ORDINARY ISO-DIFFERENTIAL EQUATIONS

VOLUME 3

SVETLIN GEORGIEV

nova publishers
New York

NOTICE TO THE READER

Library of Congress Cataloging-in-Publication Data

ISBN: 978-1-63463-323-9

Published by Nova Science Publishers, Inc. † New York

Contents

Preface

This book is intended for readers who have had a course in iso-differential calculus and it can be used for a senior undergraduate course.

Chapter 1 deals with exact iso-differential equations, while first-order iso-differential equations are studied in Chapter 2 and Chapter 3. Chapter 4 discusses iso-integral inequalities.

Many iso-differential equations cannot be solved as finite combinations of elementary functions. Therefore, it is important to know whether a given iso-differential equation has a solution and if it is unique. These aspects of the existence and uniqueness of the solutions for first-order initial value problems are considered in Chapter 5. Iso-differential inequalities are discussed in Chapter 6. Continuity and differentiability of solutions with respect to initial conditions are examined in Chapter 7. Chapter 8 extends existence-uniqueness results and continuous dependence on initial data for linear iso-differential systems. Basic properties of solutions of linear iso-differential systems are given in Chapter 9. Chapter 10 deals with the fundamental matrix solutions. In Chapter 11 necessary and sufficient conditions are provided so that a linear iso-differential system has only periodic solutions. The asymptotic behaviour of the solutions of linear systems is investigated in Chapter 12. Chapters 13 and 14 are devoted on some aspects of the stability of solutions of iso-differential systems.

The last major topic covered in this book is that of boundary value problems involving second-order iso-differential equations. After linear boundary value problems are introduced in Chapter 15, Green's function and its construction is discussed in Chapter 16.

I will be very grateful to anybody who wants to inform me about errors or just misprints, or wants to express criticism or other comments, to my e-mails svetlingeorgiev1@gmail.com, sgg2000bg@yahoo.com.

Svetlin Georgiev
Paris, France
August 15, 2014

Chapter 1

Exact Equations

Here we suppose

$$\hat{T} : \mathbb{R} \longrightarrow (0,\infty), \qquad \hat{T} \in C^1(\mathbb{R}).$$

We consider the equation

$$\hat{M}^{\wedge}(\hat{x},\hat{y})\,\hat{\times}\,\hat{d}\hat{x} + \hat{N}^{\wedge}(\hat{x},\hat{y})\,\hat{\times}\,\hat{d}\hat{y}^{\wedge\wedge} = 0, \tag{1}$$

or

$$\hat{M}^{\wedge}(\hat{x},\hat{y}) + \hat{N}^{\wedge}(\hat{x},\hat{y})\left(\hat{y}^{\wedge\wedge}\right)^{\circledast} = 0,$$

where M and N are continuous functions having continuous partial derivatives M_y and N_x in the rectangle

$$S = \Big\{(x,y) \in \mathbb{R}^2 : \quad |x-x_0| \le a, \qquad |y-y_0| \le b\Big\}, \qquad 0 < a, \quad b < \infty.$$

The equation (1) we can rewrite in the following form

$$\left(M(x,y)\left(1 - x\frac{\hat{T}'(x)}{\hat{T}(x)}\right) - yN(x,y)\frac{\hat{T}'(x)}{\hat{T}(x)}\right)dx + N(x,y)dy = 0.$$

Definition 1.0.1. *The equation* (1) *is said to be exact if there exists a function* $u(x,y)$ *such that*

$$u_x(x,y) = M(x,y)\left(1 - x\frac{\hat{T}'(x)}{\hat{T}(x)}\right) - yN(x,y)\frac{\hat{T}'(x)}{\hat{T}(x)}, \qquad u_y(x,y) = N(x,y). \tag{2}$$

The nomenclature comes from the fact that

$$\left(M(x,y)\left(1 - x\frac{\hat{T}'(x)}{\hat{T}(x)}\right) - yN(x,y)\frac{\hat{T}'(x)}{\hat{T}(x)}\right)dx + N(x,y)dy = u_x(x,y)dx + u_y(x,y)dy$$

is exactly the differential du.

Once the equation (1) is exact its implicit solution is

$$u(x,y) = C, \tag{3}$$

where C is an arbitrary constant.

If (1) is exact, then from (2) we have

$$\begin{aligned} u_{xy} &= \frac{\partial}{\partial y}\left(M(x,y)\left(1 - x\frac{\hat{T}'(x)}{\hat{T}(x)}\right) - yN(x,y)\frac{\hat{T}'(x)}{\hat{T}(x)}\right) \\ &= M_y(x,y)\left(1 - x\frac{\hat{T}'(x)}{\hat{T}(x)}\right) - N(x,y)\frac{\hat{T}'(x)}{\hat{T}(x)} - yN_y(x,y)\frac{\hat{T}'(x)}{\hat{T}*(x)}, \\ u_{yx} &= N_x(x,y). \end{aligned}$$

Since M_y and N_x are continuous, we must have

$$u_{xy}(x,y) = u_{yx}(x,y),$$

i.e., the equation (1) to be exact it is necessary to have

$$M_y(x,y)\left(1 - x\frac{\hat{T}'(x)}{\hat{T}(x)}\right) - N(x,y)\frac{\hat{T}'(x)}{\hat{T}(x)} - yN_y(x,y)\frac{\hat{T}'(x)}{\hat{T}*(x)} = N_x(x,y). \tag{4}$$

Conversely, if M and N satisfy (4) then the equation (1) is exact. To establish this we shall exhibit a function u satisfying (2). We integrate the both sides of the following equality

$$u_x(x,y) = M(x,y)\left(1 - x\frac{\hat{T}'(x)}{\hat{T}(x)}\right) - yN(x,y)\frac{\hat{T}'(x)}{\hat{T}(x)}$$

with respect to x, to obtain

$$u(x,y) = \int_{x_0}^{x}\left(M(s,y)\left(1 - s\frac{\hat{T}'(s)}{\hat{T}(s)}\right) - yN(s,y)\frac{\hat{T}'(s)}{\hat{T}(s)}\right)ds + g(y). \tag{5}$$

Here $g(y)$ is an arbitrary function of y and plays the role of the constant of integration. We shall obtain g by using the equation

$$u_y(x,y) = N(x,y).$$

Indeed, we have

$$\frac{\partial}{\partial y}\int_{x_0}^{x}\left(M(s,y)\left(1 - s\frac{\hat{T}'(s)}{\hat{T}(s)}\right) - yN(s,y)\frac{\hat{T}'(s)}{\hat{T}(s)}\right)ds + g'(y) = N(x,y), \tag{6}$$

and since

$$\begin{aligned} N_x(x,y) &= \frac{\partial^2}{\partial x \partial y}\int_{x_0}^{x}\left(M(s,y)\left(1 - s\frac{\hat{T}'(s)}{\hat{T}(s)}\right) - yN(s,y)\frac{\hat{T}'(s)}{\hat{T}(s)}\right)ds \\ &= N_x(x,y) - M_y(x,y)\left(1 - x\frac{\hat{T}'(x)}{\hat{T}(x)}\right) - N(x,y)\frac{\hat{T}'(x)}{\hat{T}(x)} - yN_y(x,y)\frac{\hat{T}'(x)}{\hat{T}(x)} \\ &= 0, \end{aligned}$$

the function

$$N(x,y) - \frac{\partial}{\partial y}\int_{x_0}^{x}\left(M(s,y)\left(1 - s\frac{\hat{T}'(s)}{\hat{T}(s)}\right) - yN(s,y)\frac{\hat{T}'(s)}{\hat{T}(s)}\right)ds$$

must depends on y alone.

Therefore g can be obtained from (6), and finally the function u, satisfying (2), is given by (5).

We summarize this important result in the following theorem.

Theorem 1.0.2. *Let the functions $M(x,y)$ and $N(x,y)$ together with their partial derivatives $M_y(x,y)$ and $N_x(x,y)$ be continuous in the rectangle S. Then the DE (1) is exact if and only if the condition (4) is satisfied.*

Obviously, in this result S may be replaced by any region which does not include any "hole".

The above proof of this theorem is, in fact, constructive, i.e., we can explicitly find a solution of (1). For this, we compute $g(y)$ from (6),

$$g(y) = \int_{y_0}^{y} N(x,t)dx - \int_{x_0}^{x}\left(M(s,y)\left(1 - s\tfrac{\hat{T}'(s)}{\hat{T}(s)}\right) - yN(s,y)\tfrac{\hat{T}'(s)}{\hat{T}(s)}\right)ds$$

$$+ \int_{x_0}^{x}\left(M(s,y_0)\left(1 - s\tfrac{\hat{T}'(s)}{\hat{T}(s)}\right) - y_0N(s,y_0)\tfrac{\hat{T}'(s)}{\hat{T}(s)}\right)ds + g(y_0).$$

Therefore, from (5), it follows that

$$u(x,y) = \int_{y_0}^{y} N(x,t)dt + \int_{x_0}^{x}\left(M(s,y_0)\left(1 - s\frac{\hat{T}'(s)}{\hat{T}(s)}\right) - y_0N(s,y_0)\frac{\hat{T}'(s)}{\hat{T}(s)}\right)ds + g(y_0),$$

and hence the solution of the exact equation (1) is given by

$$\int_{y_0}^{y} N(x,t)dt + \int_{x_0}^{x}\left(M(s,y_0)\left(1 - s\frac{\hat{T}'(s)}{\hat{T}(s)}\right) - y_0N(s,y_0)\frac{\hat{T}'(s)}{\hat{T}(s)}\right)ds = C, \qquad (7)$$

where C is an arbitrary constant.

Now we integrate the both sides of

$$u_y(x,y) = N(x,y)$$

with respect to y, to obtain

$$u(x,y) = \int_{y_0}^{y} N(x,s)ds + f(x). \qquad (8)$$

Here $f(x)$ is an arbitrary function of x and plays the role of the constant of integration. We will obtain the function f by using the equality

$$u_x(x,y) = M(x,y).$$

We have

$$\frac{\partial}{\partial x}\int_{y_0}^{y} N(x,s)ds + f'(x) = M(x,y), \tag{9}$$

and because

$$M_y(x,y) - \frac{\partial^2}{\partial x \partial y}\int_{y_0}^{y} N(x,s)ds = M_y(x,y) - N_x(x,y) = 0,$$

the function

$$M(x,y) - \frac{\partial}{\partial x}\int_{y_0}^{y} N(x,s)ds$$

must depend on x alone. Therefore, the function f can be obtained from the equality (9), from where

$$f(x) = \int_{x_0}^{x} M(s,y)ds - \int_{y_0}^{y} N(x,s)ds + \int_{y_0}^{y} N(x_0,s)ds + f(x_0).$$

Now, using (8), we get

$$u(x,y) = \int_{x_0}^{x} M(s,y)ds + \int_{y_0}^{y} N(x_0,s)ds + f(x_0),$$

whereupon a solution of the exact equation (1) is given by the following equality

$$\int_{x_0}^{x} M(s,y)ds + \int_{y_0}^{y} N(x_0,s)ds = C, \tag{10}$$

where C is an arbitrary constant.

In (7) and (10) the choice of x_0 and y_0 is at our disposal, except that these must be chosen so that the integrals remain proper.

Example 1.0.3. *Let*

$$S = \left\{(x,y) \in \mathbb{R}^2 : \left|x - \frac{1}{2}\right| \leq \frac{1}{2}, \qquad |y-1| \leq 1\right\},$$

$\hat{T}(x) = \frac{1}{x+2}$, $M(x,y) = ye^x$, $N(x,y) = 2e^x$, $(x,y) \in S$. *Then*

$$\hat{T}'(x) - -\frac{1}{(x+2)^2},$$

$$\begin{aligned} M(x,y)\left(1 - x\frac{\hat{T}'(x)}{\hat{T}(x)}\right) - yN(x,y)\frac{\hat{T}'(x)}{\hat{T}(x)} \\ = ye^x\left(1 - x\frac{-\frac{1}{(x+2)^2}}{\frac{1}{x+2}}\right) - y2e^x\frac{-\frac{1}{(x+2)^2}}{\frac{1}{x+2}} \\ = ye^x\left(1 + \frac{x}{x+2}\right) + 2ye^x\frac{1}{x+2} \\ = 2ye^x. \end{aligned}$$

The equation (1) *we can rewrite in the form*

$$2ye^x dx + 2e^x dy = 0. \tag{11}$$

Since

$$\frac{\partial}{\partial y}\left(2ye^x\right) = \frac{\partial}{\partial x}\left(2e^x\right) = 2e^x,$$

then the equation (11) *is an exact equation.*

Let u be a solution to the equation (11)*. Then*

$$u_x(x,y) = 2ye^x, \tag{12}$$

$$u_y(x,y) = 2e^x. \tag{13}$$

We integrate the equality (12) *with respect to the variable x and we get*

$$\begin{aligned} u(x,y) &= 2y\int_0^x e^s ds + g(y) \\ &= 2ye^x - 2y + g(y), \end{aligned} \tag{14}$$

from where

$$u_y(x,y) = 2e^x - 2 + g'(y).$$

From the last equality and from (13) *we obtain*

$$2e^x - 2 + g'(y) = 2e^x$$

or

$$g'(y) = 2.$$

Therefore

$$g(y) = 2\int_1^y ds + g(1) = 2y - 2 + g(1).$$

Now, using (14)*, we get*

$$u(x,y) = 2ye^x - 2 + g(1).$$

Consequently, a solution to the equation (11) *is*

$$2ye^x = C,$$

where C is an arbitrary constant.

Remark 1.0.4. *Let us consider the equation*

$$M(x,y)dx + N(x,y)dy = 0, \tag{15}$$

where M and N are the functions from the last example. Then

$$\frac{\partial}{\partial y}M(x,y) = e^x \neq 2e^x = \frac{\partial}{\partial x}N(x,y).$$

Therefore the equation (15) *is not an exact equation.*

Consequently, there are cases such that the initial equation is not an exact equation and the corresponding iso-lift is an exact equation.

Example 1.0.5. *Let*

$$S=\Big\{(x,y)\in\mathbb{R}^2:|x-1|\le 1,\qquad |y-2|\le 2\Big\},$$

$\hat{T}(x)=x^2+1$, $M(x,y)=y+2xe^y$, $N(x,y)=x\Big(1+xe^y\Big)$. *Then*

$$\begin{aligned}
&\hat{T}'(x)=2x,\\
&M(x,y)\Big(1-x\tfrac{\hat{T}'(x)}{\hat{T}(x)}\Big)-yN(x,y)\tfrac{\hat{T}'(x)}{\hat{T}(x)}\\
&=\Big(y+2xe^y\Big)\Big(1-x\tfrac{2x}{x^2+1}\Big)-yx\Big(1+xe^y\Big)\tfrac{2x}{x^2+1}\\
&=\Big(y+2xe^y\Big)\tfrac{1-x^2}{1+x^2}-2x^2y\tfrac{1+xe^y}{1+x^2},
\end{aligned}$$

the equation (1) *we can rewrite in the form*

$$\bigg(\Big(y+2xe^y\Big)\frac{1-x^2}{1+x^2}-2x^2y\frac{1+xe^y}{1+x^2}\bigg)dx+x\Big(1+xe^y\Big)dy=0.$$

We have

$$\begin{aligned}
&\tfrac{\partial}{\partial y}\bigg(\Big(y+2xe^y\Big)\tfrac{1-x^2}{1+x^2}-2x^2y\tfrac{1+xe^y}{1+x^2}\bigg)\\
&=\Big(1+2xe^y\Big)\tfrac{1-x^2}{1+x^2}-2x^2\tfrac{1+xe^y}{1+x^2}-\tfrac{2x^3ye^y}{1+x^2}\\
&=\tfrac{1-3x^2+2xe^y-4x^3e^y-2x^3ye^y}{1+x^2},\\
&\tfrac{\partial}{\partial x}\Big(x\Big(1+xe^y\Big)\Big)=\Big(1+xe^y\Big)+xe^y=1+2xe^y.
\end{aligned}$$

Consequently the condition (4) *is not satisfied.*

Therefore the considered equation is not an exact equation.

Remark 1.0.6. *If we consider the equation*

$$M(x,y)dx+N(x,y)dy=0, \tag{16}$$

where M and N are the functions from the last example, we have

$$\frac{\partial}{\partial y}M(x,y)=\frac{\partial}{\partial x}N(x,y)=1+2xe^y.$$

Therefore the equation (16) *is an exact equation.*

Consequently, there are cases such that the initial equation is an exact equation and the corresponding iso-lift is not an exact equation.

Exercise 1.0.7. *Let*

$$S=\left\{(x,y)\in\mathbb{R}^2:|x-1|\le 1,\qquad |y-2|\le 2\right\},$$

$\hat{T}(x)=e^x$, $M(x,y)=xy$, $N(x,y)=x-y$, $(x,y)\in S$. *Determine the equation* (1) *and check if it is an exact equation. If it is an exact equation, find an its solution.*

Solution. We have

$$\hat{T}'(x)=e^x,$$

$$M(x,y)\left(1-x\frac{\hat{T}'(x)}{\hat{T}(x)}\right)-yN(x,y)\frac{\hat{T}'(x)}{\hat{T}(x)}$$

$$=xy\left(1-x\frac{e^x}{e^x}\right)-y(x-y)\frac{e^x}{e^x}$$

$$=xy(1-x)-y(x-y)$$

$$=y^2-x^2y.$$

Then the equation (1) we can rewrite in the form

$$(y^2-x^2y)dx+(x-y)dy=0.$$

Since

$$\frac{\partial}{\partial y}(y^2-x^2y)=2y-x^2\neq 1=\frac{\partial}{\partial x}(x-y),$$

then the equation (1) is not an exact equation.

Exercise 1.0.8. *Let*

$$S=\left\{(x,y)\in\mathbb{R}^2:|x|\le 2,\qquad |y-1|\le 3\right\},$$

$\hat{T}(x)=e^x$, $M(x,y)=x^2+y^2$, $N(x,y)=x+y$, $(x,y)\in S$. *Determine the equation* (1) *and check if it is an exact equation. If it is an exact equation, find an its solution.*

Answer.

$$(x^2-x^3-xy^2-xy)dx+(x+y)dy=0.$$

It is not an exact equations.

Exercise 1.0.9. *Let*

$$S=\left\{(x,y)\in\mathbb{R}^2:\left|x-\frac{1}{2}\right|\le\frac{1}{3},\qquad |y|\le 2\right\},$$

$\hat{T}(x)=e^x$, $M(x,y)=\frac{y+xy+y^2}{1-x}$, $N(x,y)=x+y$, $(x,y)\in S$. *Determine the equation* (1) *and check if it is an exact equation. If it is an exact equation, find an its solution.*

Solution. We have

$$\hat{T}'(x) = e^x,$$

$$M(x,y)\left(1 - x\frac{\hat{T}'(x)}{\hat{T}(x)}\right) - yN(x,y)\frac{\hat{T}'(x)}{\hat{T}(x)}$$

$$= \frac{y+xy+y^2}{1-x}(1-x) - y(x+y)$$

$$= y + xy + y^2 - xy - y^2$$

$$= y.$$

Then the equation (1) we can rewrite in the following form

$$ydx + (x+y)dy = 0.$$

Since

$$\frac{\partial}{\partial y}(y) = 1\frac{\partial}{\partial x}(x+y),$$

then it is an exact equation.

Let u be an its solution. Then

$$u_x(x,y) = y, \tag{17}$$

$$u_y(x,y) = x + y. \tag{18}$$

We integrate the equality (18) with respect to the variable y and we get

$$u(x,y) = \int_0^y (x+s)ds + f(x)$$

$$= xy + \frac{y^2}{2} + f(x),$$

from here

$$u_x(x,y) = y + f'(x).$$

From the last equality and (17) we obtain

$$y + f'(x) = y$$

or

$$f'(x) = 0.$$

Therefore $f(x) = C_0$, where C_0 is a constant, and

$$u(x,y) = xy + \frac{y^2}{2} + C_0.$$

Consequently, a solution is

$$2xy + y^2 = C,$$

where C is a constant.

Exercise 1.0.10. *Let*

$$S=\left\{(x,y)\in\mathbb{R}^2 : |x|\leq\frac{1}{2},\qquad |y|\leq\frac{1}{2}\right\},$$

$\hat{T}(x)=e^x$, $M(x,y)=\frac{y+xy-y^2}{1-x}$, $N(x,y)=x-y$, $(x,y)\in S$. *Determine the equation* (1) *and check if it is an exact equation. If it is an exact equation, find an its solution.*

Answer.

$$ydx+(x-y)dy=0.$$

It is an exact equation. A solution is

$$2xy-y^2=C,$$

where C is an arbitrary constant.

Now we consider the equation

$$M(x,y)+N(x,y)\hat{\times}\left(\hat{y}^{\wedge\wedge}\right)^{\circledast}=0 \tag{19}$$

or

$$M(x,y)\hat{\times}\hat{d}\hat{x}+N(x,y)\hat{\times}\hat{d}\hat{y}^{\wedge\wedge}=0,$$

where M and N are continuous functions having continuous partial derivatives M_y and N_x in the rectangle S, defined in the begin of this chapter.

The equation (19) we can rewrite in the form

$$\Big(M(x,y)(\hat{T}(x)-x\hat{T}'(x))-y\hat{T}'(x)N(x,y)\Big)dx+\hat{T}(x)N(x,y)dy=0.$$

Definition 1.0.11. *The equation* (19) *is said to be exact if there exists a function* $u(x,y)$ *such that*

$$u_x(x,y)=M(x,y)(\hat{T}(x)-x\hat{T}'(x))-y\hat{T}'(x)N(x,y), \tag{20}$$

$$u_y(x,y)=\hat{T}(x)N(x,y). \tag{21}$$

The nomenclature comes from the fact that

$$\Big(M(x,y)(\hat{T}(x)-x\hat{T}'(x))-y\hat{T}'(x)N(x,y)\Big)dx+\hat{T}(x)N(x,y)dy=u_x(x,y)dx+u_y(x,y)dy$$

is exactly the differential du.

Theorem 1.0.12. *Let the functions* $M(x,y)$ *and* $N(x,y)$ *together with their partial derivatives* $M_y(x,y)$ *and* $N_x(x,y)$ *be continuous in the rectangle S. Then the differential equation* (19) *is exact if and only if the condition*

$$M_y(x,y)(\hat{T}(x)-x\hat{T}'(x))-2N(x,y)\hat{T}'(x)=yN_y(x,y)\hat{T}'(x)+\hat{T}(x)N_x(x,y) \tag{22}$$

is satisfied. In the case when the equation (19) *is an exact equation, then a solution of* (19) *is given by*

$$\int_{x_0}^{x}\Big(M(s,y)(\hat{T}(s)-s\hat{T}'(s))-y_0\hat{T}(s)M(s,y_0)\Big)ds+\hat{T}(x)\int_{y_0}^{y}N(x,t)dt=C$$

or

$$\int_{x_0}^{x}\Big(M(s,y)(\hat{T}(s)-s\hat{T}'(s))-y\hat{T}'(s)N(s,y)\Big)ds+\hat{T}(x_0)\int_{y_0}^{y}N(x_0,t)dt=C,$$

where C is a constant.

Proof. Let us suppose that (19) is an exact equation. Then there exists a function u satisfying (20) and (21). Then

$$u_{xy}(x,y) = M_y(x,y)(\hat{T}(x) - x\hat{T}'(x)) - \hat{T}'(x)N(x,y) - y\hat{T}'(x)N_y(x,y),$$

$$u_{yx}(x,y) = \hat{T}'(x)N(x,y) + \hat{T}(x)N_x(x,y).$$

Since $M_y(x,y)$ and $N_x(x,y)$ are continuous functions in the rectangle S, then we must have

$$u_{xy}(x,y) = u_{yx}(x,y),$$

i.e., for (19) to be exact it is necessary that

$$M_y(x,y)(\hat{T}(x) - x\hat{T}'(x)) - \hat{T}'(x)N(x,y) - y\hat{T}'(x)N_y(x,)$$

$$= \hat{T}'(x)N(x,y) + \hat{T}(x)N_x(x,y).$$

Conversely, if M and N satisfy (22), then we integrate the equality (20) with respect to the variable x and we get

$$u(x,y) = \int_{x0}^{x} \Big(M(s,y)(\hat{T}(s) - s\hat{T}'(s)) - y\hat{T}'(s)N(s,y)\Big)ds + g(y), \tag{23}$$

here the function g is an arbitrary function of y and plays the role of the constant of integration. Now we differentiate the last equality with respect to the variable y and using (21), we obtain

$$u_y(x,y) = \frac{\partial}{\partial y}\int_{x0}^{x} \Big(M(s,y)(\hat{T}(s) - s\hat{T}'(s)) - y\hat{T}'(s)N(s,y)\Big)ds + g'(y)$$

$$= \hat{T}(x)N(x,y),$$

from where

$$g'(y) = -\frac{\partial}{\partial y}\int_{x0}^{x} \Big(M(s,y)(\hat{T}(s) - s\hat{T}'(s)) - y\hat{T}'(s)N(s,y)\Big)ds$$

$$+\hat{T}(x)N(x,y),$$

which we integrate with respect to the variable y and we go to

$$g(y) = -\int_{x0}^{x} \Big(M(s,y)(\hat{T}(s) - s\hat{T}'(s)) - y\hat{T}(s)N(s,y)\Big)ds$$

$$+\int_{x0}^{x} \Big(M(s,y_0)(\hat{T}(s) - s\hat{T}'(s)) - y_0\hat{T}(s)N(s,y_0)\Big)ds$$

$$+\hat{T}(x)\int_{y0}^{y} N(x,t)dt + g(y_0).$$

From the last equality and from (23), we obtain

$$u(x,y) = \int_{x0}^{x} \Big(M(s,y)(\hat{T}(s) - s\hat{T}'(s)) - y\hat{T}(s)N(s,y)\Big)ds$$

$$- \int_{x0}^{x} \Big(M(s,y)(\hat{T}(s) - s\hat{T}'(s)) - y\hat{T}(s)N(s,y)\Big)ds$$

$$+ \int_{x0}^{x} \Big(M(s,y_0)(\hat{T}(s) - s\hat{T}'(s)) - y_0\hat{T}(s)M(s,y_0)\Big)ds$$

$$+g(y_0) + \hat{T}(x) \int_{y0}^{y} N(x,t)dt$$

$$= \int_{x0}^{x} \Big(M(s,y_0)(\hat{T}(s) - s\hat{T}'(s)) - y_0\hat{T}(s)M(s,y_0)\Big)ds$$

$$+\hat{T}(x) \int_{y0}^{y} N(x,t)dt + g(y_0).$$

Therefore, a solution of the exact equation (19) is given by

$$\int_{x0}^{x} \Big(M(s,y_0)(\hat{T}(s) - s\hat{T}'(s)) - y_0\hat{T}(s)M(s,y_0)\Big)ds + \hat{T}(x) \int_{y0}^{y} N(x,t)dt = C,$$

where C is a constant.

Now we integrate the equality (21) with respect to the variable y and we obtain

$$u(x,y) = \hat{T}(x) \int_{y0}^{y} N(x,t)dt + f(x). \tag{24}$$

Here the function f is an arbitrary function of x and plays the role of the constant of integration. We differentiate the last equality with respect to the variable x and using (20) we have

$$u_x(x,y) = \hat{T}'(x) \int_{y0}^{y} N(x,t)dt + \hat{T}(x)\frac{\partial}{\partial x} \int_{y0}^{y} N(x,t)dt + f'(x)$$

$$= M(x,y)(\hat{T}(x) - x\hat{T}'(x)) - y\hat{T}'(x)N(x,y)$$

or

$$f'(x) = M(x,y)(\hat{T}(x) - x\hat{T}'(x)) - y\hat{T}'(x)N(x,y)$$

$$-\hat{T}'(x) \int_{y0}^{y} N(x,t)dt - \hat{T}(x)\frac{\partial}{\partial x} \int_{y0}^{y} N(x,t)dt,$$

which we integrate with respect to the variable x and we obtain

$$f(x) = \int_{x_0}^{x} \Big(M(s,y)(\hat{T}(s) - s\hat{T}'(s)) - y\hat{T}'(s)N(s,y) \Big) ds$$

$$- \int_{x_0}^{x} \hat{T}(s) \tfrac{\partial}{\partial s} \int_{y_0}^{y} N(s,t)dtds + f(x_0) - \int_{x_0}^{x} \hat{T}'(s) \int_{y_0}^{y} N(s,t)dtds$$

$$= \int_{x_0}^{x} \Big(M(s,y)(\hat{T}(s) - s\hat{T}'(s)) - y\hat{T}'(s)N(s,y) \Big) ds - \int_{x_0}^{x} \hat{T}'(s) \int_{y_0}^{y} N(s,t)dtds$$

$$-\hat{T}(x) \int_{y_0}^{y} N(x,t)dt + \hat{T}(x_0) \int_{y_0}^{y} N(x_0,t)dt + \int_{x_0}^{x} \hat{T}'(s) \int_{y_0}^{y} N(s,t)dtds + f(x_0)$$

$$= \int_{x_0}^{x} \Big(M(s,y)(\hat{T}(s) - s\hat{T}'(s)) - y\hat{T}'(s)N(s,y) \Big) ds$$

$$-\hat{T}(x) \int_{y_0}^{y} N(x,t)dt + \hat{T}(x_0) \int_{y_0}^{y} N(x_0,t)dt + f(x_0).$$

From here and from (24), we get

$$u(x,y) = \int_{x_0}^{x} \Big(M(s,y)(\hat{T}(s) - s\hat{T}'(s)) - y\hat{T}'(s)N(s,y) \Big) ds$$

$$+\hat{T}(x_0) \int_{y_0}^{y} N(x_0,t)dt + f(x_0).$$

Consequently, a solution of the exact equation (19) is given by

$$\int_{x_0}^{x} \Big(M(s,y)(\hat{T}(s) - s\hat{T}'(s)) - y\hat{T}'(s)N(s,y) \Big) ds + \hat{T}(x_0) \int_{y_0}^{y} N(x_0,t)dt = C,$$

where C is a constant. □

Remark 1.0.13. 1. *Obviously, in this result S may be replaced by any region which does not include any "hole".*

2. *The choice of x_0 and y_0 is at our disposal, except that these must be chosen so that the integrals remain proper.*

3. *The function*

$$\frac{\partial}{\partial y} \int_{x_0}^{x} \Big(M(s,y)(\hat{T}(s) - s\hat{T}'(s)) - y\hat{T}'(s)N(s,y) \Big) ds - \hat{T}(x)N(x,y)$$

must depend on y alone.

4. *The function*

$$M(x,y)(\hat{T}(x) - x\hat{T}'(x)) - y\hat{T}'(x)N(x,y) - \hat{T}'(x) \int_{y_0}^{y} N(x,t)dt - \hat{T}(x) \tfrac{\partial}{\partial x} \int_{y_0}^{y} N(x,t)dt$$

must depend on x alone.

Example 1.0.14. *Let*

$$S=\{(x,y)\in\mathbb{R}^2 : |x-1|\leq 1, \qquad |y|\leq 1\},$$

$\hat{T}(x)=x^2+1$, $M(x,y)=x-y$, $N(x,y)=x+y$, $(x,y)\in S$. *Then*

$$\hat{T}'(x)=2x,$$

$$M(x,y)(\hat{T}(x)-x\hat{T}'(x))-y\hat{T}'(x)N(x,y)=(x-y)(x^2+1-2x^2)-2xy(x+y)$$

$$=(1-x^2)(x-y)-2xy(x+y)$$

$$=x-x^3-y+x^2y-2xy^2-2x^2y$$

$$=-x^3-x^2y-2xy^2+x-y,$$

$$\hat{T}(x)N(x,y)=(x^2+1)(x+y)$$

$$=x^3+x^2y+x+y.$$

Therefore, the equation (19) *takes the form*

$$(-x^3-x^2y-2xy^2+x-y)dx+(x^3+x^2y+x+y)dy=0.$$

Since

$$\tfrac{\partial}{\partial y}(-x^3-x^2y-2xy^2+x-y)=-x^2-4xy-1$$

$$\neq \tfrac{\partial}{\partial x}(x^3+x^2y+x+y)=3x^2+2xy+1,$$

the equation (19) *is not an exact equation.*

Example 1.0.15. *Let*

$$S=\{(x,y)\in\mathbb{R}^2 : |x-4|\leq 1, \qquad |y|\leq 0\},$$

$\hat{T}(x)=1+x^2$, $M(x,y)=\frac{5x^2y+3xy^2+y}{1-x^2}$, $N(x,y)=x+y$, $(x,y)\in S$. *Then*

$$\hat{T}'(x)=2x,$$

$$M(x,y)(\hat{T}(x)-x\hat{T}'(x))-y\hat{T}'(x)N(x,y)=\tfrac{5x^2y+3xy^2+y}{1-x^2}(1-x^2)-y(x+y)2x$$

$$=5x^2y+3xy^2+y-2x^2y-2xy^2$$

$$=3x^2y+xy^2+y,$$

$$\hat{T}(x)N(x,y)=(x^2+1)(x+y)$$

$$=x^3+x^2y+x+y.$$

Then the equation (19) *takes the form*

$$(3x^2y+xy^2+y)dx+(x^3+x^2y+x+y)dy=0.$$

Since

$$\frac{\partial}{\partial y}(3x^2y+xy^2+y)=\frac{\partial}{\partial x}(x^3+x^2y+x+y)=3x^2+2xy+1,$$

then the equation (19) *is an exact equation.*

Therefore there exists a function u such that

$$u_x(x,y)=3x^2y+xy^2+y, \tag{25}$$

$$u_y(x,y)=x^3+x^2y+x+y. \tag{26}$$

We integrate the equality (25) *with respect to the variable x and we obtain*

$$u(x,y)=\int_0^x(3s^2y+sy^2+y)ds+g(y)$$

$$=x^3y+\tfrac{x^2y^2}{2}+xy+g(y),$$

here the function g is an arbitrary function of y and plays the role of the constant of integration. From here, using (26), *we get*

$$u_y(x,y)=x^3+x^2y+x+g'(y)=x^3+x^2y+x+y$$

or

$$g'(y)=y,$$

whereupon

$$g(y)=\frac{y^2}{2}+C_0,$$

where C_0 is a constant. Then

$$u(x,y)=x^3y+\frac{x^2y^2}{2}+xy+\frac{y^2}{2}+C_0$$

and

$$2x^3y+x^2y^2+2xy+y^2=C$$

is a solution to the equation (19). *Here C is a constant.*

Exercise 1.0.16. *Let*

$$S=\{(x,y)\in\mathbb{R}^2:|x|\leq 1,\qquad |y|\leq 1\},$$

$\hat{T}(x)=e^x$, $M(x,y)=x^2-y$, $N(x,y)=xy$. *Determine the equation* (19) *and check if it is an exact equation. If it is an exact equation, find an its solution.*

Answer.

$$e^x(-x^3-xy^2+x^2-y+xy)dx+xye^xdy=0.$$

It is not an exact equation.

Exercise 1.0.17. *Let*

$$S = \{(x,y) \in \mathbb{R}^2 : |x-5| \le 1, \qquad |y| \le 1\},$$

$\hat{T}(x) = e^x$, $M(x,y) = \frac{\frac{3}{2}y^2+2xy+y}{1-x}$, $N(x,y) = x+y$, $(x,y) \in S$. *Determine the equation* (19) *and check if it is an exact equation. If it is an exact equation, find an its solution.*

Answer.

$$\left(\frac{1}{2}y^2 + xy + y\right)e^x dx + e^x(x+y)dy = 0.$$

It is an exact solution. An its solution is given by

$$2e^x xy + e^x y^2 = C,$$

where C is a constant.

Now we consider the equation

$$M(x,y^\wedge) + N(x,y^\wedge)\hat{\times}\left(y^\wedge\right)^{\circledast} = 0 \tag{27}$$

or

$$M(x,y^\wedge)\hat{\times}\hat{d}\hat{x} + N(x,y^\wedge)\hat{\times}\hat{d}y^\wedge = 0,$$

where M and N are continuous functions having continuous partial derivatives $M_{y^\wedge}$ and N_x in the rectangle S, defined in the begin of this chapter.

The equation (27) we can rewrite in the form

$$M(x,y^\wedge)(\hat{T}(x) - x\hat{T}'(x))dx + N(x,y^\wedge)\hat{T}^2(x)dy^\wedge = 0.$$

For convenience, below we will use the notation

$$M(x,y)(\hat{T}(x) - x\hat{T}'(x))dx + N(x,y)\hat{T}^2(x)dy = 0.$$

Definition 1.0.18. *The equation* (27) *is said to be exact if there exists a function* $u(x,y)$ *such that*

$$u_x(x,y) = M(x,y)(\hat{T}(x) - x\hat{T}'(x)), \qquad u_y(x,y) = N(x,y)\hat{T}^2(x).$$

The nomenclature comes from the fact that

$$M(x,y)(\hat{T}x) - x\hat{T}'(x))dx + N(x,y)\hat{T}^2x)dy = u_x(x,y)dx + u_y(x,y)dy$$

is exactly the differential du.

Once the iso-differential equation (27) is exact its implicit solution is

$$u(x,y) = C,$$

where C is a constant.

Theorem 1.0.19. *Let the functions $M(x,y)$ and $N(x,y)$ together with their partial derivatives $M_y(x,y)$ and $N_x(x,y)$ be continuous in the region S. Then the iso-differential equation* (27) *is exact if and only if the condition*

$$M_y(x,y)(\hat{T}(x)-x\hat{T}'(x)) = N_x(x,y)\hat{T}^2(x)+2N(x,y)\hat{T}(x)\hat{T}'(x) \tag{28}$$

is satisfied. A solution to the exact equation (27) *is given by*

$$\hat{T}^2(x)\int_{y_0}^{y} N(x,t)dt + \int_{x_0}^{x} M(s,y_0)(\hat{T}(s)-s\hat{T}'(s))ds = C,$$

or

$$\hat{T}^2(x_0)\int_{y_0}^{y} N(x_0,t)dt + \int_{x_0}^{x} M(s,y)(\hat{T}(s)-s\hat{T}'(s))ds = C,$$

where C is a constant.

Proof. Let us suppose that the equation (27) is an exact equation. Then there exists a function $u(x,y)$ such that

$$u_x(x,y) = M(x,y)(\hat{T}(x)-x\hat{T}'(x)), \tag{29}$$

$$u_y(x,y) = N(x,y)\hat{T}^2(x). \tag{30}$$

Then we have

$$u_{xy}(x,y) = M_y(x,y)(\hat{T}(x)-x\hat{T}'(x))$$

and

$$u_{yx}(x,y) = N_x(x,y)\hat{T}^2(x)+2N(x,y)\hat{T}(x)\hat{T}'(x).$$

Since $M_y(x,y)$ and $N_x(x,y)$ are continuous functions in the region S we must have

$$u_{xy}(x,y) = u_{yx}(x,y).$$

Consequently the condition (28) is satisfied.

Conversely, if M and N satisfy the condition (28), then we integrate the equality (29) with respect to the variable x and we get

$$u(x,y) = \int_{x_0}^{x} M(s,y)(\hat{T}(s)-s\hat{T}'(s))ds + g(y). \tag{31}$$

Here the function g is an arbitrary function of y and plays the role of the constant of integration. We differentiate the last equality with respect to the variable y and using (30), we obtain

$$u_y(x,y) = \frac{\partial}{\partial y}\int_{x_0}^{x} M(s,y)(\hat{T}(s)-s\hat{T}'(s))ds + g'(y) = N(x,y)\hat{T}^2(x),$$

whereupon

$$g'(y) = N(x,y)\hat{T}^2(x) - \frac{\partial}{\partial y}\int_{x_0}^{x} M(s,y)(\hat{T}(s)-s\hat{T}'(s))ds. \tag{32}$$

We observe that the function on the right hand of the last equality must depend on y alone. We integrate the equality (32) with respect to the variable y and we get

$$g(y) = \int_{y_0}^{y} N(x,t)\hat{T}^2(x)dt - \int_{x_0}^{x} M(s,y)(\hat{T}(s) - s\hat{T}'(s))ds$$

$$+\int_{x_0}^{x} M(s,y_0)(\hat{T}(s) - s\hat{T}'(s))ds + g(y_0),$$

from the last equality and from (31) we have

$$u(x,y) = \int_{x_0}^{x} M(s,y)(\hat{T}s) - s\hat{T}'(s))ds + \int_{y_0}^{y} N(x,t)\hat{T}^2(x)dt$$

$$-\int_{x_0}^{x} M(s,y)(\hat{T}(s) - s\hat{T}'(s))ds + \int_{x_0}^{x} M(s,y_0)(\hat{T}(s) - s\hat{T}'(s))ds + g(y_0)$$

$$= \hat{T}^2(x)\int_{y_0}^{y} N(x,t)dt + \int_{x_0}^{x} M(s,y_0)(\hat{T}(s) - s\hat{T}'(s))ds + g(y_0).$$

Therefore. a solution of the exact equation (27) is given by

$$\hat{T}^2(x)\int_{y_0}^{y} N(x,t)dt + \int_{x_0}^{x} M(s,y_0)(\hat{T}(s) - s\hat{T}'(s))ds = C$$

for some constant C.

Now we integrate the equality (30) with respect to the variable y and we obtain

$$u(x,y) = \hat{T}^2(x)\int_{y_0}^{y} N(x,t)dt + f(x), \tag{33}$$

where the function f is an arbitrary function of y and plays the role of the constant of integration. We differentiate the last equality with respect to the variable x and using (29), we get

$$u_x(x,y) = 2\hat{T}(x)\hat{T}'(x)\int_{y_0}^{y} N(x,t)dt + \hat{T}^2(x)\frac{\partial}{\partial x}\int_{y_0}^{y} N(x,t)dt + f'(x)$$

$$= M(x,y)(\hat{T}(x) - x\hat{T}'(x)),$$

from where

$$\begin{aligned} f'(x) &= -2\hat{T}(x)\hat{T}'(x)\int_{y_0}^{y} N(x,t)dt \\ &-\hat{T}^2(x)\frac{\partial}{\partial x}\int_{y_0}^{y} N(x,t)dt + M(x,y)(\hat{T}(x) - x\hat{T}'(x)). \end{aligned} \tag{34}$$

From the last equality it follows that its right hand must depend on x alone. Now we integrate the equality (34) with respect to the variable x and we go to

$$f(x) = \ -2\int_{x_0}^{x}\hat{T}(s)\hat{T}'(s)\int_{y_0}^{y}N(s,t)dtds - \int_{x_0}^{x}\hat{T}^2(s)\frac{\partial}{\partial s}\int_{y_0}^{y}N(s,t)dtds$$

$$+\int_{x_0}^{x}M(s,y)\hat{T}(s) - s\hat{T}'(s))ds + f(x_0)$$

$$= -2\int_{x_0}^{x}\hat{T}(s)\hat{T}'(s)\int_{y_0}^{y}N(s,t)dtds - \hat{T}^2(x)\int_{y_0}^{y}N(x,t)dt$$

$$+\hat{T}^2(x_0)\int_{y_0}^{y}N(x_0,t)dt + 2\int_{x_0}^{x}\hat{T}(s)\hat{T}'(s)\int_{y_0}^{y}N(s,t)dtds$$

$$+\int_{x_0}^{x}M(s,y)(\hat{T}s) - s\hat{T}'(s))ds + f(x_0)$$

$$= -\hat{T}^2(x)\int_{y_0}^{y}N(x,t)dt + \hat{T}^2(x_0)\int_{y_0}^{y}N(x_0,t)dt$$

$$+\int_{x_0}^{x}M(s,y)(\hat{T}(s) - s\hat{T}'(s))ds + f(x_0).$$

From the last equality and (33) we have

$$u(x,y) = \hat{T}^2(x)\int_{y_0}^{y}N(x,t)dt - \hat{T}^2(x)\int_{y_0}^{y}N(x,t)dt + \hat{T}^2(x_0)\int_{y_0}^{y}N(x_0,t)dt$$

$$+\int_{x_0}^{x}M(s,y)(\hat{T}(s) - s\hat{T}'(s))ds + f(x_0)$$

$$= \hat{T}^2(x_0)\int_{y_0}^{y}N(x_0,t)dt + \int_{x_0}^{x}M(s,y)(\hat{T}(s) - s\hat{T}'(s))ds + f(x_0).$$

Therefore, a solution of the exact equation (27) is given by

$$\hat{T}^2(x_0)\int_{y_0}^{y}N(x_0,t)dt + \int_{x_0}^{x}M(s,y)(\hat{T}(s) - s\hat{T}'(s))ds = C$$

for some constant C.

We note that in this result S can be replaced by any region which does not include any "hole". Also, the choice of x_0 and y_0 is at our disposal, except that these must be chosen so that the integrals remain proper. □

Example 1.0.20. *Let*

$$S = \{(x,y) \in \mathbb{R}^2 : |x| \le 1, \qquad |y| \le 1\},$$

$\hat{T}(x) = x^2 + 1$, $M(x,y) = x - y$, $N(x,y) = x + y$, $(x,y) \in S$. *Then*

$$\hat{T}'(x) = 2x + 1,$$

$$M(x,y)(\hat{T}(x) - x\hat{T}'(x)) = (x-y)(x^2 + 1 - 2x^2)$$

$$= (x-y)(1-x^2)$$

$$= -x^3 + x^2y + x - y,$$

$$N(x,y)\hat{T}(x) = (x+y)(x^2+1)$$

$$= x^3 + x^2y + x + y.$$

The equation (27) *takes the form*

$$(-x^3+x^2y+x+y)dx+(x^3+x^2y+x+y)dy=0.$$

Since

$$\frac{\partial}{\partial y}(-x^3+x^2y+x+y)=x^2-1$$

$$\neq \frac{\partial}{\partial x}(x^3+x^2y+x+y)=3x^2+2xy+1,$$

then it is not an exact equation.

Exercise 1.0.21. *Let*

$$S=\{(x,y)\in\mathbb{R}^2 : |x-1|\leq 3, \qquad |y|\leq 4\},$$

$\hat{T}(x)=e^x$, $M(x,y)=xy$, $N(x,y)=x+y$, $(x,y)\in S$. *Determine the equation* (27) *and check if it is an exact equation. If it is an exact equation, find an its solution.*

Answer.

$$(xy-x^2y)e^x dx+e^{2x}(x+y)dy=0.$$

It is not an exact equation.

Example 1.0.22. *Let*

$$S=\{(x,y)\in\mathbb{R}^2 : |x-5|\leq 1, \qquad |y|\leq 2\},$$

$\hat{T}(x)=x^2+1$, $M(x,y)=\frac{5x^4y+2x^3y^2+2xy^2+6x^2y+y}{1-x^2}$, $N(x,y)=x+y$, $(x,y)\in S$. *Then*

$$\hat{T}'(x)=2x,$$

$$M(x,y)(\hat{T}(x)-x\hat{T}'(x))=\frac{5x^4y+2x^3y^2+2xy^2+6x^2y+y}{1-x^2}(1-x^2)$$

$$=5x^4y+2x^3y^2+2xy^2+6x^2y+y,$$

$$N(x,y)\hat{T}(x)^2=(x+y)(x^2+1)^2$$

$$=(x+y)(x^4+2x^2+1)$$

$$=x^5+x^4y+2x^3+2x^2y+x+y,$$

then the equation (27) *takes the form*

$$(5x^4y+2x^3y^2+2xy^2+6x^2y+y)dx+(x^5+x^4y+2x^3+2x^2y+x+y)dy=0.$$

Since

$$\frac{\partial}{\partial y}(5x^4y+2x^3y^2+2xy^2+6x^2y+y)=5x^4+4x^3y+4xy+6x^2+1$$

$$=\frac{\partial}{\partial x}(x^5+x^4y+2x^3+2x^2y+x+y),$$

then the equation (27) *is an exact equation. Therefore, there exists a function* $u(x,y)$ *such that*

$$u_x(x,y) = 5x^4y + 2x^3y^2 + 2xy^2 + 6x^2y + y, \tag{35}$$

$$u_y(x,y) = x^5 + x^4y + 2x^3 + 2x^2y + x + y. \tag{36}$$

We integrate the equality (35) *with respect to the variable x and we get*

$$u(x,y) = \int_0^x (5s^4y + 2s^3y^2 + 2sy^2 + 6s^2y + y)ds + g(y)$$

$$= x^5y + \frac{x^4}{2}y^2 + x^2y^2 + 2x^3y + xy + g(y),$$

where the function g is an arbitrary function of y and plays the role of the constant of integration.

We differentiate the last equality with respect to the variable y and using (36), *we obtain*

$$x^5 + x^4y + 2x^2y + 2x^3 + x + g'(y) = x^5 + x^4y + 2x^3 + 2x^2y + x + y$$

or

$$g'(y) = y,$$

from where

$$g(y) = \frac{y^2}{2} + C_0,$$

where C_0 *is a constant.*

From here,

$$u(x,y) = x^5y + \frac{x^4}{2}y^2 + x^2y^2 + 2x^3y + xy + \frac{y^2}{2} + C_0,$$

and a solution is given by

$$2x^5y + x^4y^2 + 2x^2y^2 + 4x^3y + 2xy + y^2 = C,$$

where C is a constant.

Exercise 1.0.23. *Let*

$$S = \{(x,y) \in \mathbb{R}^2 : |x-4| \leq 1, \qquad |y| \leq 2\},$$

$\hat{T}(x) = e^x$, $M(x,y) = e^x\frac{y+2xy-y^2}{1-x}$, $N(x,y) = x-y$, $(x,y) \in S$. *Determine the equation* (27) *and check if it is an exact equation. If it is an exact equation, find an its solution.*

Answer.

$$e^{2x}(y + 2xy - y^2)dx + e^{2x}(x-y)dy = 0.$$

It is an exact equation. A solution is given by

$$e^{2x}(2xy - y^2) = C$$

for some constant C.

Now we consider the equation

$$\hat{M}^{\wedge}(\hat{x},\hat{y})+\hat{N}^{\wedge}(\hat{x},\hat{y})\hat{\times}\left(y^{\vee}\right)^{\circledast}=0 \tag{37}$$

or

$$\hat{M}^{\wedge}(\hat{x},\hat{y})\hat{\times}\hat{d}\hat{x}+\hat{N}^{\wedge}(\hat{x},\hat{y})\hat{\times}\hat{d}y^{\vee}=0,$$

where M and N are continuous functions having continuous partial derivatives $M_{y^{\vee}}$ and N_x in the region S, defined in the begin of this chapter.

The equation (37) we can rewrite in the form

$$M(x,y^{\vee})\left(1-x\frac{\hat{T}'(x)}{\hat{T}(x)}\right)dx+N(x,y^{\vee})\hat{T}(x)dy^{\vee}=0.$$

For convenience, below we will use the notation

$$M(x,y)\left(1-x\frac{\hat{T}'(x)}{\hat{T}(x)}\right)dx+N(x,y)\hat{T}(x)dy=0.$$

Definition 1.0.24. *The equation* (37) *is said to be exact if there exists a function* $u(x,y)$ *such that*

$$u_x(x,y)=M(x,y)\left(1-x\frac{\hat{T}'(x)}{\hat{T}(x)}\right),\qquad u_y(x,y)=N(x,y)\hat{T}(x).$$

The nomenclature comes from the fact that

$$M(x,y)\left(1-x\frac{\hat{T}'(x)}{\hat{T}(x)}\right)dx+M(x,y)\hat{T}(x)dy=u_x(x,y)dx+u_y(x,y)dy$$

is exactly the differential du.

Once the iso-differential equation is an exact equation its implicit solution is

$$u(x,y)=C,$$

where C is a constant.

Theorem 1.0.25. *Let the functions* $M(x,y)$ *and* $N(x,y)$ *together with their partial derivatives* $M_y(x,y)$ *and* $N_x(x,y)$ *be continuous in the rectangle S. Then the iso-differential equation* (37) *is exact if and only if the condition*

$$M_y(x,y)\left(1-x\frac{\hat{T}'(x)}{\hat{T}(x)}\right)=N_x(x,y)\hat{T}(x)+N(x,y)\hat{T}'(x) \tag{38}$$

is satisfied. A solution of the exact equation (37) *is given by*

$$\int_{x_0}^{x}M(s,y_0)\left(1-s\frac{\hat{T}'(s)}{\hat{T}(s)}\right)ds+\hat{T}(x)\int_{y_0}^{y}N(x,s)ds+\hat{T}'(x)\int_{y_0}^{y}N_x(x,s)ds=C,$$

or

$$\hat{T}(x_0)\int_{y_0}^{y}M(x_0,t)dt+\int_{x_0}^{x}M(s,t)\left(1-s\frac{\hat{T}'(s)}{\hat{T}(s)}\right)ds=C,$$

where C is a constant.

Proof. Let us suppose that the equation (37) is an exact equation. Then there exists a function $u(x,y)$ such that

$$u_x(x,y) = M(x,y)\left(1 - x\frac{\hat{T}'(x)}{\hat{T}(x)}\right), \tag{39}$$

$$u_y(x,y) = N(x,y)\hat{T}(x). \tag{40}$$

Then we have

$$u_{xy}(x,y) = M_y(x,y)\left(1 - x\frac{\hat{T}'(x)}{\hat{T}(x)}\right)$$

and

$$u_{yx}(x,y) = N_x(x,y)\hat{T}(x) + N(x,y)\hat{T}'(x).$$

Since $M_y(x,y)$ and $N_x(x,y)$ are continuous functions in S, then we must have

$$u_{xy}(x,y) = u_{yx}(x,y)$$

or the condition (38) is satisfied.

Conversely, if M and N satisfy (38), then we integrate the equality (39) with respect to the variable x and we get

$$u(x,y) = \int_{x_0}^{x} M(s,y)\left(1 - s\frac{\hat{T}'(s)}{\hat{T}(s)}\right)ds + g(y), \tag{41}$$

where the function g is an arbitrary function of y and plays the role of the constant of integration. We differentiate the last equality with respect to the variable y and using (40), we get

$$\frac{\partial}{\partial y}\int_{x_0}^{x} M(s,y)\left(1 - s\frac{\hat{T}'(s)}{\hat{T}(s)}\right)ds + g'(y) = N_x(x,y)\hat{T}(x) + N(x,y)\hat{T}'(x)$$

or

$$g'(y) = N_x(x,y)\hat{T}(x) + N(x,y)\hat{T}'(x) - \frac{\partial}{\partial y}\int_{x_0}^{x} M(s,y)\left(1 - s\frac{\hat{T}'(s)}{\hat{T}(s)}\right)ds.$$

From the last equality it follows that its right hand must depend on y alone and after we integrate the last equality with respect to the variable y we obtain

$$g(y) = \hat{T}(x)\int_{y_0}^{y} N_x(x,s)ds + \hat{T}'(x)\int_{y_0}^{y} N(x,s)ds - \int_{x_0}^{x} M(s,y)\left(1 - s\frac{\hat{T}'(s)}{\hat{T}(s)}\right)ds$$

$$+ \int_{x_0}^{x} M(s,y_0)\left(1 - s\frac{\hat{T}'(s)}{\hat{T}(s)}\right)ds + g(y_0).$$

From here and (41), we have

$$u(x,y) = \int_{x_0}^{x} M(s,y)\left(1 - s\frac{\hat{T}'(s)}{\hat{T}(s)}\right)ds + \hat{T}(x)\int_{y_0}^{y} N_x(x,s)ds$$

$$+\hat{T}'(x)\int_{y_0}^{y} N(x,s)ds - \int_{x_0}^{x} M(s,y)\left(1 - s\frac{\hat{T}'(s)}{\hat{T}(s)}\right)ds$$

$$+ \int_{x_0}^{x} M(s,y_0)\left(1 - s\frac{\hat{T}'(s)}{\hat{T}(s)}\right)ds + g(y_0)$$

$$= \int_{x_0}^{x} M(s,y_0)\left(1 - s\frac{\hat{T}'(s)}{\hat{T}(s)}\right)ds + \hat{T}(x)\int_{y_0}^{y} N_x(x,s)ds + \hat{T}'(x)\int_{y_0}^{y} N(x,s)ds + g(y_0).$$

Consequently, a solution of the exact equation (37) is given by

$$\int_{x_0}^{x} M(s,y_0)\left(1-s\frac{\hat{T}'(s)}{\hat{T}(s)}\right)ds+\hat{T}(x)\int_{y_0}^{y} N_x(x,s)ds+\hat{T}'(x)\int_{y_0}^{y} N(x,s)ds=C,$$

where C is a constant.

Now we integrate the equality (40) with respect to the variable y and we obtain

$$u(x,y)=\hat{T}(x)\int_{y_0}^{y} N(x,t)dt+f(x), \tag{42}$$

where f is an arbitrary function of x and plays the role of the constant of integration. We differentiate the last equality with respect to the variable x and using (39), we get

$$u_x(x,y)=\hat{T}'(x)\int_{y_0}^{y} N(x,t)dt+\hat{T}(x)\frac{\partial}{\partial x}\int_{y_0}^{y} N(x,t)dt+f'(x)$$

$$=M(x,y)\left(1-x\frac{\hat{T}'(x)}{\hat{T}(x)}\right)$$

or

$$f'(x)=-\hat{T}'(x)\int_{y_0}^{y} N(x,t)dt-\hat{T}(x)\frac{\partial}{\partial x}\int_{y_0}^{y} N(x,t)dt+M(x,y)\left(1-x\frac{\hat{T}'(x)}{\hat{T}(x)}\right).$$

We observe that the right hand of the last equality must depend on x alone and we integrate it with respect to the variable x, we go to

$$f(x)=-\int_{x_0}^{x}\hat{T}'(s)\int_{y_0}^{y} N(s,t)dtds-\int_{x_0}^{x}\hat{T}(x)\frac{\partial}{\partial s}\int_{y_0}^{y} N(s,t)dtds$$

$$+\int_{x_0}^{x} M(s,y)\left(1-s\frac{\hat{T}'(s)}{\hat{T}(s)}\right)ds+f(x_0)$$

$$=-\hat{T}(x)\int_{y_0}^{y} N9x,t)dt+\hat{T}(x_0)\int_{y_0}^{y} N(x_0,t)dt$$

$$-\int_{x_0}^{x}\hat{T}(s)\frac{\partial}{\partial s}\int_{y_0}^{y} N(s,t)dtds+\int_{x_0}^{x}\hat{T}(s)\frac{\partial}{\partial s}\int_{y_0}^{y} N9s,t)dtds$$

$$+\int_{x_0}^{x} M(s,y)\left(1-s\frac{\hat{T}'(s)}{\hat{T}(s)}\right)ds+f(x_0)$$

$$=-\hat{T}(x)\int_{y_0}^{y} N(x,t)dt+\hat{T}(x_0)\int_{y_0}^{y} N(x_0,t)dt+\int_{x_0}^{x} M(s,y)\left(1-s\frac{\hat{T}'(s)}{\hat{T}(s)}\right)ds.$$

From here and (42), it follows that

$$u(x,y)=\hat{T}(x)\int_{y_0}^{y} N(x,t)dt-\hat{T}(x)\int_{y_0}^{y} N(x,t)dt+\hat{T}(x_0)\int_{y_0}^{y} N(x_0,t)dt$$

$$-\int_{x_0}^{x} M(s,y)\left(1-s\frac{\hat{T}'(s)}{\hat{T}(s)}\right)ds+f(x_0)$$

$$=\hat{T}(x_0)\int_{y_0}^{y} N(x_0,t)dt+\int_{x_0}^{x} M(s,y)\left(1-s\frac{\hat{T}'(s)}{\hat{T}(s)}\right)ds+fx_0).$$

Consequently, a solution of the exact equation (37) is given by

$$\hat{T}(x_0)\int_{y_0}^{y} N(x_0,t)dt+\int_{x_0}^{x} M(s,y)\left(1-s\frac{\hat{T}'(s)}{\hat{T}(s)}\right)ds=C,$$

where C is a constant.

We will note that the rectangle S can be replaced by any region which does not include any "hole". Also, the choice of x_0 and y_0 is at our disposal, except that these must be chosen so that the integrals remain proper. □

Example 1.0.26. *Let*

$$S = \{(x,y) \in R^2 : |x-1| \leq 1, \qquad |y| \leq 1\},$$

$\hat{T}(x) = e^x$, $M(x,y) = x+y$, $N(x,y) = x-y$, $(x,y) \in S$. *Then*

$$M(x,y)\left(1 - x\frac{\hat{T}'(x)}{\hat{T}(x)}\right) = (x+y)\left(1 - x\frac{e^x}{e^x}\right)$$

$$= (x+y)(1-x)$$

$$= -x^2 - xy + x + y,$$

$$N(x,y)\hat{T}(x) = (x-y)e^x.$$

The equation (37) *takes the form*

$$(-x^2 - xy + x + y)dx + (x-y)e^y dy = 0.$$

Since

$$\frac{\partial}{\partial y}(-x^2 - xy + x + y) = -x + y$$

$$\neq \frac{\partial}{\partial x}\left((x-y)e^y\right) = e^y,$$

then it is not an exact equation.

Exercise 1.0.27. *Let*

$$S = \{(x,y) \in \mathbb{R}^2 : |x| \leq 1, \qquad |y| \leq 1\},$$

$\hat{T}(x) = x^2 + 1$, $M(x,y) = x^2 - y$, $N(x,y) = xy$, $(x,y) \in S$. *Determine the equation* (37) *and check if it is an exact equation.*

Answer.

$$\frac{-x^4 + x^2y + x^2 - y}{1 + x^2}dx + (x^3y + xy)dy = 0.$$

It is not an exact equation.

Example 1.0.28. *Let*

$$S = \{(x,y) \in \mathbb{R}^2 : |x-5| \leq 1, \qquad |y| \leq 1\},$$

$\hat{T}(x) = e^x$, $M(x,y) = \frac{e^x(y+xy)}{1-x}$, $N(x,y) = x$, $(x,y) \in S$. *Then*

$$M(x,y)\left(1 - x\frac{\hat{T}'(x)}{\hat{T}(x)}\right) = \frac{e^x(y+xy)}{1-x}(1-x)$$

$$= e^x(y+xy),$$

$$N(x,y)\hat{T}(x) = xe^x.$$

The equation (37) *takes the form*

$$e^x(y+xy)dx + xe^x dy = 0.$$

Since

$$\frac{\partial}{\partial y}\left(e^x(y+xy)\right) = e^x(x+1) = \frac{\partial}{\partial x}\left(xe^x\right),$$

then it is an exact equation. Then there exists a function $u(x,y)$ *such that*

$$u_x(x,y) = e^x(y+xy), \tag{43}$$

$$u_y(x,y) = xe^x. \tag{44}$$

We integrate the equality (44) *with respect to the variable* y *and we get*

$$u(x,y) = xe^x \int_0^y dt + f(x) = xye^x + f(x), \tag{45}$$

where f *is an arbitrary function of* x *and plays the role of the constant of integration. Now we differentiate the last equality with respect to the variable* x *and using* (43), *we have*

$$(x+1)ye^x + f'(x) = e^x(y+xy)$$

or

$$f'(x) = 0,$$

whereupon $f(x) = C_0$, *where* C_0 *is a constant. From here and* (45) *we obtain that*

$$u(x,y) = xye^x + C_0.$$

Consequently, a solution is given by

$$xye^x = C$$

for some constant C.

Exercise 1.0.29. *Let*

$$S = \{(x,y) \in \mathbb{R}^2 : |x-6| \leq 2, \qquad |y| \leq 2\},$$

$\hat{T}(x) = e^x$, $M(x,y) = e^x\frac{y+xy+\frac{y^2}{2}}{1-x}$, $N(x,y) = x+y$, $(x,y) \in S$. *Determine the equation* (37) *and check if it is an exact equation. If it is an exact equation, find an its solution.*

Answer.

$$e^x\left(y+xy+\frac{y^2}{2}\right)dx+(x+y)e^x dy=0.$$

It is an exact equation. A solution is given by

$$(2xy+y^2)e^x=C$$

for some constant C.

Definition 1.0.30. *For the equation* (1) *a non-zero function* $\mu(x,y)$ *is called an integrating factor if the equivalent iso-differential equation*

$$\mu(x,y)\left(M(x,y)\left(1-x\frac{\hat{T}'(x)}{\hat{T}(x)}\right)-yN(x,y)\frac{\hat{T}'(x)}{\hat{T}(x)}\right)dx+\mu(x,y)N(x,y)dy=0 \tag{46}$$

is an exact equation.

If $u(x,y)=C$, C is a constant, is a solution of the equation (1), then y' computed from (1) and the equality

$$u_x(x,y)+u_x(x,y)y'=0 \tag{46'}$$

must be the same, i.e.,

$$y'=-\frac{u_x(x,y)}{u_y(x,y)}=-\frac{M(x,y)\left(1-x\frac{\hat{T}'(x)}{\hat{T}(x)}\right)-yN(x,y)\frac{\hat{T}'(x)}{\hat{T}(x)}}{N(x,y)}$$

or

$$\frac{u_x(x,y)}{M(x,y)\left(1-x\frac{\hat{T}'(x)}{\hat{T}(x)}\right)-yN(x,y)\frac{\hat{T}'(x)}{\hat{T}(x)}}=\frac{u_y(x,y)}{N(x,y)}=\mu(x,y), \tag{47}$$

where μ is some function of x and y. Thus we have

$$\begin{aligned}&\mu(x,y)\left(\left(M(x,y)\left(1-x\tfrac{\hat{T}'(x)}{\hat{T}(x)}\right)-yN(x,y)\tfrac{\hat{T}'(x)}{\hat{T}(x)}\right)+N(x,y)y'\right)\\&=u_x(x,y)+u_y(x,y)y'=\tfrac{du}{dx}\end{aligned} \tag{48}$$

and from here the iso-differential equation (46) is exact, and an integrating factor μ of (1) is given by (47).

Theorem 1.0.31. *If the iso-differential equation* (1) *has* $u(x,y)=C$*,* C *is a constant, as its solution, it admits an infinite number of integrating factors.*

Proof. Let $\phi(u)$ be any function of u. Since $u(x,y)=C$ we have (48). From here

$$\begin{aligned}&\mu(x,y)\phi(u)\left(M(x,y)\left(1-x\tfrac{\hat{T}'(x)}{\hat{T}(x)}\right)-yN(x,y)\hat{T}'(x)+N(x,y)y'\right)\\&=\phi(u)\tfrac{du}{dx}=\tfrac{d}{dx}\int_0^u\phi(s)ds.\end{aligned}$$

Hence, $\mu(x,y)\phi(u)$ is an integrating factor of (1). Since ϕ is an arbitrary function, we have established the result. □

The function $\mu(x,y)$ is an integrating factor of (1) provided (46) is exact, i.e., if and only if

$$\frac{\partial}{\partial y}\left(\mu(x,y)\left(M(x,y)\left(1-x\frac{\hat{T}'(x)}{\hat{T}(x)}\right)\right)-yN(x,y)\hat{T}'(x)\right)=\frac{\partial}{\partial x}(\mu(x,y)N(x,y)).$$

This implies that the integrating factor of (1) must satisfy the equation

$$\begin{aligned}&N(x,y)\mu_x(x,y)-\mu_y(x,y)\left(M(x,y)\left(1-x\tfrac{\hat{T}'(x)}{\hat{T}(x)}\right)-yN(x,y)\hat{T}'(x)\right)\\&=\mu(x,y)\left(\left(M_y(x,y)\left(1-x\tfrac{\hat{T}'(x)}{\hat{T}(x)}\right)-N(x,y)\hat{T}'(x)-yN_y(x,y)\hat{T}'(x)\right)-N_x(x,y)\right).\end{aligned}\tag{49}$$

A solution of (49) gives an integrating factor of (1), but it is not easy to be found a solution of the partial differential equation (49). However, a particular non-zero solution of (49) is all we need for the solution of (46).

If we assume that

$$\mu(x,y)=X(x)Y(y),$$

then from (49) we have

$$\begin{aligned}&N(x,y)\tfrac{1}{X}\tfrac{dX}{dx}-\left(M(x,y)\left(1-x\tfrac{\hat{T}'(x)}{\hat{T}(x)}\right)-yN(x,y)\hat{T}'(x)\right)\tfrac{1}{Y}\tfrac{dY}{dy}\\&=M_y(x,y)\left(1-x\tfrac{\hat{T}'(x)}{\hat{T}(x)}\right)-N(x,y)\hat{T}'(x)-yN_y(x,y)\hat{T}'(x).\end{aligned}\tag{50}$$

Hence, if

$$\begin{aligned}&M_y(x,y)\left(1-x\tfrac{\hat{T}'(x)}{\hat{T}(x)}\right)-N(x,y)\hat{T}'(x)-yN_y(x,y)\hat{T}'(x)\\&=N(x,y)g(x)-\left(M(x,y)\left(1-x\tfrac{\hat{T}'(x)}{\hat{T}(x)}\right)-yN(x,y)\hat{T}'(x)\right)h(y),\end{aligned}$$

then (50) is satisfied provided

$$\frac{1}{X}\frac{dX}{dx}=g(x)\quad\text{and}\quad\frac{1}{Y}\frac{dY}{dy}=h(y),\tag{50'}$$

i.e.,

$$X(x)=e^{\int g(x)dx},\qquad Y(y)=e^{\int h(y)dy}.\tag{50''}$$

We will illustrate this in the following example.

Example 1.0.32. *Let*

$$S=\{(x,y)\in\mathbb{R}^2:|x-5|\le 1,\qquad |y-7|\le 2\},$$

$\hat{T}(x)=e^x$, $M(x,y)=\frac{xy+y-y^2}{1-x}$, $N(x,y)=x$, $(x,y)\in S$. *We will search an integration factor* $\mu(x,y)$ *in the form*

$$\mu(x,y)=x^my^n,\qquad m,n\in\mathbb{R},\quad (m,n)\neq(0,0).$$

In this case the equation (1) *takes the form*

$$\left(\frac{xy+y-y^2}{1-x}(1-x)-yx\right)dx+xdy=0,$$

or

$$(y-y^2)dx+xdy=0. \qquad (51)$$

We multiply the equation (51) *with* $x^m y^n$ *and we obtain*

$$x^m y^n(y-y^2)dx+x^{m+1}y^n dy=0.$$

We will find m and n by the equation

$$\frac{\partial}{\partial y}\left(x^m y^n(y-y^2)\right)=\frac{\partial}{\partial x}\left(x^{m+1}y^n\right)$$

or

$$nx^m y^{n-1}(y-y^2)+x^m y^n(1-2y)=(m+1)x^m y^n,$$

or

$$(n+1)x^m y^n-(n+2)x^m y^{n+1}=(m+1)x^m y^n,$$

whereupon

$$n+1=m+1, \qquad n+2=0,$$

i.e.,

$$m=n=-2.$$

Consequently

$$\mu(x,y)=\frac{1}{x^2y^2}.$$

Then the considered equation takes the form

$$\frac{1-y}{x^2y}dx+\frac{1}{xy^2}dy=0,$$

which is an exact equation. Therefore there exists a function $u(x,y)$ *such that*

$$u_x(x,y)=\frac{1-y}{x^2y} \qquad (52)$$

and

$$u_y(x,y)=\frac{1}{xy^2}, \qquad (53)$$

from where, after integrating (53) *with respect to the variable y, we have*

$$u(x,y)=-\frac{1}{xy}+f(x),$$

where f *is an arbitrary function of x and plays the role of the constant of integration. We differentiate the last equality with respect to the variable x and using* (52), *we obtain*

$$\frac{1}{x^2y}+f'(x)=\frac{1}{x^2y}-\frac{1}{x^2}$$

or

$$f'(x) = -\frac{1}{x^2}.$$

From here, after we integrate with respect to the variable x, we get

$$f(x) = \frac{1}{x} + C_0$$

for some constant C_0. Hence,

$$u(x,y) = \frac{1}{x} - \frac{1}{xy} + C_0 = \frac{y-1}{xy} + C_0$$

and a solution is given by

$$y(1-Cx) = 1$$

for some constant C.

One may also look for an integrating factor of the form $\mu = \mu(v)$, where v is an function of x and y. Then (49) takes the form

$$\frac{1}{\mu}\frac{d\mu}{dv} = \frac{M_y(x,y)\left(1 - x\frac{\hat{T}'(x)}{\hat{T}(x)}\right) - N(x,y)\hat{T}'(x) - yN_y(x,y)\hat{T}(x) - N_x(x,y)}{N(x,y)v_x - \left(M(x,y)\left(1 - x\frac{\hat{T}'(x)}{\hat{T}(x)}\right) - yN(x,y)\hat{T}'(x)\right)v_y}. \tag{54}$$

Thus if the expression on the right side of (54) is a function of v alone, say, $\phi(v)$ then the integrating factor is given by

$$\mu = e^{\int \phi(v)dv}. \tag{55}$$

Below, some special classes of v and the corresponding function $\phi(v)$ are given.

1. $v = x$,

$$\phi(x) = \frac{M_y(x,y)\left(1 - x\frac{\hat{T}'(x)}{\hat{T}(x)}\right) - N(x,y)\hat{T}'(x) - yN_y(x,y)\hat{T}(x) - N_x(x,y)}{N(x,y)}.$$

2. $v = y$,

$$\phi(v) = \frac{M_y(x,y)\left(1 - x\frac{\hat{T}'(x)}{\hat{T}(x)}\right) - N(x,y)\hat{T}'(x) - yN_y(x,y)\hat{T}(x) - N_x(x,y)}{-M(x,y)\left(1 - x\frac{\hat{T}'(x)}{\hat{T}(x)}\right) + yN(x,y)\hat{T}'(x)}.$$

3. $v = x - y$,

$$\phi(v) = \frac{M_y(x,y)\left(1 - x\frac{\hat{T}'(x)}{\hat{T}(x)}\right) - N(x,y)\hat{T}'(x) - yN_y(x,y)\hat{T}(x) - N_x(x,y)}{N(x,y) + \left(M(x,y)\left(1 - x\frac{\hat{T}'(x)}{\hat{T}(x)}\right) - yN(x,y)\hat{T}'(x)\right)}.$$

4. $v = xy$,

$$\phi(v) = \frac{M_y(x,y)\left(1 - x\frac{\hat{T}'(x)}{\hat{T}(x)}\right) - N(x,y)\hat{T}'(x) - yN_y(x,y)\hat{T}(x) - N_x(x,y)}{yN(x,y) - \left(M(x,y)\left(x - x^2\frac{\hat{T}'(x)}{\hat{T}(x)}\right) - xyN(x,y)\hat{T}'(x)\right)}.$$

5. $v = \frac{x}{y}$,

$$\phi(v) = y^2 \frac{M_y(x,y)\left(1 - x\frac{\hat{T}'(x)}{\hat{T}(x)}\right) - N(x,y)\hat{T}'(x) - yN_y(x,y)\hat{T}(x) - N_x(x,y)}{yN(x,y) + \left(M(x,y)\left(x - x^2\frac{\hat{T}'(x)}{\hat{T}(x)}\right) - xyN(x,y)\hat{T}'(x)\right)}.$$

6. $v = x^2 + y^2$,

$$\phi(v) = \frac{M_y(x,y)\left(1 - x\frac{\hat{T}'(x)}{\hat{T}(x)}\right) - N(x,y)\hat{T}'(x) - yN_y(x,y)\hat{T}(x) - N_x(x,y)}{2xN(x,y) - 2y\left(M(x,y)\left(1 - x\frac{\hat{T}'(x)}{\hat{T}(x)}\right) - yN(x,y)\hat{T}'(x)\right)}.$$

If in the above equalities the right hand is a function of v, then the equation (1) has integrating factor μ given by (55).

Lemma 1.0.33. *Suppose* (1) *is exact and has an integrating factor* $\mu(x,y)$ ($\neq$ *const*), *then* $\mu(x,y) = C$ *is a solution to the equation* (1).

Proof. In view of the hypothesis, the condition (49) implies that

$$N(x,y)\mu_x(x,y) - \mu_y(x,y)\left(M(x,y)\left(1 - x\frac{\hat{T}'(x)}{\hat{T}(x)}\right) - yN(x,y)\hat{T}'(x)\right) = 0.$$

Multiplying the equation (1) with $\mu_y(x,y)$ we get

$$\left(M(x,y)\mu_y(x,y)\left(1 - x\frac{\hat{T}'(x)}{\hat{T}(x)}\right) - y\mu_y(x,y)N(x,y)\frac{\hat{T}'(x)}{\hat{T}(x)}\right)dx + N(x,y)\mu_y(x,y)dy$$

$$= N(x,y)(\mu_x(x,y) + \mu_y(x,y)dy)$$

$$= N(x,y)d\mu(x,y) = 0$$

and this implies the lemma. □

Theorem 1.0.34. *If* $\mu_1(x,y)$ *and* $\mu_2(x,y)$ *are two integrating factors of the iso-differential equation* (1) *such that their ratio is not a constant, then* $\mu_1(x,y) = C\mu_2(x,y)$ *is a solution of* (1).

Proof. Clearly, the iso-differential equations

$$\left(\mu_1(x,y)M(x,y)\left(1 - x\frac{\hat{T}'(x)}{\hat{T}(x)}\right) - y\mu_1(x,y)N(x,y)\frac{\hat{T}'(x)}{\hat{T}(x)}\right)dx + \mu_1(x,y)N(x,y)dy = 0, \tag{56}$$

$$\left(\mu_2(x,y)M(x,y)\left(1-x\frac{\hat{T}'(x)}{\hat{T}(x)}\right)-y\mu_2(x,y)N(x,y)\frac{\hat{T}'(x)}{\hat{T}(x)}\right)dx+\mu_2(x,y)N(x,y)dy=0, \tag{57}$$

are exact.

Multiplying (57) by $\frac{\mu_1(x,y)}{\mu_2(x,y)}$ converts it to the exact equation (56) and the last lemma implies that $\frac{\mu_1(x,y)}{\mu_2(x,y)} = C$ is a solution of (57), i.e., of (1). □

Exercise 1.0.35. *Let*

$$S = \{(x,y) \in \mathbb{R}^2 : |x-7| \le 2, \qquad |y-10| \le 3\},$$

$\hat{T}(x) = e^x$, $M(x,y) = \frac{x^3y+x^2y+y+xy}{1-x}$, $N(x,y) = x+x^3$, $(x,y) \in S$. *Determine the equation* (1)*, find an its integrating factor and an its solution.*

Answer. The equation (1) is

$$(x^2y+y+1)dx+(x+x^3)dy=0,$$

an its integrating factor is

$$\mu(x,y) = \frac{1}{1+x^2},$$

an its solution is

$$xy+\tan^{-1}(x) = C$$

for some constant C.

Exercise 1.0.36. *Suppose that the equation* (19) *is an exact equation. Prove that* $\hat{T}(x)$ *is an integrating factor for the equation* (1).

Definition 1.0.37. *For the equation* (19) *a non-zero function* $\mu(x,y)$ *is called an integrating factor if the equivalent equation*

$$\mu(x,y)\Big(M(x,y)(\hat{T}(x)-x\hat{T}'(x))-y\hat{T}'(x)N(x,y)\Big)dx+\mu(x,y)\hat{T}(x)N(x,y)dy=0 \tag{57'}$$

is an exact equation.

If $u(x,y) = C$, for some constant C, is a solution to the equation (19), then y' computed from the equation (19) and (46′) must be the same, i.e.,

$$y' = -\frac{u_x(x,y)}{u_y(x,y)} = -\frac{M(x,y)(\hat{T}(x)-x\hat{T}'(x))-y\hat{T}'(x)N(x,y)}{\hat{T}(x)N(x,y)}$$

or

$$\frac{u_x(x,y)}{M(x,y)(\hat{T}(x)-x\hat{T}'(x))-y\hat{T}'(x)N(x,y)} = \frac{u_y(x,y)}{\hat{T}(x)N(x,y)} = \mu(x,y), \tag{58}$$

where $\mu(x,y)$ is some function of x and y. Thus we have to have

$$\mu(x,y)\Big(M(x,y)(\hat{T}(x)-x\hat{T}'(x))-y\hat{T}'(x)N(x,y)\Big)+\mu(x,y)\hat{T}(x)N(x,y)y'$$
$$= u_x(x,y)+u_y(x,y)y' = \frac{du(x,y)}{dx} \tag{59}$$

and from here the equation (57′) is an exact equation, and an integrating factor $\mu(x,y)$ of the equation (19) is given by (58).

Theorem 1.0.38. *If the iso-differential equation* (19) *has* $u(x,y) = C$, *C is a constant, as its solution, then it admits an infinite number of integrating factors.*

Proof. Let $\phi(u)$ be any continuous function of u. Since $u(x,y) = C$ we have (59). From here,

$$\mu(x,y)\phi(u)\Big(M(x,y)(\hat{T}(x) - x\hat{T}'(x)) - y\hat{T}'(x)N(x,y)\Big) + \mu(x,y)\phi(u)\hat{T}(x)N(x,y)$$

$$= \phi(u)\frac{du(x,y)}{dx}$$

$$= \frac{d}{dx}\int_0^u \phi(s)ds.$$

Hence, $\mu(x,y)\phi(u)$ is an integrating factor of (19). Since ϕ is an arbitrary function, then we have established the result. □

The function $\mu(x,y)$ is an integrating factor of (19) provided (57) is exact, i.e., if and only if

$$\frac{\partial}{\partial y}\Big(\mu(x,y)\Big(M(x,y)(\hat{T}(x) - x\hat{T}'(x)) - y\hat{T}'(x)N(x,y)\Big)\Big) = \frac{\partial}{\partial x}\Big(\mu(x,y)\hat{T}(x)N(x,y)\Big).$$

This implies that an integrating factor of (19) must satisfy the equation

$$\mu_x(x,y)\hat{T}(x)N(x,y) - \mu_y(x,y)\Big(M(x,y)\Big(\hat{T}(x) - x\hat{T}'(x)\Big) - y\hat{T}'(x)N(x,y)\Big)$$

$$= \mu(x,y)\Big(M_y(x,y)\Big(\hat{T}(x) - x\hat{T}'(x)\Big) - \hat{T}'(x)N(x,y) - y\hat{T}'(x)N_y(x,y)$$

$$-\hat{T}'(x)N(x,y) - \hat{T}(x)N_x(x,y)\Big) \tag{60}$$

$$= \mu(x,y)\Big(M_y(x,y)\Big(\hat{T}(x) - x\hat{T}'(x)\Big) - 2\hat{T}'(x)N(x,y)$$

$$-y\hat{T}'(x)N_y(x,y) - \hat{T}(x)N_x(x,y)\Big).$$

A solution of (60) gives an integrating factor of (19), but it is not easy to be found a solution of the partial differential equation (60). However, a particular non-zero solution of (60) is all we need for the solution of (19).

If we assume that $\mu(x,y) = X(x)Y(y)$, then from (60) we have

$$\hat{T}(x)N(x,y)\frac{1}{X(x)}\frac{dX(x)}{dx} - \Big(M(x,y)\Big(\hat{T}(x) - x\hat{T}'(x)\Big) - y\hat{T}'(x)N(x,y)\Big)\frac{1}{Y(y)}\frac{dY(y)}{dy}$$

$$= M_y(x,y)\Big(\hat{T}(x) - x\hat{T}'(x)\Big) - 2\hat{T}'(x)N(x,y) - y\hat{T}'(x)N_y(x,y) - \hat{T}(x)N_x(x,y). \tag{61}$$

Hence,

$$\hat{T}(x)N(x,y)g(x) - \Big(M(x,y)\Big(\hat{T}(x) - x\hat{T}'(x)\Big) - y\hat{T}'(x)N(x,y)\Big)h(y)$$

$$= M_y(x,y)\Big(\hat{T}(x) - x\hat{T}'(x)\Big) - 2\hat{T}'(x)N(x,y) - y\hat{T}'(x)N_y(x,y) - \hat{T}(x)N_x(x,y),$$

then (61) satisfied provided (50′) and (50″).

Example 1.0.39. *Let*

$$S=\{(x,y)\in\mathbb{R}^2 : |x-7|\le 1, \qquad |y-10|\le 3\},$$

$\hat{T}(x)=x^2+1$, $M(x,y)=\frac{4x^2y^2+y^2}{1-x^4}$, $N(x,y)=\frac{3xy}{2x^2+2}$. *Then*

$$\begin{aligned} &M(x,y)\left(\hat{T}(x)-x\hat{T}'(x)\right)-y\hat{T}'(x)N(x,y)\\ &=\tfrac{4x^2y^2+y^2}{1-x^4}(1-x^2)-2xy\tfrac{3xy}{2x^2+2}\\ &=\tfrac{4x^2y^2+y^2}{1+x^2}-\tfrac{3x^2y^2}{x^2+1}\\ &=\tfrac{x^2y^2+y^2}{1+x^2}\\ &=y^2,\\ &\hat{T}(x)N(x,y)=(x^2+1)\tfrac{3xy}{2x^2+2}=\tfrac{3}{2}xy. \end{aligned}$$

Then the equation (19) *takes the form*

$$y^2dx+\frac{3}{2}xydy=0. \tag{62}$$

Since

$$\frac{\partial}{\partial y}(y^2)=2y\neq\frac{\partial}{\partial x}\left(\frac{3}{2}xy\right)=\frac{3}{2}y,$$

then the equation (19) *is not an exact equation.*

We will search an integrating factor of the equation (62) *of the form* $\mu(x,y)=X(x)Y(y)$. *We multiply* (62) *with* $\mu(x,y)$ *and we get*

$$y^2X(x)Y(y)dx+\frac{3}{2}xyX(x)Y(y)dy=0.$$

We want

$$\begin{aligned} &\tfrac{\partial}{\partial y}\left(y^2X(x)Y(y)\right)=\tfrac{\partial}{\partial x}\left(\tfrac{3}{2}xyX(x)Y(y)\right) &&\iff\\ &2yX(x)Y(y)+y^2X(x)Y'(y)=\tfrac{3}{2}yX(x)Y(y)+\tfrac{3}{2}xyX'(x)Y(y) &&\Longrightarrow\\ &\tfrac{y}{2}+y^2\tfrac{1}{Y(y)}\tfrac{dY(y)}{dy}=\tfrac{3}{2}xy\tfrac{1}{X(x)}\tfrac{dX(x)}{dx} &&\Longrightarrow\\ &\tfrac{1}{2}+y\tfrac{1}{Y(y)}\tfrac{dY(y)}{dy}=\tfrac{3}{2}x\tfrac{1}{X(x)}\tfrac{dX(x)}{dx}. \end{aligned}$$

From here, it follows that there exists $\lambda \in \mathbb{R}$ such that

$$\frac{1}{2} + y\frac{1}{Y(y)}\frac{dY(y)}{dy} = \lambda,$$

$$\frac{3}{2}x\frac{1}{X(x)}\frac{dX(x)}{dx} = \lambda,$$

whereupon

$$X(x) = C_1|x|^{\frac{2}{3}\lambda}, \qquad Y(y) = C_2|y|^{\frac{2\lambda-1}{2}}.$$

In particular, for $\lambda = \frac{3}{2}$, we get an integrating factor

$$\mu(x,y) = xy.$$

One may also look for an integrating factor of the form $\mu = \mu(v)$, where v is known function of x and y, then (60) leads to

$$\frac{1}{\mu(v)}\frac{d\mu}{dv} = \frac{M_y(x,y)\left(\hat{T}(x)-x\hat{T}'(x)\right)-2\hat{T}'(x)N(x,y)-y\hat{T}'(x)N_y(x,y)-\hat{T}(x)N_x(x,y)}{v_x\hat{T}(x)N(x,y)-v_y\left(M(x,y)(\hat{T}(x)-x\hat{T}'(x))-y\hat{T}'(x)N(x,y)\right)}. \tag{63}$$

Thus, if the expression in (63) is a function of v alone, say, $\phi(v)$, then the integrating factor is given by (55).

Some special classes of v and the corresponding $\phi(v)$ are given in the following table.

1. $v = x$,

$$\phi(v) = \frac{M_y(x,y)\left(\hat{T}(x)-x\hat{T}'(x)\right)-2\hat{T}'(x)N(x,y)-y\hat{T}'(x)N_y(x,y)-\hat{T}(x)N_x(x,y)}{\hat{T}(x)N(x,y)}.$$

2. $v = y$,

$$\phi(v) = -\frac{M_y(x,y)\left(\hat{T}(x)-x\hat{T}'(x)\right)-2\hat{T}'(x)N(x,y)-y\hat{T}'(x)N_y(x,y)-\hat{T}(x)N_x(x,y)}{M(x,y)\left(\hat{T}(x)-x\hat{T}'(x)\right)-y\hat{T}'(x)N(x,y)}.$$

3. $v = x-y$,

$$\phi(v) = \frac{M_y(x,y)\left(\hat{T}(x)-x\hat{T}'(x)\right)-2\hat{T}'(x)N(x,y)-y\hat{T}'(x)N_y(x,y)-\hat{T}(x)N_x(x,y)}{\hat{T}(x)N(x,y)+M(x,y)\left(\hat{T}(x)-x\hat{T}'(x)\right)-y\hat{T}'(x)N(x,y)}.$$

4. $v = xy$,

$$\phi(v) = \frac{M_y(x,y)\left(\hat{T}(x)-x\hat{T}'(x)\right)-2\hat{T}'(x)N(x,y)-y\hat{T}'(x)N_y(x,y)-\hat{T}(x)N_x(x,y)}{y\hat{T}(x)N(x,y)-x\left(M(x,y)\left(\hat{T}(x)-x\hat{T}'(x)\right)-y\hat{T}'(x)N(x,y)\right)}.$$

5. $v = \frac{x}{y}$,

$$\phi(v) = y^2 \frac{M_y(x,y)\left(\hat{T}(x) - x\hat{T}'(x)\right) - 2\hat{T}'(x)N(x,y) - y\hat{T}'(x)N_y(x,y) - \hat{T}(x)N_x(x,y)}{y\hat{T}(x)N(x,y) - x\left(M(x,y)\left(\hat{T}(x) - x\hat{T}'(x)\right) - y\hat{T}'(x)N(x,y)\right)}.$$

6. $v = x^2 + y^2$,

$$\phi(v) = \frac{M_y(x,y)\left(\hat{T}(x) - x\hat{T}'(x)\right) - 2\hat{T}'(x)N(x,y) - y\hat{T}'(x)N_y(x,y) - \hat{T}(x)N_x(x,y)}{2x\hat{T}(x)N(x,y) - 2y\left(M(x,y)\left(\hat{T}(x) - x\hat{T}'(x)\right) - y\hat{T}'(x)N(x,y)\right)}.$$

If the expressions in the right hand are functions of v, then the equation (19) has integrating factors given by (55).

Lemma 1.0.40. *Suppose* (19) *is exact and has an integrating factor* $\mu(x,y)(\neq$ *const), then* $\mu(x,y) = C$, *C is a constant, is a solution of the equation* (19).

Proof. In view of hypothesis, the condition (60) implies

$$\mu_x(x,y)\hat{T}(x)N(x,y) - \mu_y(x,y)\left(M(x,y)\left(\hat{T}(x) - x\hat{T}'(x)\right) - y\hat{T}'(x)N(x,y)\right) = 0.$$

We multiply the equation (19) with $\mu_y(x,y)$ and using the last equality, we get

$$0 = \mu_y(x,y)\left(M(x,y)\left(\hat{T}(x) - x\hat{T}'(x)\right) - y\hat{T}'(x)N(x,y)\right) + \mu_y(x,y)\hat{T}(x)N(x,y)y'$$

$$= \mu_x(x,y)\hat{T}(x)N(x,y) + \mu_y(x,y)\hat{T}(x)N(x,y)y'$$

$$= \hat{T}(x)N(x,y)\left(\mu_x(x,y) + \mu_y(x,y)y'\right)$$

$$= \hat{T}(x)N(x,y)\frac{d\mu}{dx},$$

i.e., $\mu(x,y) = C$, C is a constant, is a solution of (19). □

Theorem 1.0.41. *If* $\mu_1(x,y)$ *and* $\mu_2(x,y)$ *are two integrating factors of the equation* (19) *such that their ratio is not a constant, then* $\mu_1(x,y) = C\mu_2(x,y)$ *is a solution of* (19).

Proof. We have that the equations

$$\mu_1(x,y)\left(M(x,y)\left(\hat{T}(x) - x\hat{T}'(x)\right) - y\hat{T}'(x)N(x,y)\right) + \hat{T}(x)\mu_1(x,y)N(x,y)y' = 0, \qquad (63')$$

$$\mu_2(x,y)\left(M(x,y)\left(\hat{T}(x) - x\hat{T}'(x)\right) - y\hat{T}'(x)N(x,y)\right) + \hat{T}(x)\mu_2(x,y)N(x,y)y' = 0, \qquad (64)$$

are exact. Multiplying the equation $(63')$ with $\frac{\mu_2(x,y)}{\mu_1(x,y)}$ we obtain the exact equation (64). From here and the last Lemma it follows that $\mu_1(x,y) = C\mu_2(x,y)$, C is a constant, is a solution of $(63')$, i.e., of (19). □

Exercise 1.0.42. *Suppose that* (1) *is an exact equation. Prove that* $\frac{1}{\hat{T}(x)}$ *is an integrating factor of* (19).

Definition 1.0.43. *For the equation* (27) *a non-zero function* $\mu(x,y)$ *is called an integrating factor if the equivalent iso-differential equation*

$$\mu(x,y)M(x,y)\left(\hat{T}(x)-x\hat{T}'(x)\right)dx+\mu(x,y)N(x,y)\hat{T}^2(x)dy=0 \tag{65}$$

is an exact equation.

If $u(x,y)=C$, C is a constant, is a solution to the equation (27), then y' computed from (27) and (46′) must be the same, i.e.,

$$y'=-\frac{u_x(x,y)}{u_y(x,y)}=-\frac{M(x,y)\left(\hat{T}(x)-x\hat{T}'(x)\right)}{N(x,y)\hat{T}^2(x)}$$

or

$$\frac{u_x(x,y)}{M(x,y)\left(\hat{T}(x)-x\hat{T}'(x)\right)}=\frac{u_y(x,y)}{N(x,y)\hat{T}^2(x)}=\mu(x,y), \tag{66}$$

where $\mu(x,y)$ is a function of x and y. Thus we have

$$\begin{aligned}
&\mu(x,y)\left(M(x,y)\left(\hat{T}(x)-x\hat{T}'(x)\right)+N(x,y)\hat{T}^2(x)y'\right)\\
&=\mu(x,y)\left(\tfrac{u_x(x,y)}{\mu(x,y)}+\tfrac{u_y(x,y)}{\mu(x,y)}y'\right)\\
&=u_x(x,y)+u_y(x,y)y'\\
&=\tfrac{du(x,y)}{dx},
\end{aligned} \tag{67}$$

from here, the equation (65) is an exact equation and its integrating factor is given by (66).

Theorem 1.0.44. *If the iso-differential equation* (27) *has* $u(x,y)=C$, C *is a constant, as its solution, then it admits an infinite number of integrating factors.*

Proof. Let $\phi(u)$ be any continuous function of u. Since $u(x,y)=C$ we have the equation (67). From here,

$$\begin{aligned}
&\mu(x,y)\phi(u)\left(M(x,y)\left(\hat{T}(x)-x\hat{T}'(x)\right)+M(x,y)\hat{T}^2(x)y'\right)\\
&=\phi(u)\tfrac{du(x,y)}{dx}\\
&=\tfrac{d}{dx}\int_0\phi(s)ds.
\end{aligned}$$

Hence, $\mu(x,y)\phi(u)$ is an integrating factor of the equation (27) and since $\phi(u)$ is an arbitrary continuous function of u, we have established the result. □

The function $\mu(x,y)$ is an integrating factor of (27) provided (65) is exact, i.e., if and only if

$$\frac{\partial}{\partial y}\left(\mu(x,y)M(x,y)\left(\hat{T}(x)-x\hat{T}'(x)\right)\right)=\frac{\partial}{\partial x}\left(\mu(x,y)N(x,y)\hat{T}^2(x)\right)\quad\Longleftrightarrow$$

$$\mu_y(x,y)M(x,y)\left(\hat{T}(x)-x\hat{T}'(x)\right)+\mu(x,y)M_y(x,y)\left(\hat{T}(x)-x\hat{T}'(x)\right)$$

$$=\mu_x(x,y)N(x,y)\hat{T}^2(x)+\mu(x,y)N_x(x,y)\hat{T}^2(x)+2\mu(x,y)N(x,y)\hat{T}(x)\hat{T}'(x).$$

Thus implies that an integrating factor of (27) must satisfy the equation

$$\begin{aligned}&\mu_x(x,y)N(x,y)\hat{T}^2(x)-\mu_y(x,y)M(x,y)\left(\hat{T}(x)-x\hat{T}'(x)\right)\\&=\mu(x,y)\left(M_y(x,y)\left(\hat{T}(x)-x\hat{T}'(x)\right)-N_x(x,y)\hat{T}^2(x)-2N(x,y)\hat{T}(x)\hat{T}'(x)\right).\end{aligned}\tag{68}$$

A solution of (68) gives an integrating factor of (27), but it is not easy to be found a solution of the partial differential equation (68). However, a particular non-zero solution of (68) is all we need for the solution of (27).

If we assume that $\mu(x,y)=X(x)Y(y)$, then from (68) we have

$$\begin{aligned}&N(x,y)\hat{T}^2(x)\tfrac{1}{X(x)}\tfrac{dX(x)}{dx}-M(x,y)\left(\hat{T}(x)-x\hat{T}'(x)\right)\tfrac{1}{Y(y)}\tfrac{dY(y)}{dy}\\&=M_y(x,y)\left(\hat{T}(x)-x\hat{T}'(x)\right)-N_x(x,y)\hat{T}^2(x)-2N(x,y)\hat{T}(x)\hat{T}'(x).\end{aligned}\tag{69}$$

Hence, if

$$N(x,y)\hat{T}^2(x)g(x)-M(x,y)\left(\hat{T}(x)-x\hat{T}'(x)\right)h(y)$$

$$=M_y(x,y)\left(\hat{T}(x)-x\hat{T}'(x)\right)-N_x(x,y)\hat{T}^2(x)-2N(x,y)\hat{T}(x)\hat{T}'(x),$$

then (69) is satisfied provided (50′) and (50″).

Example 1.0.45. *Let*

$$S=\{(x,y)\in\mathbb{R}^2:|x-10|\le 1,\qquad |y-6|\le 2\},$$

$\hat{T}(x)=x^2+1$, $M(x,y)=\frac{1}{1-x^2}$, $N(x,y)=\frac{x}{(x^2+1)^2}$, $(x,y)\in S$. *Then*

$$M(x,y)\left(\hat{T}(x)-x\hat{T}'(x)\right)=\tfrac{1}{1-x^2}(1+x^2-2x^2)$$

$$=\tfrac{1}{1-x^2}(1-x^2)$$

$$=1,$$

$$N(x,y)\hat{T}^2(x)=\tfrac{x}{(x^2+1)^2}(x^2+1)^2$$

$$=x,$$

and the equation (27) *takes the form*

$$dx + xdy = 0. \tag{70}$$

Since

$$\frac{\partial}{\partial y}(1) = 0 \neq 1 = \frac{\partial}{\partial x}(x),$$

then it is not an exact equation. We will search an integrating factor of the form $\mu(x,y) = X(x)Y(y)$. *We multiply* (70) *with* $\mu(x,y)$ *and we get*

$$X(x)Y(y)dx + xX(x)Y(y)dy = 0.$$

We want

$$\frac{\partial}{\partial y}\Big(X(x)Y(y)\Big) = \frac{\partial}{\partial x}\Big(xX(x)Y(y)\Big) \iff$$

$$X(x)Y'(y) = X(x)Y(y) + xX'(x)Y(y) \implies$$

$$\frac{1}{Y(y)}\frac{dY(y)}{dy} = 1 + x\frac{1}{X(x)}\frac{dX(x)}{dx},$$

from where it follows that there exists $\lambda \in \mathbb{R}$ *such that*

$$1 + x\frac{1}{X(x)}\frac{dX(x)}{dx} = \lambda, \qquad \frac{1}{Y(y)}\frac{dY(y)}{dy} = \lambda,$$

whereupon

$$X(x) = C_1|x|^{\lambda-1}, \qquad Y(y) = C_2e^{\lambda y},$$

C_1 *and* C_2 *are constant. In particular, for* $\lambda = 2$, *we have* $\mu(x,y) = xe^{2y}$ *is an integrating factor of* (70).

One may also look for an integrating factor of the form $\mu = \mu(v)$, where v is known function of x and y. Then

$$\frac{1}{\mu(v)}\frac{d\mu(v)}{dv} = \frac{M_y(x,y)\Big(\hat{T}(x) - x\hat{T}'(x)\Big) - N_x(x,y)\hat{T}^2(x) - 2N(x,y)\hat{T}(x)\hat{T}'(x)}{v_xN(x,y)\hat{T}^2(x) - v_yM(x,y)\Big(\hat{T}(x) - x\hat{T}'(x)\Big)}. \tag{71}$$

Thus, if the expression in the right side of (71) is a function of v alone, say $\phi(v)$, then the integrating factor is given by (55).

Some special classes of v and the corresponding $\phi(v)$ are given in the following table.

1. $v = x$,

$$\phi(v) = \frac{M_y(x,y)\Big(\hat{T}(x) - x\hat{T}'(x)\Big) - N_x(x,y)\hat{T}^2(x) - 2N(x,y)\hat{T}(x)\hat{T}'(x)}{N(x,y)\hat{T}^2(x)}.$$

2. $v = y$,

$$\phi(v) = -\frac{M_y(x,y)\Big(\hat{T}(x) - x\hat{T}'(x)\Big) - N_x(x,y)\hat{T}^2(x) - 2N(x,y)\hat{T}(x)\hat{T}'(x)}{M(x,y)\Big(\hat{T}(x) - x\hat{T}'(x)\Big)}.$$

3. $v = x - y$,

$$\phi(v) = \frac{M_y(x,y)\left(\hat{T}(x) - x\hat{T}'(x)\right) - N_x(x,y)\hat{T}^2(x) - 2N(x,y)\hat{T}(x)\hat{T}'(x)}{N(x,y)\hat{T}^2(x) + M(x,y)\left(\hat{T}(x) - x\hat{T}'(x)\right)}.$$

4. $v = xy$,

$$\phi(v) = \frac{M_y(x,y)\left(\hat{T}(x) - x\hat{T}'(x)\right) - N_x(x,y)\hat{T}^2(x) - 2N(x,y)\hat{T}(x)\hat{T}'(x)}{yN(x,y)\hat{T}^2(x) - xM(x,y)\left(\hat{T}(x) - x\hat{T}'(x)\right)}.$$

5. $v = \frac{x}{y}$,

$$\phi(v) = y^2 \frac{M_y(x,y)\left(\hat{T}(x) - x\hat{T}'(x)\right) - N_x(x,y)\hat{T}^2(x) - 2N(x,y)\hat{T}(x)\hat{T}'(x)}{yN(x,y)\hat{T}^2(x) + xM(x,y)\left(\hat{T}(x) - x\hat{T}'(x)\right)}.$$

6. $v = x^2 + y^2$,

$$\phi(v) = \frac{M_y(x,y)\left(\hat{T}(x) - x\hat{T}'(x)\right) - N_x(x,y)\hat{T}^2(x) - 2N(x,y)\hat{T}(x)\hat{T}'(x)}{2xN(x,y)\hat{T}^2(x) - 2yM(x,y)\left(\hat{T}(x) - x\hat{T}'(x)\right)}.$$

If the expressions of the right hand are functions of v alone, then (27) has integrating factors given by (55).

Lemma 1.0.46. *Suppose* (27) *is exact and has an integrating factor* $\mu(x,y)$ *(*$\neq$ *const), then* $\mu(x,y) = C$*, C is a constant, is a solution to the equation* (27).

Proof. In view of the hypothesis, the condition (68) implies to

$$\mu_x(x,y)N(x,y)\hat{T}^2(x) - \mu_y(x,y)M(x,y)\left(\hat{T}(x) - x\hat{T}'(x)\right) = 0$$

or

$$\mu_y(x,y) = \frac{\mu_x(x,y)N(x,y)\hat{T}^2(x)}{M(x,y)\left(\hat{T}(x) - x\hat{T}'(x)\right)}.$$

Multiplying the equation (27) with $\mu_y(x,y)$, we find

$$0 = \mu_y(x,y)M(x,y)\left(\hat{T}(x) - x\hat{T}'(x)\right)dx + N(x,y)\hat{T}^2(x)\mu_y(x,y)dy$$

$$= \mu_x(x,y)N(x,y)\hat{T}^2(x)dx + N(x,y)\hat{T}^2(x)\mu_y(x,y)dy$$

$$= N(x,y)\hat{T}^2(x)\left(\mu_x(x,y)dx + \mu_y(x,y)dy\right)$$

$$= N(x,y)\hat{T}^2(x)d\mu(x,y),$$

and this implies the lemma. □

Theorem 1.0.47. *If $\mu_1(x,y)$ and $\mu_2(x,y)$ are two integrating factors of (27) such that their ration is not a constant, then $\mu_1(x,y) = C\mu_2(x,y)$, C is a constant, is a solution of (27).*

Proof. Clearly, the iso-differential equations

$$\mu_1(x,y)M(x,y)\left(\hat{T}(x) - x\hat{T}'(x)\right)dx + \mu_1(x,y)N(x,y)\hat{T}^2(x)dy = 0, \tag{72}$$

$$\mu_2(x,y)M(x,y)\left(\hat{T}(x) - x\hat{T}'(x)\right)dx + \mu_2(x,y)N(x,y)\hat{T}^2(x)dy = 0, \tag{73}$$

are exact. Multiplying (73) by $\frac{\mu_1(x,y)}{\mu_2(x,y)}$ converts it to the exact equation (72). Thus, the exact equation (73) admits an integrating factor $\frac{\mu_1(x,y)}{\mu_2(x,y)}$. From here and the last Lemma, it follows that $\mu_1(x,y) = C\mu_2(x,y)$ is a solution of (73), i.e., of (27), C is an arbitrary constant. □

Exercise 1.0.48. *Suppose that (37) is an exact equation. Prove that $\frac{1}{\hat{T}(x)}$ is an integrating factor of (27).*

Definition 1.0.49. *For the equation (37) a non-zero function $\mu(x,y)$ is called an integrating factor if the equivalent iso-differential equation*

$$\mu(x,y)M(x,y)\left(1 - x\frac{\hat{T}'(x)}{\hat{T}(x)}\right)dx + \mu(x,y)N(x,y)\hat{T}(x)dy = 0 \tag{74}$$

is an exact equation.

If $u(x,y) = C$, C is a constant, is a solution of the equation (37), then y' computed from (37) and (46′) must be the same, i.e.,

$$y' = -\frac{u_x(x,y)}{u_y(x,y)} = -\frac{M(x,y)\left(1 - x\frac{\hat{T}'(x)}{\hat{T}(x)}\right)}{N(x,y)\hat{T}(x)}$$

or

$$\frac{u_x(x,y)}{M(x,y)\left(1 - x\frac{\hat{T}'(x)}{\hat{T}(x)}\right)} = \frac{u_y(x,y)}{N(x,y)\hat{T}(x)} = \mu(x,y) \tag{75}$$

for some function $\mu(x,y)$. Thus we have

$$\begin{aligned} &\mu(x,y)M(x,y)\left(1 - x\frac{\hat{T}'(x)}{\hat{T}(x)}\right)dx + \mu(x,y)N(x,y)\hat{T}(x)dy \\ &= u_x(x,y)dx + u_y(x,y)dy \\ &= du(x,y) \end{aligned} \tag{76}$$

and from here, the equation (74) is an exact equation and the integrating factor of (37) is given by (75).

Theorem 1.0.50. *If the iso-differential equation (37) has $u(x,y) = C$, C is a constant, as its solution, then it admits an infinite number of integrating factors.*

Proof. Let $\phi(u)$ be an arbitrary continuous function of u. Since $u(x,y)=C$ we have (76). From here

$$\mu(x,y)\phi(u)\left(M(x,y)\left(1-x\frac{\hat{T}'(x)}{\hat{T}(x)}\right)+N(x,y)\hat{T}(x)y'\right)$$

$$=\phi(u)\frac{du(x,y)}{dx}$$

$$=\frac{d}{dx}\int_0 \phi(s)ds.$$

Hence, $\mu(x,y)\phi(u)$ is an integrating factor of (37). Since $\phi(u)$ is an arbitrary continuous function, we have established the theorem. □

The function $\mu(x,y)$ is an integrating factor of (37) provided (74) is exact, i.e., if and only if

$$\frac{\partial}{\partial y}\left(\mu(x,y)M(x,y)\left(1-x\frac{\hat{T}'(x)}{\hat{T}(x)}\right)\right)=\frac{\partial}{\partial x}\left(\mu(x,y)N(x,y)\hat{T}(x)\right) \iff$$

$$\mu_y(x,y)M(x,y)\left(1-x\frac{\hat{T}'(x)}{\hat{T}(x)}\right)+\mu(x,y)M_y(x,y)\left(1-x\frac{\hat{T}'(x)}{\hat{T}(x)}\right)$$

$$=\mu_x(x,y)N(x,y)\hat{T}(x)+\mu(x,y)N_x(x,y)\hat{T}(x)+\mu(x,y)N(x,y)\hat{T}'(x).$$

This implies that an integrating factor of (37) must satisfy the equation

$$\begin{aligned}&\mu_x(x,y)N(x,y)\hat{T}(x)-\mu_y(x,y)M(x,y)\left(1-x\frac{\hat{T}'(x)}{\hat{T}(x)}\right)\\&=\mu(x,y)\left(M_y(x,y)\left(1-x\frac{\hat{T}'(x)}{\hat{T}(x)}\right)-N_x(x,y)\hat{T}(x)-N(x,y)\hat{T}'(x)\right).\end{aligned} \tag{77}$$

A solution of (77) gives an integrating factor of (37), but it is not easy to be found a solution to the partial differential equation (77). However, a particular non-zero solution of (77) is all we need for the solution of (37).

If we assume that $\mu(x,y)=X(x)Y(y)$, then from (77) we have

$$\begin{aligned}&\frac{1}{X(x)}\frac{dX(x)}{dx}N(x,y)\hat{T}(x)-\frac{1}{Y(y)}\frac{dY(y)}{dy}M(x,y)\left(1-x\frac{\hat{T}'(x)}{\hat{T}(x)}\right)\\&=M_y(x,y)\left(1-x\frac{\hat{T}'(x)}{\hat{T}(x)}\right)-N_x(x,y)\hat{T}(x)-N(x,y)\hat{T}'(x),\end{aligned} \tag{78}$$

then (78) is satisfied provided (50′) and (50″).

Example 1.0.51. *Let*

$$S=\{(x,y)\in\mathbb{R}^2:|x-10|\leq 5, \qquad |y-10|\leq 1\},$$

$\hat{T}(x) = e^x$, $M(x,y) = \frac{y}{1-x}e^{-y}$, $N(x,y) = \frac{x^2+2y}{2x}e^{-x-y}$, $(x,y) \in S$. *Then*

$$M(x,y)\left(1 - x\frac{\hat{T}'(x)}{\hat{T}(x)}\right)$$

$$= \frac{y}{1-x}e^{-y}(1-x)$$

$$= ye^{-y},$$

$$N(x,y)\hat{T}(x) = \frac{x^2+2y}{2x}e^{-x-y}e^x$$

$$= \frac{x^2+2y}{2x}e^{-y},$$

the equation (37) *takes the form*

$$ye^{-y}dx + \frac{x^2+2y}{2x}e^{-y}dy = 0. \tag{79}$$

Since

$$\frac{\partial}{\partial y}\left(ye^{-y}\right) = (1-y)e^{-y}$$

$$\neq \frac{\partial}{\partial x}\left(\left(\frac{x^2+y}{2x}\right)e^{-y}\right) = \left(\frac{1}{2} - \frac{y}{x^2}\right)e^{-y},$$

then it is not an exact equation. We will search an integrating factor of (79) *of the form* $\mu(x,y) = X(x)Y(y)$. *We multiply* (79) *with* $\mu(x,y)$ *and we find*

$$ye^{-y}X(x)Y(y)dx + \frac{x^2+2y}{2x}e^{-y}X(x)Y(y)dy = 0.$$

We want

$$\frac{\partial}{\partial y}\left(ye^{-y}X(x)Y(y)\right) = \frac{\partial}{\partial x}\left(\frac{x^2+2y}{2x}e^{-y}X(x)Y(y)\right) \iff$$

$$(1-y)e^{-y}X(x)Y(y) + ye^{-y}X(x)Y'(y)$$

$$= \left(\frac{1}{2} - \frac{y}{x^2}\right)e^{-y}X(x)Y(y) + \frac{x^2+2y}{2x}e^{-y}X'(x)Y(y) \implies$$

$$\frac{x^2+2y}{2x^2}X(x)Y(y) - yX(x)Y(y) + yX(x)Y'(y) = \frac{x^2+2y}{2x}X'(x)Y(y),$$

and

$$\frac{x^2+2y}{2x^2}X(x)Y(y) = \frac{x^2+2y}{2x}X'(x)Y(y),$$

$$yX(x)Y(y) = yX(x)Y'(y),$$

whereupon

$$X(x) = C_1x, \qquad Y(y) = C_2e^y,$$

C_1 *and* C_2 *are constants. In particular,* $\mu(x,y) = xe^y$ *is an integrating factor of* (79).

One may also look for an integrating factor of the form $\mu = \mu(v)$, where v is known function of x and y. Then (77) leads to

$$\frac{1}{\mu(v)}\frac{d\mu}{dv} = \frac{M_y(x,y)\left(1 - x\frac{\hat{T}'(x)}{\hat{T}(x)}\right) - N_x(x,y)\hat{T}(x) - N(x,y)\hat{T}'(x)}{v_x N(x,y)\hat{T}(x) - v_y M(x,y)\left(1 - x\frac{\hat{T}'(x)}{\hat{T}(x)}\right)}. \tag{80}$$

Thus, if the expression in the right side of (80) is a function of v alone, say, $\phi(v)$, then the integrating factor is given by (55).

Some classes of v and the corresponding $\phi(v)$ are given in the following table.

1. $v = x$,

$$\phi(v) = \frac{M_y(x,y)\left(1 - x\frac{\hat{T}'(x)}{\hat{T}(x)}\right) - N_x(x,y)\hat{T}(x) - N(x,y)\hat{T}'(x)}{N(x,y)\hat{T}(x)}.$$

2. $v = y$,

$$\phi(v) = -\frac{M_y(x,y)\left(1 - x\frac{\hat{T}'(x)}{\hat{T}(x)}\right) - N_x(x,y)\hat{T}(x) - N(x,y)\hat{T}'(x)}{M(x,y)\left(1 - x\frac{\hat{T}'(x)}{\hat{T}(x)}\right)}.$$

3. $v = x - y$,

$$\phi(v) = \frac{M_y(x,y)\left(1 - x\frac{\hat{T}'(x)}{\hat{T}(x)}\right) - N_x(x,y)\hat{T}(x) - N(x,y)\hat{T}'(x)}{N(x,y)\hat{T}(x) + M(x,y)\left(1 - x\frac{\hat{T}'(x)}{\hat{T}(x)}\right)}.$$

4. $v = xy$,

$$\phi(v) = \frac{M_y(x,y)\left(1 - x\frac{\hat{T}'(x)}{\hat{T}(x)}\right) - N_x(x,y)\hat{T}(x) - N(x,y)\hat{T}'(x)}{yN(x,y)\hat{T}(x) - xM(x,y)\left(1 - x\frac{\hat{T}'(x)}{\hat{T}(x)}\right)}.$$

5. $v = \frac{x}{y}$,

$$\phi(v) = y^2\frac{M_y(x,y)\left(1 - x\frac{\hat{T}'(x)}{\hat{T}(x)}\right) - N_x(x,y)\hat{T}(x) - N(x,y)\hat{T}'(x)}{yN(x,y)\hat{T}(x) + xM(x,y)\left(1 - x\frac{\hat{T}'(x)}{\hat{T}(x)}\right)}.$$

6. $v = x^2 + y^2$,

$$\phi(v) = \frac{M_y(x,y)\left(1 - x\frac{\hat{T}'(x)}{\hat{T}(x)}\right) - N_x(x,y)\hat{T}(x) - N(x,y)\hat{T}'(x)}{2xN(x,y)\hat{T}(x) - 2yM(x,y)\left(1 - x\frac{\hat{T}'(x)}{\hat{T}(x)}\right)}.$$

If the expressions in the right hand are functions of v alone, then (37) has integrating factors given by (55).

Lemma 1.0.52. *Suppose* (37) *is exact and has an integrating factor* $\mu(x,y)$ ($\neq$ *const), then* $\mu(x,y) = C$, *C is a constant, is a solution to the equation* (37).

Proof. In view of hypothesis, the condition (77) implies that

$$\mu_x(x,y)N(x,y)\hat{T}(x)-\mu_y(x,y)M(x,y)\left(1-x\frac{\hat{T}'(x)}{\hat{T}(x)}\right)=0.$$

Multiplying the equation (37) by $\mu_y(x,y)$ we find

$$\begin{aligned}
&\mu_y(x,y)M(x,y)\left(1-x\frac{\hat{T}'(x)}{\hat{T}(x)}\right)dx+\mu_y(x,y)N(x,y)\hat{T}(x)dy\\
&=\mu_x(x,y)N(x,y)\hat{T}(x)dx+\mu_y(x,y)N(x,y)\hat{T}(x)dy\\
&=N(x,y)\hat{T}(x)\left(\mu_x(x,y)dx+\mu_y(x,y)dy\right)\\
&=N(x,y)\hat{T}(x)d\mu(x,y)\\
&=0
\end{aligned}$$

and this implies the Lemma. □

Theorem 1.0.53. *If $\mu_1(x,y)$ and $\mu_2(x,y)$ are two integrating factors of (37) such that their ratio is not a constant, then $\mu_1(x,y)=C\mu_2(x,y)$, C is a constant, is a solution of (37).*

Proof. Clearly, the iso-differential equations

$$\mu_1(x,y)M(x,y)\left(1-x\frac{\hat{T}'(x)}{\hat{T}(x)}\right)dx+\mu_1(x,y)N(x,y)\hat{T}(x)dy=0, \tag{81}$$

$$\mu_2(x,y)M(x,y)\left(1-x\frac{\hat{T}'(x)}{\hat{T}(x)}\right)dx+\mu_2(x,y)N(x,y)\hat{T}(x)dy=0, \tag{82}$$

are exact. Multiplying of (81) by $\frac{\mu_2(x,y)}{\mu_1(x,y)}$ converts it to the exact equation (82). In other words, the exact equation (81) admits an integrating factor $\frac{\mu_2(x,y)}{\mu_1(x,y)}$ and the last lemma implies that $\mu_1(x,y)=C\mu_2(x,y)$ is a solution of (81), i.e., of (37). □

Exercise 1.0.54. *Suppose that (27) is an exact equation. Prove that $\hat{T}(x)$ is an integrating factor of (37).*

Advanced Practical Exercises

Problem 1.0.55. *Let*

$$S=\left\{(x,y)\in\mathbb{R}^2:|x|\leq 1,\qquad |y|\leq 1\right\},$$

$\hat{T}(x)=x+3$, $M(x,y)=x$, $N(x,y)=y$. Determine the equation (1) and check if it is an exact equation. If it is an exact equation, find an its solution.

Answer.

$$\frac{x-y^2}{x+1}dx+ydy=0.$$

It is not an exact equation.

Problem 1.0.56. *Let*

$$S=\left\{(x,y)\in\mathbb{R}^2:|x|\leq\frac{1}{4},\qquad |y-1|\leq 1\right\},$$

$\hat{T}(x)=e^x$, $M(x,y)=\frac{x^2y+y^3+2xy}{1-x}$, $N(x,y)=x^2+y^2$, $(x,y)\in S$. *Determine the equation* (1) *and check if it is an exact equation. If it is an exact equation, find an its solution.*

Answer.

$$2xydx+(x^2+y^2)dy=0.$$

It is an exact equation.

A solution is

$$3x^2y+y^3=C,$$

where C is a constant.

Problem 1.0.57. *Let*

$$S=\left\{(x,y)\in\mathbb{R}^2:|x-a|\leq b,\qquad |y-c|\leq q\right\},$$

where a, b, c, $q\in[0,\infty)$, *such that* $a>b+1$, $c>q$. *Let also,* $\hat{T}(x)=e^x$ *and* $M(x,y)$, $M_y(x,y)$ *are continuous functions on S. Find a continuous function N, determined in S, so that* $N_y(x,y)$ *exists in S and it is a continuous function in S, and*

$$M_y(x,y)(1-x)=N_x(x,y),$$

$$N(x,y)=-yN_y(x,y),\qquad (x,y)\in S.$$

For such functions determine if the equation (1) *is an exact equation.*

Answer.

$$N(x,y)=\frac{f(x)}{y},$$

where f is a continuous function on $|x-a|\leq b$. The equation (1) is an exact equation.

Problem 1.0.58. *Let*

$$S=\{(x,y)\in\mathbb{R}^2:|x-1|\leq 2,\qquad |y-3|\leq 4\},$$

$\hat{T}(x)=e^x$, $M(x,y)=x-y$, $N(x,y)=xy$, $(x,y)\in S$. *Determine the equation* (19) *and check if it is an exact equation. If it is an exact equation, find an its solution.*

Answer.

$$(-x^3-2x^2y^2+x^2y+x-y)dx+(xy+x^3y)dy=0.$$

It is not an exact equation.

Problem 1.0.59. *Let*

$$S = \{(x,y) \in \mathbb{R}^2 : |x-4| \leq 1, \qquad |y-1| \leq 1\},$$

$\hat{T}(x) = e^x$, $M(x,y) = \frac{y+2xy-\frac{3}{2}y^2}{1-x}$, $N(x,y) = x-y$, $(x,y) \in S$. *Determine the equation* (19) *and check if it is an exact equation. If it is an exact equation, find an its solution.*

Answer.

$$e^x\left(-\frac{1}{2}y^2 + xy + y\right)dx + e^x(x-y)dy = 0.$$

It is an exact equation. An its solution is given by

$$2e^x xy - y^2 e^x = C,$$

where C is a constant.

Problem 1.0.60. *Let*

$$S = \{(x,y) \in \mathbb{R}^2 : |x| \leq 2, \qquad |y| \leq 3\},$$

$\hat{T}(x) = x^2+1$, $M(x,y) = xy$, $N(x,y) = x-2y$, $(x,y) \in S$. *Determine the equation* (27) *and check if it is an exact equation. If it is an exact equation, find an its solution.*

Answer.

$$(xy - x^3 y)dx + (x^5 - 2x^4 y + 2x^3 - 4x^2 y + x - 2y)dy = 0.$$

It is not an exact equation.

Problem 1.0.61. *Let*

$$S = \{(x,y) \in \mathbb{R}^2 : |x-5| \leq 3, |y| \leq 1\},$$

$\hat{T}(x) = e^{2x}$, $M(x,y) = e^{2x}\frac{y+4xy+2y^2}{1-2x}$, $N(x,y) = x+y$, $(x,y) \in S$. *Determine the equation* (27) *and check if it is an exact equation. If it is an exact equation, find an its solution.*

Answer.

$$e^{4x}(y + 4xy + 2y^2)dx + e^{4x}(x+y)dy = 0.$$

It is an exact equation. A solution is given by

$$e^{4x}(2xy + y^2) = C,$$

where C is a constant.

Problem 1.0.62. *Let*

$$S = \{(x,y) \in \mathbb{R}^2 : |x-2| \leq 2, \qquad |y| \leq 2\},$$

$\hat{T}(x) = e^x$, $M(x,y) = x^2 - y^2$, $N(x,y) = xy$, $(x,y) \in S$. *Determine the equation* (37) *and check if it is an exact equation. If it is an exact equation, find an its solution.*

Answer.

$$(-x^3+xy^2+x^2-y^2)dx+xye^x dy=0.$$

It is not an exact equation.

Problem 1.0.63. *Let*

$$S=\{(x,y)\in\mathbb{R}^2:|x-5|\le 1, \qquad |y|\le 3\},$$

$\hat{T}(x)=e^x$, $M(x,y)=e^x\frac{y^2}{2(1-x)}$, $N(x,y)=y$, $(x,y)\in S$. *Determine the equation* (37) *and check if it is an exact equation. If it is an exact equation, find an its solution.*

Answer.

$$\frac{y^2}{2}e^x dx+ye^x dy=0.$$

It is an exact equation. A solution is given by

$$y^2e^x=C$$

for some constant C.

Problem 1.0.64. *Let*

$$S=\{(x,y)\in\mathbb{R}^2:|x-10|\le 2, \qquad |y-4|\le 1\},$$

$\hat{T}(x)=e^x$, $M(x,y)=\frac{2y^3+2x^2y^2-y^2+x^2y^3+2x^3y^2-2x^2y}{1-x}$, $N(x,y)=x^2y^2+2x^3y-2x^2$, $(x,y)\in S$. *Determine the equation* (1), *find an its integrating factor and an its solution.*

Answer. The equation (1) takes the form

$$(xy^3+2x^2y^2-y^2)dx+(x^2y^2+2x^3y-2x^2)dy=0,$$

an its integrating factor is

$$\mu(x,y)=\frac{1}{x^2y^2}e^{xy},$$

an its solution is given by

$$e^{xy}(y+2x)=Cxy$$

for some constant C.

Problem 1.0.65. *Show that* $u(x,y)=C$, C *is a constant, is the general solution of the equation* (1) *if and only if*

$$\left(M(x,y)\left(1-x\frac{\hat{T}'(x)}{\hat{T}(x)}\right)-yN(x,y)\frac{\hat{T}'(x)}{\hat{T}(x)}\right)u_y(x,y)=N(x,y)u_x(x,y).$$

Problem 1.0.66. *Show that* $u(x,y)=C$, C *is a constant, is the general solution of the equation* (19) *if and only if*

$$\left(M(x,y)\left(\hat{T}(x)-x\hat{T}'(x)\right)-y\hat{T}'(x)N(x,y)\right)u_y(x,y)=\hat{T}(x)N(x,y)u_x(x,y).$$

Problem 1.0.67. *Show that* $u(x,y) = C$, *C is a constant, is the general solution of the equation* (27) *if and only if*

$$M9x,y)\left(\hat{T}(x) - x\hat{T}'(x)\right)u_y(x,y) = N(x,y)\hat{T}^2(x)u_x(x,y).$$

Problem 1.0.68. *Show that* $u(x,y) = C$, *C is a constant, is the general solution of the equation* (37) *if and only if*

$$M(x,y)\left(1 - x\frac{\hat{T}'(x)}{\hat{T}(x)}\right)u_y(x,y) = N(x,y)\hat{T}(x)u_x(x,y).$$

Chapter 2

Elementary First-Order Equations

We suppose that $\hat{T}:\mathbb{R}\longrightarrow(0,\infty)$, $\hat{T}\in\mathcal{C}^1(\mathbb{R})$. Let also

$$S=\{(x,y)\in\mathbb{R}^2: |x-x_0|\le a, \qquad |y-y_0|\le b\},$$

where x_0, $y_0\in\mathbb{R}$, a, $b\in\mathbb{R}$, $a>0$, $b>0$.

We consider the equation

$$\hat{X}_1^{\wedge}(\hat{x})\hat{\times}\hat{Y}_1(\hat{y})+\hat{X}_2^{\wedge}(\hat{x})\hat{\times}\hat{Y}_2(\hat{y})\hat{\times}\left(\hat{y}^{\wedge}(\hat{x})\right)^{\circledast}=0. \tag{1}$$

The equation (1) we can rewrite in the form

$$\tfrac{X_1(x)}{\hat{T}(x)}\hat{T}(x)\frac{Y_1\left(\frac{y}{\hat{T}(x)}\right)}{\hat{T}(x)}+\tfrac{X_2(x)}{\hat{T}(x)}\hat{T}(x)\frac{Y_2\left(\frac{y}{\hat{T}(x)}\right)}{\hat{T}(x)}\hat{T}(x)\frac{y'(x)\hat{T}(x)-y(x)\hat{T}'(x)}{\hat{T}(x)\left(\hat{T}(x)-x\hat{T}'(x)\right)}=0 \qquad \Longrightarrow$$

$$\tfrac{X_1(x)}{\hat{T}(x)}Y_1\left(\tfrac{y}{\hat{T}(x)}\right)+\tfrac{X_2(x)\hat{T}(x)}{\hat{T}(x)-x\hat{T}'(x)}\left(\tfrac{y}{\hat{T}(x)}\right)'Y_2\left(\tfrac{y}{\hat{T}(x)}\right)=0.$$

We put

$$z=\frac{y}{\hat{T}(x)}.$$

Then

$$z'=\frac{dz}{dx}$$

and the equation (1) admits the following representation

$$\frac{X_1(x)}{\hat{T}(x)}Y_1(z)+\frac{X_2(x)\hat{T}(x)}{\hat{T}(x)-x\hat{T}'(x)}Y_2(z)z'=0.$$

We observe that in the last equation the variables are separated.

Definition 2.0.69. *The iso-differential equation* (1) *is said to be separable.*

If $X_2(x)Y_1(z) \neq 0$ in S, then the solution of this exact equation is given by

$$\int \frac{X_1(x)\left(\hat{T}(x) - x\hat{T}'(x)\right)}{X_2(x)\hat{T}^2(x)} dx + \int \frac{Y_2(z)}{Y_1(z)} dz = C, \tag{2}$$

where C is a constant. Here the integrals are indefinite and constants of integration have been absorbed in C.

The equation (2) contains all solutions of (1) for which $\left(\hat{T}(x) - x\hat{T}'(x)\right)X_2(x)Y_1(z) \neq 0$. In fact, when we divide (1) by $\left(\hat{T}(x) - x\hat{T}'(x)\right)$, $X_2(x)$, $Y_1(z)$ we may have lost some solutions, and the ones which are not in (2) for some constant C must be coupled with (2) to obtain all solutions of (1).

Example 2.0.70. *Let $\hat{T}(x) = e^x$, $X_1(x) = e^x$, $X_2(x) = x(1-x)$, $Y_1(y) = y(y-1)$, $Y_2(y) = 1$. Then*

$$\begin{aligned} \tfrac{X_1(x)}{\hat{T}(x)} Y_1(z) &= \tfrac{e^x}{e^x} z(z-1) \\ &= z(z-1), \\ \tfrac{X_2(x)\hat{T}(x)}{\hat{T}(x) - x\hat{T}'(x)} Y_2(z) &= \tfrac{x(1-x)e^x}{(1-x)e^x} \\ &= x. \end{aligned}$$

Then the equation (1) *takes the form*

$$z(z-1) + xz' = 0 \qquad \Big| \cdot \frac{1}{xz(z-1)}, \qquad x \neq, z \neq 0, 1,$$

from where

$$\frac{1}{x} dx + \frac{1}{z(z-1)} dz = 0, \qquad x \neq 0, z \neq 0, 1.$$

Consequently

$$\int \frac{dx}{x} + \int \frac{dz}{z(z-1)} = C$$

or

$$x(z-1) = Cz.$$

Consequently the solutions of the considered equation are

$$x\left(y - e^x\right) = Cy, \qquad y = 0.$$

Here C is a constant.

Exercise 2.0.71. *Let*

$$\hat{T}(x) = e^x, \quad X_1(x) = xe^x, \quad X_2(x) = (1-x)(x^2+1), \quad Y_1(y) = \sin y, \quad Y_2(y) = \cos y.$$

Determine the equation (1) *and find its solutions.*

Solution. We have

$$\frac{X_1(x)}{\hat{T}(x)}Y_1(z) = \frac{xe^x}{e^x}\sin z$$

$$= x\sin z,$$

$$\frac{X_2(x)\hat{T}(x)}{\hat{T}(x)-x\hat{T}'(x)}Y_2(z) = \frac{(1-x)(x^2+1)e^x}{e^x-xe^x}\cos z$$

$$= \frac{(1-x)(x^2+1)e^x}{(1-x)e^x}\cos z$$

$$= (x^2+1)\cos z.$$

Then the equation (1) takes the form

$$x\sin z + (x^2+1)\cos(z)z' = 0 \qquad \Big| \cdot \frac{1}{(x^2+1)\sin z}, \qquad z \neq k\pi, \qquad k \in \mathbb{Z}.$$

From here,

$$\frac{x}{x^2+1}dx + \frac{\cos z}{\sin z}dz = 0 \qquad \Longrightarrow$$

$$\int \frac{x}{x^2+1}dx + \int \frac{\cos z}{\sin z}dz = C \qquad \Longrightarrow$$

$$\frac{1}{2}\int \frac{d(x^2+1)}{x^2+1} + \int \frac{d\sin z}{\sin z} = C \qquad \Longrightarrow$$

$$\log\sqrt{x^2+1} + \log|\sin z| = C \qquad \Longrightarrow$$

$$\sqrt{x^2+1}\sin z = C \tag{3}$$

or

$$\sqrt{x^2+1}\sin\left(ye^{-x}\right) = C. \tag{4}$$

Here C is a constant. Since $z = k\pi$, $k \in \mathbb{Z}$, satisfy the equality (3) for $C = 0$, then the solutions of the considered equations are given by (4).

Exercise 2.0.72. *Let*

$$\hat{T}(x) = e^x, \quad X_1(x) = x^3e^x, \quad X_2(x) = (1-x)(x^4+1)e^x, \quad Y_1(y) = \cos y, \quad Y_2(y) = \sin y.$$

Determine the equation (1) *and find its solutions.*

Answer. The equation (1) takes the form

$$x^3\cos z + (x^4+1)\sin(z)z' = 0.$$

Its solutions are given by

$$\sqrt[4]{x^4+1} = C\cos\left(ye^{-x}\right), \quad C = \text{const}, \quad y = (2k+1)\frac{\pi}{2}e^x, \quad k \in \mathbb{Z}.$$

We consider the equation

$$\hat{X}_1^{\wedge}(x)\,\hat{\times}\,\hat{Y}_1(\hat{y}) + \hat{X}_2^{\wedge}(x)\,\hat{\times}\,\hat{Y}_2(\hat{y})\,\hat{\times}\left(\hat{y}^{\wedge}(\hat{x})\right)^{\circledast} = 0, \tag{5}$$

which we can rewrite in the form

$$\frac{X_1\left(x\hat{T}(x)\right)}{\hat{T}(x)}\hat{T}(x)\frac{Y_1\left(\frac{y}{\hat{T}(x)}\right)}{\hat{T}(x)}+\frac{X_2(x\hat{T}(x))}{\hat{T}(x)}\hat{T}(x)\frac{Y_2\left(\frac{y}{\hat{T}(x)}\right)}{\hat{T}(x)}\hat{T}(x)\frac{y'\hat{T}(x)-y\hat{T}'(x)}{\hat{T}(x)\left(\hat{T}(x)-x\hat{T}'(x)\right)}=0 \qquad \Longrightarrow$$

$$\frac{X_1\left(x\hat{T}(x)\right)}{\hat{T}(x)}Y_1\left(\frac{y}{\hat{T}_1(x)}\right)+X_2\left(x\hat{T}(x)\right)Y_2\left(\frac{y}{\hat{T}(x)}\right)\frac{\hat{T}(x)}{\hat{T}(x)-x\hat{T}'(x)}\left(\frac{y}{\hat{T}(x)}\right)'=0.$$

We put $z=\frac{y}{\hat{T}(x)}$. Then we get

$$\frac{X_1\left(x\hat{T}(x)\right)}{\hat{T}(x)}Y_1(z)+X_2\left(x\hat{T}(x)\right)\frac{\hat{T}(x)}{\hat{T}(x)-x\hat{T}'(x)}Y_2(z)z'=0.$$

In the last equation the variables are separated.

Definition 2.0.73. *The equation* (5) *is said to be separable.*

If $X_2\left(x\hat{T}(x)\right)Y_1(z)\neq 0$ in S, then the solution of this exact equation is given by

$$\int\frac{X_1\left(x\hat{T}(x)\right)\left(\hat{T}(x)-x\hat{T}'(x)\right)}{X_2\left(x\hat{T}(x)\right)\hat{T}^2(x)}dx+\int\frac{Y_2(z)}{Y_1(z)}dz=C, \tag{6}$$

where C is a constant. Here both integrals are indefinite and constants of integration have been absorbed in C.

The equality (6) contains all the solutions of the equation (5) for which $\left(\hat{T}(x)-x\hat{T}'(x)\right)X_2\left(x\hat{T}(x)\right)Y_1(z)\neq 0$. In fact, when we divide by $\left(\hat{T}(x)-x\hat{T}'(x)\right)$, $X_2\left(x\hat{T}(x)\right)$, $Y_1(z)$ we might have lost some solutions, and the ones which are not in (6) for some constant C must be coupled with (6) to obtain all solutions of (5).

Example 2.0.74. *Let*

$$\hat{T}(x)=e^x,\quad X_1(x)=x^2,\quad X_2(x)=x,\quad Y_1(y)=y^2+1,\quad Y_2(y)=y.$$

Then

$$X_1\left(x\hat{T}(x)\right) = x^2\hat{T}^2(x)$$

$$= x^2e^{2x},$$

$$X_2\left(x\hat{T}(x)\right) = x\hat{T}(x)$$

$$= xe^x,$$

$$\frac{X_1\left(x\hat{T}(x)\right)}{\hat{T}(x)}Y_1(z) = \frac{x^2e^{2x}}{e^x}(z^2+1)$$

$$= x^2(z^2+1)e^x,$$

$$X_2\left(x\hat{T}(x)\right)\frac{\hat{T}(x)}{\hat{T}(x)-x\hat{T}'(x)}Y_2(z) = xe^x\frac{e^x}{e^x-xe^x}z$$

$$= \frac{xz}{1-x}e^x.$$

Then the equation (5) *takes the form*

$$x^2(z^2+1)e^x + \frac{x}{1-x}e^x zz' = 0 \qquad \Big| \cdot \frac{1-x}{xe^x(z^2+1)}, x \neq 0, \qquad \Longrightarrow$$

$$x(1-x)dx + \frac{z}{z^2+1}dz = 0 \qquad \Longrightarrow$$

$$\int x(1-x)dx + \int \frac{z}{1+z^2}dz = C \qquad \Longrightarrow$$

$$\frac{1}{2}x^2 - \frac{1}{3}x^3 + \log\sqrt{z^2+1} = C \qquad z = \frac{y}{e^x} \qquad \Longrightarrow$$

$$-x + \frac{1}{2}x^2 - \frac{1}{3}x^3 + \log\sqrt{y^2+e^{2x}} = C \qquad \Longrightarrow$$

$$-6x + 3x^2 - 2x^3 + \log\left(y^2+e^{2x}\right)^3 = C,$$

where C is a constant. Also, $x = 0$ is a solution to the considered equation.

Exercise 2.0.75. *Let*

$$\hat{T}(x) = x^2+1, \quad X_1(x) = x^3, \quad X_2(x) = x, \quad Y_1(y) = y^4+1, \quad Y_2(y) = y^3.$$

Determine the equation (5) *and find its solutions.*

Solution. We have

$$X_1\left(x\hat{T}(x)\right) = x^3\hat{T}^3(x)$$

$$= x^3(x^2+1)^3,$$

$$X_2\left(x\hat{T}(x)\right) = x\hat{T}(x)$$

$$= x(x^2+1),$$

$$\frac{X_1\left(x\hat{T}(x)\right)}{\hat{T}(x)}Y_1(z) = \frac{x^3(x^2+1)^3}{x^2+1}(z^4+1)$$

$$= x^3(x^2+1)^2(z^4+1),$$

$$X_2\left(x\hat{T}(x)\right)\frac{\hat{T}(x)}{\hat{T}(x)-x\hat{T}'(x)}Y_2(z) = x(x^2+1)^2\frac{x^2+1}{x^2+1-2x^2}z^3$$

$$= \frac{x(x^2+1)^2}{1-x^2}z^3.$$

Then the equation (5) takes the form

$$x^3(x^2+1)^2(z^4+1) + \frac{x(x^2+1)^2}{1-x^2}z^2z' = 0 \qquad \Big|\cdot\frac{1-x^2}{x(x^2+1)^2(z^4+1)}, x \neq 0, \qquad \Longrightarrow$$

$$x^2(1-x^2) + \frac{z^3}{1+z^4}z' = 0 \quad \Longrightarrow$$

$$\int x^2(1-x^2)dx + \int \frac{z^3}{1+z^4}dz = C \quad \Longrightarrow$$

$$\frac{1}{3}x^3 - \frac{1}{5}x^5 + \frac{1}{4}\log(z^4+1) = C, \quad z = \frac{y}{x^2+1}, \quad \Longrightarrow$$

$$20x^3 - 12x^5 + 15\log\frac{y^4+(x^2+1)^4}{(x^2+1)^4} = C,$$

where C is a constant. Also, $x = 0$ is a solution.

Exercise 2.0.76. *Let*

$$\hat{T}(x) = x^2+1, \quad X_1(x) = x, \quad X_2(x) = 2x, \quad Y_1(y) = e^y, \quad Y_2(y) = e^{2y}.$$

Determine the equation (5) *and find its solutions.*

Answer. The equation (5) takes the form

$$xe^z + 2\frac{x(x^2+1)}{1-x^2}e^{2z}z' = 0.$$

Its solutions are given by

$$2\arctan x - x + 2e^{\frac{y}{x^2+1}} = C,$$

where C is a constant. Also, $x = 0$ is a solution.

We consider the equation

$$\hat{X}_1(\hat{x})\hat{\times}\hat{Y}_1(\hat{y}) + \hat{X}_2(\hat{x})\hat{\times}\hat{Y}_2(\hat{y})\hat{\times}\left(\hat{y}^{\wedge}(\hat{x})\right)^{\circledast} = 0, \tag{7}$$

which we can rewrite into the following form

$$\frac{X_1\left(\frac{x}{\hat{T}(x)}\right)}{\hat{T}(x)}\hat{T}(x)\frac{Y_1\left(\frac{y}{\hat{T}(x)}\right)}{\hat{T}(x)}$$

$$+\frac{X_2\left(\frac{x}{\hat{T}(x)}\right)}{\hat{T}(x)}\hat{T}(x)\frac{Y_2\left(\frac{y}{\hat{T}(x)}\right)}{\hat{T}(x)}\hat{T}(x)\frac{\hat{T}(x)}{\hat{T}(x)-x\hat{T}'(x)}\frac{y'\hat{T}(x)-y\hat{T}'(x)}{\hat{T}^2(x)} = 0 \quad \Longrightarrow$$

$$\frac{1}{\hat{T}(x)}X_1\left(\frac{x}{\hat{T}(x)}\right)Y_1\left(\frac{y}{\hat{T}(x)}\right) + \frac{\hat{T}(x)}{\hat{T}(x)-x\hat{T}'(x)}X_2\left(\frac{x}{\hat{T}(x)}\right)Y_2\left(\frac{y}{\hat{T}(x)}\right)\left(\frac{y}{\hat{T}(x)}\right)' = 0.$$

We put $\frac{y}{\hat{T}(x)} = z$ and we get

$$\frac{1}{\hat{T}(x)}X_1\left(\frac{x}{\hat{T}(x)}\right)Y_1(z) + \frac{\hat{T}(x)}{\hat{T}(x)-x\hat{T}'(x)}X_2\left(\frac{x}{\hat{T}(x)}\right)Y_2(z)z' = 0,$$

in which the variables are separable.

Definition 2.0.77. *The iso-differential equation* (7) *is said to be called separable.*

If $X_2\left(\frac{x}{\hat{T}(x)}\right)Y_1(z) \neq 0$ in S, then the solution of this exact equation is given by

$$\int \frac{X_1\left(\frac{x}{\hat{T}(x)}\right)}{X_2\left(\frac{x}{\hat{T}(x)}\right)}\frac{1}{\hat{T}^2(x)}\left(\hat{T}(x)-x\hat{T}'(x)\right)dx + \int \frac{Y_2(z)}{Y_1(z)}dz = C \tag{8}$$

where C is a constant. The equation (8) contains all the solutions of the equation (7) for which $X_2\left(\frac{x}{\hat{T}(x)}\right)\left(\hat{T}(x)-x\hat{T}'(x)\right)Y_1(z) = 0$. In fact, when we divide the equation (7) by $X_2\left(\frac{x}{\hat{T}(x)}\right)$, $\hat{T}(x)-x\hat{T}'(x)$, $Y_1(z)$ we might have lost some solutions, and the ones which are not in (8) for some constant C must be coupled with (8) to obtain all solutions of (7).

Example 2.0.78. *Let* $\hat{T}(x) = e^x$, $X_1(x) = X_2(x) = x$, $Y_1(y) = \cos y$, $Y_2(y) = \sin y$. *Then*

$$X_1\left(\frac{x}{\hat{T}(x)}\right) = X_2\left(\frac{x}{\hat{T}(x)}\right)$$

$$= \frac{x}{\hat{T}(x)}$$

$$= xe^{-x},$$

$$\frac{1}{\hat{T}(x)}X_1\left(\frac{x}{\hat{T}(x)}\right)Y_1(z) = \frac{1}{\hat{T}(x)}\frac{x}{\hat{T}(x)}\cos z$$

$$= xe^{-2x}\cos z,$$

$$\frac{\hat{T}(x)}{\hat{T}(x)-x\hat{T}'(x)}X_2\left(\frac{x}{\hat{T}(x)}\right)Y_2(z) = \frac{\hat{T}(x)}{\hat{T}(x)\quad x\hat{T}'(x)}\frac{x}{\hat{T}(x)}\sin z$$

$$= \frac{x}{\hat{T}(x)-x\hat{T}'(x)}\sin z$$

$$= \frac{x}{e^x - xe^x}\sin z$$

$$= \frac{xe^{-x}}{1-x}\sin z.$$

The equation (7) *takes the form*

$$xe^{-2x}\cos z + \frac{xe^{-x}}{1-x}\sin z z' = 0 \quad \Bigg| \frac{(1-x)e^x}{x\cos z}, \quad x \neq 0, 1, z \neq (2k+1)\frac{\pi}{2}, \quad k \in \mathbb{Z}, \quad \Longrightarrow$$

$$e^{-x}(1-x)dx + \frac{\sin z}{\cos z}dz = 0.$$

Therefore, when $x \neq 0, 1$, $z \neq (2k+1)\frac{\pi}{2}$, $k \in \mathbb{Z}$, *the solution is given by*

$$\int e^{-x}(1-x)dx + \int \frac{\sin z}{\cos z}dz = C \quad \Longrightarrow$$

$$-\int (1-x)de^{-x} - \int \frac{d(\cos z)}{\cos z} = C \quad \Longrightarrow$$

$$-(1-x)e^{-x} - \int e^{-x}dx - \log|\cos z| = C \quad \Longrightarrow$$

$$xe^{-x} - \log\left|\cos\left(ye^{-x}\right)\right| = C,$$

where C *is a constant. We note, that* $x = 0$, $y = (2k+1)\frac{\pi}{2}e^x$, $k \in \mathbb{Z}$, *are solutions to the considered equation.*

Exercise 2.0.79. *Let* $\hat{T}(x) = e^x$, $X_1(x) = x^2$, $X_2(x) = x$, $Y_1(y) = y^2 + 1$, $Y_2(y) = 2y$. *Determine the equation* (7) *and find its solutions.*

Answer. The equation (7) takes the form

$$x^2(z^2+1)e^{-3x} + 2\frac{xz}{1-x}e^{-x}z' = 0.$$

Its solutions are given by

$$x^2e^{-2x}+\log\left(y^2+e^{2x}\right)^2=C,$$

where C is a constant. Also, $x=0$ is a solution.

Exercise 2.0.80. *Prove that the solutions of the equation*

$$X_1^{\wedge}(x)\hat{\times}\hat{Y}_1(\hat{y})+X_2^{\wedge}(x)\hat{\times}\hat{Y}_2(\hat{y})\hat{\times}\left(\hat{y}^{\wedge}(\hat{x})\right)^{\circledast}=0$$

are given by (6) *in the case when* $X_2\left(x\hat{T}(x)\right)Y_1(z)\left(\hat{T}(x)-x\hat{T}'(x)\right)\neq 0.$

Exercise 2.0.81. *Prove that the solutions of the equation*

$$X_1^{\vee}(x)\hat{\times}\hat{Y}_1(\hat{y})+X_2^{\vee}(x)\hat{\times}\hat{Y}_2(\hat{y})\hat{\times}\left(\hat{y}^{\wedge}(\hat{x})\right)^{\circledast}=0$$

are given by (8) *in the case when* $X_2\left(\frac{x}{\hat{T}(x)}\right)Y_1(z)\left(\hat{T}(x)-x\hat{T}'(x)\right)\neq 0.$

We consider the equation

$$\hat{X}_1^{\wedge}(\hat{x})\hat{\times}\hat{Y}_1(\hat{y}^{\wedge})+\hat{X}_2^{\wedge}(\hat{x})\hat{\times}\hat{Y}_2(\hat{y}^{\wedge})\hat{\times}\left(\hat{y}^{\wedge}(x)\right)^{\circledast}=0, \tag{9}$$

which we can rewrite in the form

$$\frac{X_1(x)}{\hat{T}(x)}\hat{T}(x)^{\frac{Y_1\left(\frac{y\left(x\hat{T}(x)\right)}{\hat{T}(x)}\right)}{\hat{T}(x)}}$$

$$+\frac{X_2(x)}{\hat{T}(x)}\hat{T}(x)^{\frac{Y_2\left(\frac{y\left(x\hat{T}(x)\right)}{\hat{T}(x)}\right)}{\hat{T}(x)}}\frac{y'\left(x\hat{T}(x)\right)\left(\hat{T}(x)+x\hat{T}'(x)\right)\hat{T}(x)-y\left(x\hat{T}(x)\right)\hat{T}'(x)}{\hat{T}(x)\left(\hat{T}(x)-x\hat{T}'(x)\right)}=0$$

or

$$\frac{X_1(x)}{\hat{T}(x)}Y_1\left(\frac{y\left(x\hat{T}(x)\right)}{\hat{T}(x)}\right)+X_2(x)\frac{\hat{T}(x)}{\hat{T}(x)-x\hat{T}'(x)}Y_2\left(\frac{y\left(x\hat{T}(x)\right)}{\hat{T}(x)}\right)\frac{d}{dx}\left(\frac{y\left(x\hat{T}(x)\right)}{\hat{T}(x)}\right)=0.$$

We put $z=\frac{y\left(x\hat{T}(x)\right)}{\hat{T}(x)}$. Then the equation (9) admits the form

$$\frac{X_1(x)}{\hat{T}(x)}Y_1(z)+X_2(x)\frac{\hat{T}(x)}{\hat{T}(x)-x\hat{T}'(x)}Y_2(z)dx=0,$$

in which the variables are separated.

Definition 2.0.82. *The iso-differential equation* (9) *is said to be separable.*

If $\hat{T}(x)-x\hat{T}'(x)\neq 0$, $X_2(x)\neq 0$, $Y_1(z)\neq 0$ in S, then the solutions of the equation (9) are given by

$$\int \frac{X_1(x)\left(\hat{T}(x)-x\hat{T}'(x)\right)}{X_2(x)\hat{T}^2(x)}dx+\int \frac{Y_2(z)}{Y_1(z)}dz=C, \tag{10}$$

where C is a constant. Here both integrals are indefinite and constants of integration have been absorbed in C.

In fact, when we divide by $\hat{T}(x)-x\hat{T}'(x)$, $X_2(x)$ and $Y_1(z)$ might have lost some solutions of (9), and the ones which are not in (10) for some constant C must be coupled with (10) to obtain all solutions of (9).

Exercise 2.0.83. *Prove that the solutions of the following iso-differential equation*

$$\hat{X}_1^{\wedge}(\hat{x})\hat{\times}Y_1(\hat{y}^{\wedge})+\hat{X}_2^{\wedge}(\hat{x})\hat{\times}\hat{Y}_2(\hat{y}^{\wedge})\hat{\times}\left(\hat{y}^{\wedge}(x)\right)^{\circledast}=0$$

are given by

$$\int \frac{X_1(x)\left(\hat{T}(x)-x\hat{T}'(x)\right)}{X_2(x)\hat{T}(x)}dx+\int \frac{Y_2(z)}{Y_1(z)}dz=C,$$

where C is a constant, in the case when $X_2(x)\neq 0$, $\hat{T}(x)-x\hat{T}'(x)\neq 0$, $Y_1(z)\neq 0$ in S.

We consider the equation

$$\hat{X}_1^{\wedge}(\hat{x})\hat{\times}\hat{Y}_1(y^{\wedge})+\hat{X}_2^{\wedge}(\hat{x})\hat{\times}\hat{Y}_2(y^{\wedge})\hat{\times}\left(y^{\wedge}(x)\right)^{\circledast}=0, \tag{11}$$

which we can rewrite in the form

$$\frac{X_1(x)}{\hat{T}(x)}\hat{T}(x)\frac{Y_1\left(x\hat{T}(x)\right)}{\hat{T}(x)}$$

$$+\frac{X_2(x)}{\hat{T}(x)}\hat{T}(x)\frac{Y_2\left(y\left(x\hat{T}(x)\right)\right)}{\hat{T}(x)}\hat{T}(x)\frac{y'\left(x\hat{T}(x)\right)\hat{T}(x)\left(\hat{T}(x)+x\hat{T}'(x)\right)}{\hat{T}(x)-x\hat{T}'(x)}=0$$

or

$$\frac{X_1(x)}{\hat{T}(x)}Y_1\left(y\left(x\hat{T}(x)\right)\right)+X_2(x)\frac{\hat{T}(x)\left(\hat{T}(x)+x\hat{T}'(x)\right)}{\hat{T}(x)-x\hat{T}'(x)}Y_2\left(y\left(x\hat{T}(x)\right)\right)y'\left(x\hat{T}(x)\right)=0.$$

We put $y\left(x\hat{T}(x)\right)=z$. Then the equation (11) admits the following representation

$$\frac{X_1(x)}{\hat{T}(x)}Y_1(z)dx+\frac{X_2(x)\hat{T}(x)}{\hat{T}(x)-x\hat{T}'(x)}Y_2(z)dz=0,$$

in which the variables are separated.

Definition 2.0.84. *The iso-differential equation (11) is said to be separable.*

When $\hat{T}(x) - x\hat{T}'(x) \neq 0$, $X_2(x) \neq 0$, $Y_1(z) \neq 0$ in S, then the solutions of the iso-differential equation are given by

$$\int \frac{X_1(x)\left(\hat{T}(x) - x\hat{T}'(x)\right)}{X_2(x)\hat{T}^2(x)}dx + \int \frac{Y_2(z)}{Y_1(z)}dz = C, \tag{12}$$

where C is a constant. In fact, when we divide by $\hat{T}(x) - x\hat{T}'(x)$, $X_2(x)$ and $Y_1(z)$ we might have lost some solutions of the equation (11), and the ones which are not in (12) for some constant C must be coupled with (12) to obtain all solutions of (11).

Exercise 2.0.85. *Prove that the solutions of the equation*

$$\hat{X}_1^{\wedge}(\hat{x})\hat{\times}Y_1(y^{\wedge}) + P\hat{X}_2^{\wedge}(\hat{x})\hat{\times}Y_2(y^{\wedge})\hat{\times}\left(y^{\wedge}(x)\right)^{\circledast} = 0$$

are given by

$$\int \frac{X_1(x)\left(\hat{T}(x) - x\hat{T}'(x)\right)}{X_2(x)\hat{T}(x)}dx + \int \frac{Y_2(z)}{Y_1(z)}dz = C,$$

where C is a constant, in the case when $\hat{T}(x) - x\hat{T}'(x) \neq 0$, $X_2(x) \neq 0$ and $Y_1(z) \neq 0$ in S.

We consider the following iso-differential equation

$$\hat{X}_1^{\wedge}(\hat{x})\hat{\times}\hat{Y}_1(y^{\vee})\hat{\times}\hat{Y}_1(y^{\vee}) + \hat{X}_2^{\wedge}(\hat{x})\hat{\times}\hat{Y}_2(y^{\vee})\left(y^{\vee}(x)\right)^{\circledast} = 0, \tag{13}$$

which we can rewrite in the form

$$\tfrac{X_1(x)}{\hat{T}(x)}\hat{T}(x)\tfrac{Y_1(y^{\vee})}{\hat{T}(x)} + \tfrac{X_2(x)}{\hat{T}(x)}\hat{T}(x)\tfrac{Y_2(y^{\vee})}{\hat{T}(x)}\hat{d}y^{\vee}(x) \nearrow \hat{d}\hat{x} = 0,$$

or

$$\tfrac{X_1(x)}{\hat{T}(x)}Y_1(y^{\vee}) + \tfrac{X_2(x)}{\hat{T}(x)}Y_2(y^{\vee})\tfrac{1}{\hat{T}(x)}\tfrac{\hat{d}y^{\vee}(x)}{\hat{d}\hat{x}} = 0,$$

or

$$\tfrac{X_1(x)}{\hat{T}(x)}Y_1(y^{\vee}) + \tfrac{X_2(x)}{\hat{T}(x)}Y_2(y^{\vee})\tfrac{1}{\hat{T}(x)}\tfrac{\hat{T}(x)dy^{\vee}(x)}{\hat{T}(x)d\hat{x}} = 0,$$

or

$$\hat{T}(x)X_1(x)Y_1(y^{\vee}) + X_2(x)Y_2(y^{\vee})\tfrac{dy^{\vee}(x)}{d\hat{x}} = 0,$$

or

$$\hat{T}(x)X_1(x)Y_1(y^{\vee})d\hat{x} + X_2(x)Y_2(y^{\vee})dy^{\vee}(x) = 0,$$

or

$$X_1(x)\frac{\hat{T}(x) - x\hat{T}'(x)}{\hat{T}(x)}Y_1(y^{\vee})dx + X_2(x)Y_2(y^{\vee})dy^{\vee} = 0.$$

We put $z = y^{\vee}(x)$. Then we obtain the following representation of the iso-differential equation (13)

$$X_1(x)\frac{\hat{T}(x) - x\hat{T}'(x)}{\hat{T}(x)}Y_1(z)dx + X_2(x)Y_2(z)dz = 0,$$

in which the variables are separated.

Definition 2.0.86. *The equation* (13) *is said to be separable.*

In the case when $\hat{T}(x) - x\hat{T}'(x) \neq 0$, $X_2(x) \neq 0$ and $Y_1(z) \neq 0$ in S, the solutions of the equation (13) are given by

$$\int \frac{X_1(x)\left(\hat{T}(x) - x\hat{T}'(x)\right)}{X_2(x)\hat{T}(x)}dx + \int \frac{Y_2(z)}{Y_1(z)}dz = C, \tag{14}$$

where C is a constant. Here both integrals are indefinite and constants of integration have been absorbed in C.

In fact, when we divide by $\hat{T}(x) - x\hat{T}'(x)$, $X_2(x)$ and $Y_1(z)$ we might have lost some of the solutions of (13), and the ones which are not in (14) for some constant C should be coupled with (14) to obtain all solutions of (13).

Exercise 2.0.87. *Prove that the solutions of the equation*

$$\hat{X}_1(\hat{x})\hat{Y}_1(y^{\vee}) + \hat{X}_2(\hat{x})\hat{\times}\hat{Y}_2(y^{\vee})\left(y^{\vee}(x)\right)^{\circledast} = 0$$

are given by

$$\int \frac{X_1(x)\left(\hat{T}(x) - x\hat{T}'(x)\right)}{X_2(x)\hat{T}^2(x)}dx + \int \frac{Y_2(z)}{Y_1(z)}dz = C,$$

where C *is a constant, in the case when* $\hat{T}(x) - x\hat{T}'(x) \neq 0$, $X_2(x) \neq 0$, $Y_1(z) \neq 0$ *in* S.

Now we consider the equation

$$\hat{X}_1^{\wedge}(\hat{x})\hat{\times}\hat{Y}_1(\hat{y}(\hat{x})) + \hat{X}_2^{\wedge}(\hat{x})\hat{\times}\hat{Y}_2(\hat{y}(\hat{x}))\hat{\times}\left(\hat{y}(\hat{x})\right)^{\circledast} = 0, \tag{15}$$

which we can rewrite in the following form

$$\frac{X_1(x)}{\hat{T}(x)}\frac{Y_1(\hat{y}(\hat{x}))}{\hat{T}(x)} + \frac{X_2(x)}{\hat{T}(x)}\hat{T}(x)\frac{Y_2(\hat{y}(\hat{x}))}{\hat{T}(x)}\hat{T}(x)d\hat{y}(\hat{x}) \nearrow d\hat{x} = 0,$$

or

$$\frac{X_1(x)}{\hat{T}(x)}Y_1(\hat{y}(\hat{x})) + \frac{X_2(x)}{\hat{T}(x)}Y_2(\hat{y}(\hat{x}))\frac{1}{\hat{T}(x)}\frac{d\hat{y}(\hat{x})}{d\hat{x}} = 0,$$

or

$$\frac{X_1(x)}{\hat{T}(x)}Y_1(\hat{y}(\hat{x}))d\hat{x} + X_2(x)Y_2(\hat{y}(\hat{x}))d\hat{y}(\hat{x}) = 0,$$

or

$$X_1(x)\frac{\hat{T}(x) - x\hat{T}'(x)}{\hat{T}(x)}Y_1(\hat{y}(\hat{x}))dx + X_2(x)Y_2(\hat{y}(\hat{x}))d\hat{y}(\hat{x}) = 0.$$

We put $z = \hat{y}(\hat{x})$. Then the equation (15) admits the following representation

$$X_1(x)\frac{\hat{T}(x)-x\hat{T}'(x)}{\hat{T}(x)}Y_1(z)dx+X_2(x)Y_2(z)dz=0,$$

which is an equation with separated variables.

Definition 2.0.88. *The equation* (15) *is said to be separable.*

When $\hat{T}(x)-x\hat{T}'(x)\neq 0$, $X_2(x)\neq 0$ and $Y_1(z)\neq 0$ in S, then the solutions of the iso-differential equation (15) are given by

$$\int\frac{X_1(x)\left(\hat{T}(x)-x\hat{T}'(x)\right)}{X_2(x)\hat{T}(x)}dx+\int\frac{Y_2(z)}{Y_1(z)}dz=C, \tag{16}$$

where C is a constant. Here both integrals are indefinite integrals and constants of integration have been absorbed in C.

In fact, when we divide by $\hat{T}(x)-x\hat{T}'(x)$, $X_2(x)$ and $Y_1(z)$ we might have lost some of the solutions of (15), and the ones which are not (16) for some constant C must be coupled with (16) to obtain all solutions of (15).

Exercise 2.0.89. *Prove that the solutions of the equation*

$$\hat{X}_1^{\wedge}(\hat{x})\hat{Y}_1(\hat{y}(\hat{x}))+\hat{X}_2^{\wedge}(\hat{x})\hat{\times}\hat{Y}_2(\hat{y}(\hat{x}))\hat{\times}\left(\hat{y}(\hat{x})\right)^{\circledast}=0$$

are given by

$$\int\frac{X_1(x)\left(\hat{T}(x)-x\hat{T}'(x)\right)}{X_2(x)\hat{T}^2(x)}dx+\int\frac{Y_2(z)}{Y_1(z)}dz=C,$$

where C is a constant, in the case when $\hat{T}(x)-x\hat{T}'(x)\neq 0$, $X_2(x)\neq 0$ and $Y_1(z)\neq 0$ in S.

Now we consider the equation

$$M(x,\hat{y})+\hat{N}(x,\hat{y})\hat{\times}(\hat{x})'\hat{\times}\left(\hat{y}^{\wedge\wedge}\right)^{\circledast}=0, \tag{17}$$

where $M(x,\hat{y})$ and $N(x,\hat{y})$ are homogeneous functions of the same degree, say, n.

Definition 2.0.90. *The iso-differential equation* (17) *is said to be homogeneous.*

The equation (17) we can rewrite in the following form

$$M(x,\hat{y})+\frac{N(x,\hat{y})}{\hat{T}(x)}\hat{T}(x)\frac{\hat{T}(x)-x\hat{T}'(x)}{\hat{T}^2(x)}\left(\hat{y}\right)'\frac{\hat{T}(x)}{\hat{T}(x)-x\hat{T}'(x)}=0$$

or

$$M(x,\hat{y})+N(x,\hat{y})\left(\hat{y}\right)'=0.$$

We put $\hat{y}=z$ and we get

$$M(x,z)+N(x,z)z'=0.$$

Using that $M(x,z)$ and $N(x,z)$ are homogeneous functions of degree n, from the last equation we find

$$x^n M\left(1,\frac{z}{x}\right)+x^n N\left(1,\frac{z}{x}\right)z'=0. \tag{18}$$

In (18) we use the substitution $z(x)=xv(x)$ and we find

$$x^n M(1,v)+x^n N(1,v)(v+xv')=0$$

or

$$x^n\Big(M(1,v)+vN(1,v)\Big)+x^{n+1}N(1,v)v'=0. \tag{19}$$

In this way we reduce the equation (17) to a separable equation. The equation (19) admits the integrating factor

$$\mu=\frac{1}{x^{n+1}\Big(M(1,v)+vN(1,v)\Big)}=\frac{1}{xM(x,\hat{y})+yN(x,\hat{y})} \tag{20}$$

provided $xM(x,\hat{y})+yN(x,\hat{y})\neq 0$.

Exercise 2.0.91. *Prove that* (20) *is an integrating factor for the equation* (19).

The vanishing of $xM(1,v)+yN(1,v)$ implies that (19) is simply

$$x^{n+1}N(1,v)v'=xN(x,\hat{y})v'=0,$$

for which the integrating factor is $\frac{1}{xN(x,\hat{y})}$. Thus, in this case the general solution of (17) is

$$\hat{y}(x)=Cx \qquad \text{or} \qquad y(x)=Cx\hat{T}(x),$$

where C is a constant.

Example 2.0.92. *Let* $\hat{T}(x)=e^x$, $M(x,y)=-x^2-xy-y^2$, $N(x,y)=x^2$, $x\neq 0$. *Then*

$$M(x,\hat{y})=-x^2-x\hat{y}-\hat{y}^2, \qquad N(x,\hat{y})=x^2.$$

The equation (17) *takes the form*

$$-x^2-x\hat{y}-\hat{y}^2+x^2\left(\hat{y}\right)'=0.$$

We put $z=\hat{y}$ *and we get*

$$-x^2-xz-z^2+x^2z'=0$$

or

$$z'=\frac{x^2+xz+z^2}{x^2}=1+\frac{z}{x}+\left(\frac{z}{x}\right)^2.$$

Let $v=\frac{z}{x}$. *Therefore*

$$z=xv, \qquad z'=xv'+v$$

and

$$v'x+v=1+v+v^2 \quad \Longrightarrow$$

$$xv'=1+v^2 \quad \Longrightarrow$$

$$\frac{dv}{1+v^2}=\frac{dx}{x} \quad \Longrightarrow$$

$$\arctan v=\log|x|+C \quad \Longrightarrow$$

$$\arctan\left(\frac{y}{xe^x}\right)=\log|x|+C$$

is the general solution of the considered equation. Here C is a constant.

Exercise 2.0.93. *Let* $\hat{T}(x)=e^x$, $M(x,y)=-y-xe^{\frac{y}{x}}$, $N(x,y)=x$, $x\neq 0$. *Find the general solution of the equation* (17).

Answer.

$$e^{-\frac{y}{xe^x}}+\log|x|=C,$$

where C is a constant.

Now we suppose that the functions $M(x,y)$ and $N(x,y)$ in (17) satisfy the condition

$$-\frac{M(x,y)}{N(x,y)}=f\Big(\frac{a_1x+b_1y+c_1}{a_2x+b_2y+c_2}\Big)$$

in which a_1, a_2, b_1, b_2, c_1 and c_2 are constants. If c_1 and c_2 are not both zero, then it can be converted to a homogeneous equation by means of the transformation

$$x=u+h, \qquad y=v+k,$$

where h and k are the solutions of the system of simultaneous linear equations

$$\begin{aligned} a_1h+b_1k+c_1&=0 \\ a_2h+b_2k+c_2&=0. \end{aligned} \tag{20'}$$

and the resulting homogeneous equation

$$\frac{dv}{du}=f\Big(\frac{a_1u+b_1v}{a_2u+b_2v}\Big)$$

can be solved easily.

However, the system (20′) can be solved for h and k provided

$$\Delta=a_1b_2-a_2b_1\neq 0.$$

If $\Delta=0$, then a_1x+b_1y is proportional to a_2x+b_2y, and hence (17) is of the form

$$z'=f(\alpha x+\beta z),$$

which can be solved easily by using the substitution

$$\alpha x+\beta z=q.$$

Example 2.0.94. *Let* $\hat{T}(x) = x^2 + 1$,

$$\frac{M(x,y)}{N(x,y)} = -\frac{2x+3y}{3x+2y+4}.$$

Then the equation (17) *we can represent in the form*

$$z' = \frac{2x+3z}{3x+2z+4}. \tag{21}$$

We put

$$x = u+h, \qquad z = v+k,$$

where h and k satisfy the system

$$\begin{array}{ccccc} 2h+3k=0 & & h=-\frac{3}{2}k & & h=-\frac{13}{5} \\ & \Longrightarrow & & \Longrightarrow & \\ 3h+2k+4=0 & & 3h+2k+4=0 & & k=\frac{8}{5}. \end{array}$$

From here,

$$\begin{array}{ccc} x = u-\frac{13}{5} & & u = x+\frac{13}{5} \\ & \Longrightarrow & \\ z = v+\frac{8}{5} & & v = z-\frac{8}{5} = \frac{y}{x^2+1}-\frac{8}{5}. \end{array}$$

Then the equation (21) *admits the representation*

$$v' = \frac{2u+3v}{3u+2v}.$$

We set

$$\frac{v}{u} = q \qquad \text{or} \qquad v = qu,$$

whereupon

$$v' = q'u+q$$

and

$$q'u+q = \frac{2+3q}{3+2q}$$

or

$$q'u = \frac{2-2q^2}{3+2q}. \tag{22}$$

Therefore

$$\frac{3+2q}{2(1-q^2)}dq = \frac{du}{u}, \qquad q \neq \pm 1.$$

From here, for $q \neq \pm 1$,

$$\int \frac{3+2q}{2(1-q^2)}dq = \int \frac{du}{u} + C_1.$$

Consequently, for $q \neq \pm 1$,

$$-\tfrac{5}{4}\log|1-q|+\tfrac{1}{4}\log|1+q| = \log|u|+C_1 \qquad \Longrightarrow$$

$$\log\tfrac{1}{|1-q|^5}+\log|1+q| = \log u^4 + C_2, \qquad C_2 = 4C_1, \qquad \Longrightarrow$$

$$1+q = C_3 u^4(1-q)^5, \qquad C_3 = e^{C_2}\mathrm{sign}(1-q), \qquad \Longrightarrow$$

$$u+v = C_3(u-v)^5 \qquad \Longrightarrow$$

$$x+1+\tfrac{y}{x^2+1} = C_3\left(x+\tfrac{21}{5}-\tfrac{y}{x^2+1}\right)^5.$$

Here C_1 is a constant.

Also, $q=\pm 1$ are solutions of (22), therefore

$$x+1+\frac{y}{x^2+1} = 0, \qquad x-\frac{y}{x^2+1}+\frac{21}{5} = 0$$

are solutions of the considered equation (17).

Consequently the solutions of the considered equation (17) are given by

$$x+1+\tfrac{y}{x^2+1} = C_3\left(x+\tfrac{21}{5}-\tfrac{y}{x^2+1}\right)^5,$$

$$x+1+\tfrac{y}{x^2+1} = 0, \qquad x-\tfrac{y}{x^2+1}+\tfrac{21}{5} = 0.$$

Example 2.0.95. *Let $\hat{T}(x) = e^x$,*

$$\frac{M(x,y)}{N(x,y)} = -\frac{x+y}{3x+3y+2}.$$

Then the equation (17) we can represent in the following form

$$z' = \frac{x+z}{3x+3z+2}.$$

Since

$$(x,z) = 3(x,z),$$

we set

$$u = x+z,$$

whereupon

$$z' = u'-1$$

and

$$u'-1 = \frac{u}{3u+1}$$

or

$$u' = \frac{1+4u}{1+3u}. \tag{23}$$

For $u \neq -\frac{1}{4}$ *we get*

$$\frac{1+3u}{1+4u}du = dx \qquad \Longrightarrow$$

$$\int \frac{1+3u}{1+4u}du = \int dx + C_1 \qquad \Longrightarrow$$

$$\frac{3}{4}u + \frac{1}{16}\log|1+4u| = x + C_1 \qquad \Longrightarrow$$

$$12u + \log|1+4u| = 16x + C_2, \qquad C_2 = 16C_1$$

or

$$12\left(x+ye^{-x}\right) + \log\left|1+4\left(x+ye^{-x}\right)\right| = 16x + C_2$$

is a solution of the equation (17). *Here* C_1 *is a constant.*

Also, $u = -\frac{1}{4}$ *is a solution to the equation* (23), *therefore*

$$1+4\left(x+ye^{-x}\right) = 0$$

is a solution of the equation (17).

Consequently, the solutions of the equation (17) *are given by*

$$12\left(x+ye^{-x}\right) + \log\left|1+4\left(x+ye^{-x}\right)\right| = 16x + C, \qquad 1+4\left(x+ye^{-x}\right) = 0,$$

where C *is a constant.*

Exercise 2.0.96. *Let* $\hat{T}(x) = e^x$,

$$\frac{M(x,y)}{N(x,y)} = -\frac{1}{2}\left(\frac{x+y-1}{x+2}\right)^2.$$

Find the general solution of the equation (17).

Answer.

$$2\tan^{-1}\frac{y-3e^x}{xe^x+2e^x} = \log|x+2| + C,$$

where C is a constant.

Exercise 2.0.97. *Let* $\hat{T}(x) = e^x$,

$$\frac{M(x,y)}{N(x,y)} = -\frac{x+y+1}{2x+2y+1}.$$

Find the general solution of (17).

Answer.

$$x + 2ye^{-x} + \log|x+ye^{-x}| = C, \qquad x+ye^{-x} = 0,$$

where C is a constant.

Sometimes, it is possible to introduce a new set of variables given by the equations

$$u = \phi(x,y), \qquad v = \psi(x,y), \tag{24}$$

which convert a given iso-differential equation into a form that can be solved rather easily. We assume that

$$\frac{\partial(u,v)}{\partial(x,y)} \neq 0$$

over a region in $\mathbb{R}^2$, which implies that there is no functional relationship between u and v. Thus, if (u_0, v_0) is the image of (x_0, y_0) under the transformation (24), then it can be uniquely solved for x and y in a neighborhood of the point (x_0, y_0). This leads to the inverse transformation

$$x = x(u,v), \qquad y = y(u,v). \tag{25}$$

The relations (24) and (25) can be used to convert some iso-differential equation in terms of u and v, which hopefully can be solved explicitly. Finally, replacement of u and v in terms of x and y by using (24) leads to an implicit solution of the considered iso-differential equation.

Example 2.0.98. *Let* $\hat{T}(x) = e^x$,

$$\frac{M(x,y)}{N(x,y)} = -\frac{3x^5}{y(y^2 - x^3)}.$$

The equation (17) *takes the following form*

$$z' = \frac{3x^5}{z(z^2 - x^3)}. \tag{26}$$

Let

$$u = x^3, \qquad v = z^2.$$

We have

$$\frac{\partial(u,v)}{\partial(x,z)} = \det\begin{pmatrix} \frac{\partial u}{\partial x} & \frac{\partial v}{\partial x} \\ \frac{\partial u}{\partial z} & \frac{\partial v}{\partial z} \end{pmatrix} = \det\begin{pmatrix} 3x^2 & 0 \\ 0 & 2z \end{pmatrix} \neq 0 \quad \text{in} \quad \mathbb{R}^2 \backslash \{(0,0)\}.$$

Also, we have

$$v'(u) = \frac{dv}{du} = \frac{dz^2}{dx^3} = \frac{2zdz}{3x^2dx} = \frac{2z}{3x^2} z',$$

$$z' = \frac{3x^2}{2z} v'.$$

From here and (26) *we obtain*

$$\frac{2z}{3x^2} z' = \frac{2x^3}{z^2 - x^3}$$

or

$$v' = \frac{2u}{v - u} = \frac{2}{\frac{v}{u} - 1}. \tag{27}$$

We set

$$\frac{v}{u} = q \qquad \text{or} \qquad v = qu,$$

whereupon

$$v' = q'u + q.$$

From the last expression and (27) *we get*

$$q'u+q=\frac{2}{q-1},$$

and

$$q'u=-\frac{q^2-q-2}{q-1} \tag{28}$$

and for $u\neq 0$, $q\neq 1,2$,

$$\frac{q-1}{(q-2)(q+1)}dq=-\frac{du}{u} \qquad \Longrightarrow$$

$$\int\frac{q-1}{(q-2)(q+1)}dq=-\int\frac{du}{u}+C_1 \qquad \Longrightarrow$$

$$\frac{2}{3}\log|q+1|+\frac{1}{3}\log|q-2|=-\log|u|+C_1 \qquad \Longrightarrow$$

$$\log\left((q+1)^2|q-2|\right)=-3\log|u|+C_2, \qquad C_2=3C_1, \qquad \Longrightarrow$$

$$(q+1)^2(q-2)u^3=C_3, \qquad C_3=e^{C_2}\mathrm{sign}((q-2)u) \qquad \Longrightarrow$$

$$\left(y^2e^{-2x}+x^3\right)^2\left(y^2e^{-2x}-2x^3\right)=C_3$$

is a solution of (17), *where* C_1 *is a constant.*

Also, $q=-1,2$ *are solutions of* (28), *from where*

$$y^2e^{-2x}+x^3=0, \qquad y^2e^{-2x}-2x^3=0$$

are solutions of the equation (17).

Consequently, the solutions of the equation (17) *are given by*

$$\left(y^2e^{-2x}+x^3\right)^2\left(y^2e^{-2x}-2x^3\right)=C,$$

where C *is a constant.*

Exercise 2.0.99. *Let* $\hat{T}(x)=e^x$,

$$\frac{M(x,y)}{N(x,y)}=\frac{2x+y}{x+5y}.$$

Find the solutions of (17) *using the transformation*

$$u=x-z, \qquad v=x+2z.$$

Answer.

$$2x^2+2xye^{-x}+5y^2e^{-2x}=C,$$

where C is a constant.

Exercise 2.0.100. *Let* $\hat{T}(x) = e^x$,

$$\frac{M(x,y)}{N(x,y)} = \frac{x+2y}{y-2x}.$$

Find the solutions of the equation (17) *using the transformation*

$$x = r\cos\theta, \qquad z = r\sin\theta,$$

where $z = \hat{y}$.

Answer.

$$\sqrt{x^2+y^2e^{-2x}} = Ce^{2\tan^{-1}\frac{ye^{-x}}{x}},$$

where C is a constant.

Advanced Practical Exercises

Problem 2.0.101. *Let*

$$\hat{T}(x) = e^x, \quad X_1(x) = (x^2-1)(x+2)e^x, \quad X_2(x) = -(x-1)^2, \quad Y_1(y) = y^2+1, \quad Y_2(y) = y.$$

Determine the equation (1) *and find its solutions.*

Answer. The equation (1) takes the form

$$(x^2-1)(x+2)(z^2+1)+(x-1)zz' = 0.$$

Its solutions are given by

$$2x^3+9x^2+6\log\left(y^2+e^{2x}\right) = C, \quad x = 1,$$

where C is a constant.

Problem 2.0.102. *Let*

$$\hat{T}(x) = e^x, \quad X_1(x) = X_2(x) = x, \quad Y_1(y) = y^6+1, \quad Y_2(y) = y^5.$$

Determine the equation (5) *and find its solution.*

Answer. The equation (5) takes the form

$$x(z^6+1)+\frac{xe^x}{1-x}z^5z' = 0.$$

Its solutions are given by

$$6x\left(e^{-x}-1\right)+\log\left(y^6+e^{6x}\right) = C,$$

where C is a constant. Also, $x = 0$ is a solution.

Problem 2.0.103. *Let $\hat{T}(x)=e^x$, $X_1(x)=X_2(x)=x$, $Y_1(y)=y^3+1$, $Y_2(y)=\frac{1}{3}y^2$. Determine the equation* (7) *and find its solutions.*

Answer. The equation (7) takes the form

$$x(z^3+1)e^{-2x}+\frac{1}{3}\frac{xz^2}{1-x}e^{-x}z'=0.$$

Its solutions are given by

$$xe^{-x}+\log\left|y^3+e^{3x}\right|=C,$$

where C is a constant. Also, $x=0$ and $y=-e^x$ are its solutions.

Problem 2.0.104. *Prove that the solutions of the equation*

$$\hat{X}_1^{\wedge}(\hat{x})\hat{\times}\hat{Y}_1(\hat{y})+\hat{X}_2^{\wedge}(\hat{x})\hat{\times}Y_2(\hat{y})\hat{\times}\left(\hat{y}^{\wedge}(\hat{x})\right)^{\circledast}=0$$

are given by

$$\int\frac{X_1(x)}{X_2(x)}\frac{\hat{T}(x)-x\hat{T}'(x)}{\hat{T}^3(x)}dx+\int\frac{Y_2(z)}{Y_1(z)}dz=C,$$

where C is a constant, in the case when $X_2(x)\left(\hat{T}(x)-x\hat{T}'(x)\right)Y_1(z)\neq 0$.

Problem 2.0.105. *Prove that the solutions of the equation*

$$\hat{X}_1(\hat{x})\hat{Y}_1(\hat{y})+X_2(x)\hat{\times}Y_2(\hat{y})\hat{\times}\left(\hat{y}^{\wedge}(\hat{x})\right)^{\circledast}=0$$

are given by

$$\int\frac{X_1\left(\frac{x}{\hat{T}(x)}\right)}{X_2(x)}\frac{\hat{T}(x)-x\hat{T}'(x)}{\hat{T}^5(x)}dx+\int\frac{Y_2(z)}{Y_1(z)}dz=C,$$

where C is a constant, in the case when $X_2(x)\left(\hat{T}(x)-x\hat{T}'(x)\right)Y_1(z)\neq 0$.

Problem 2.0.106. *Prove that the solutions of the following iso-differential equation*

$$\hat{X}_1^{\wedge}(\hat{x})Y_1(\hat{y}^{\wedge})+\hat{X}_2^{\wedge}(\hat{x})\hat{\times}\hat{Y}_2(\hat{y}^{\wedge})\hat{\times}\left(\hat{y}^{\wedge}(x)\right)^{\circledast}=0$$

are given by

$$\int\frac{X_1(x)\left(\hat{T}(x)-x\hat{T}'(x)\right)}{X_2(x)\hat{T}^2(x)}dx+\int\frac{Y_2(z)}{Y_1(z)}dz=C,$$

where C is a constant, in the case when $X_2(x)\neq 0$, $\hat{T}(x)-x\hat{T}'(x)\neq 0$, $Y_1(z)\neq 0$ in S. Here $z=\hat{y}^{\wedge}(x)$.

Problem 2.0.107. *Prove that the solutions of the equation*

$$\hat{X}_1^{\wedge}(\hat{x})Y_1(y^{\wedge})+\hat{X}_2^{\wedge}(\hat{x})\hat{\times}Y_2(y^{\wedge})\hat{\times}\left(y^{\wedge}(x)\right)^{\circledast}=0$$

are given by

$$\int\frac{X_1(x)\left(\hat{T}(x)-x\hat{T}'(x)\right)}{X_2(x)\hat{T}^2(x)}dx+\int\frac{Y_2(z)}{Y_1(z)}dz=C,$$

where C is a constant, $\hat{T}(x)-x\hat{T}'(x)\neq 0$, $X_2(x)\neq 0$ *and* $Y_1(z)\neq 0$ *in S. Here* $z=y^{\wedge}(x)$.

Problem 2.0.108. *Prove that the solutions of the equation*

$$X_1^{\vee}(\hat{x})\hat{Y}_1(y^{\vee})+\hat{X}_2^{\vee}(\hat{x})\hat{\times}\hat{Y}_2(y^{\vee})\left(y^{\vee}(x)\right)^{\circledast}=0$$

are given by

$$\int\frac{X_1(x)(\left(\hat{T}(x)-x\hat{T}'(x)\right)}{X_2(x)\hat{T}(x)}dx+\int\frac{Y_2(z)}{Y_1(z)}dz=C,$$

where C is a constant, in the case when $\hat{T}(x)-x\hat{T}'(x)\neq 0$, $X_2(x)\neq 0$ *and* $Y_1(z)\neq 0$ *in S. Here* $z=y^{\vee}(x)$.

Problem 2.0.109. *Prove that the solutions of the equation*

$$\hat{X}_1^{\wedge}(x)\hat{Y}_1(\hat{y}(\hat{x}))+\hat{X}_2^{\wedge}(\hat{x})\hat{\times}\hat{Y}_2(\hat{y}(\hat{x}))\hat{\times}\left(\hat{y}(\hat{x})\right)^{\circledast}=0$$

are given by

$$\int\frac{X_1\left(x\hat{T}(x)\right)\left(\hat{T}(x)-x\hat{T}'(x)\right)}{X_2(x)\hat{T}^2(x)}dx+\int\frac{Y_2(z)}{Y_1(z)}dz=C,$$

where C is a constant, in the case when $\hat{T}(x)-x\hat{T}'(x)\neq 0$, $X_2(x)\neq 0$ *and* $Y_1(z)\neq 0$ *in S. Here* $z=\hat{y}(\hat{x})$.

Problem 2.0.110. *Let* $\hat{T}(x)=x^2+1$, $M(x,y)=-y-x\sin\frac{y-x}{x}$, $N(x,y)=x$, $x\neq 0$. *Find the general solution of the equation* (17).

Answer.

$$\tan\frac{y-x^3-x}{2x^3+2x}=Cx,$$

where C is a constant.

Problem 2.0.111. *Let* $\hat{T}(x)=e^x$,

$$\frac{M(x,y)}{N(x,y)}=-\frac{3x-y-5}{-x+3y+7}.$$

Find the general solution of (17).

Answer.

$$\left(x+ye^{-x}+1\right)^2\left(ye^{-x}-x+3\right)=C,$$

where C is a constant.

Problem 2.0.112. *Let* $\hat{T}(x)=e^x$,

$$\frac{M(x,y)}{N(x,y)}=-\frac{(2x^2+3y^2-7)x}{(3x^2+2y^2-8)y}.$$

Find the solutions of the equation (17) *using the transformation*

$$u=x^2, \qquad v=z^2,$$

where $z=\hat{y}$.

Answer.

$$\left(x^2-y^2e^{-2x}-1\right)^5=C\left(x^2+y^2e^{-2x}-3\right),$$

where C is a constant.

Chapter 3

First-Order Linear Equations

In this chapter we will suppose that

$$\hat{T} \in C^1(\mathbb{R}), \qquad \hat{T}(x) > 0 \qquad \text{for} \qquad \forall x \in \mathbb{R}. \tag{1}$$

Let also, J be an interval in $\mathbb{R}$.

We consider the equation

$$\left(\hat{y}^{\wedge\wedge}\right)^{\circledast} = \hat{a}^{\wedge}(\hat{x})\hat{\times}\hat{y}^{\wedge\wedge} + \hat{b}^{\wedge}(\hat{x}), \tag{2}$$

where the functions $a, b \in C(J)$.

The equation (2) we can rewrite in the following form

$$\frac{\frac{y'(x)\hat{T}(x)-y(x)\hat{T}'(x)}{\hat{T}(x)\left(\hat{T}(x)-x\hat{T}'(x)\right)}} {} = \frac{a(x)}{\hat{T}(x)}\hat{T}(x)\frac{y(x)}{\hat{T}(x)} + \frac{b(x)}{\hat{T}(x)},$$

or

$$\frac{y'(x)\hat{T}(x)-y(x)\hat{T}'(x)}{\hat{T}(x)-x\hat{T}'(x)} = a(x)y(x)+b(x),$$

or

$$y'(x)\hat{T}(x)-y(x)\hat{T}'(x) = a(x)\left(\hat{T}(x)-x\hat{T}'(x)\right)y(x)+b(x)\left(\hat{T}(x)-x\hat{T}'(x)\right),$$

or

$$y'(x)\hat{T}(x) = \left(a(x)\left(\hat{T}(x)-x\hat{T}'(x)\right)+\hat{T}'(x)\right)y(x)+b(x)\left(\hat{T}(x)-x\hat{T}'(x)\right),$$

or

$$y'(x) = \left(a(x)\frac{\hat{T}(x)-x\hat{T}'(x)}{\hat{T}(x)}+\frac{\hat{T}'(x)}{\hat{T}(x)}\right)y(x)+b(x)\frac{\hat{T}(x)-x\hat{T}'(x)}{\hat{T}(x)}. \tag{3}$$

Definition 3.0.113. *The equation* (2) *will be called first-order linear iso-differential equation.*

The corresponding homogeneous equation

$$y'(x) = \left(a(x)\frac{\hat{T}(x)-x\hat{T}'(x)}{\hat{T}(x)}+\frac{\hat{T}'(x)}{\hat{T}(x)}\right)y(x) \tag{4}$$

obtained by taking $b(x)=0$ in (2)can be solved by separating the variables, i.e.,

$$\frac{dy}{y}=\left(a(x)\frac{\hat{T}(x)-x\hat{T}'(x)}{\hat{T}(x)}+\frac{\hat{T}'(x)}{\hat{T}(x)}\right)dx$$

and now integrating it to obtain

$$y(x)=Ce^{\int\left(a(x)\frac{\hat{T}(x)-x\hat{T}'(x)}{\hat{T}(x)}+\frac{\hat{T}'(x)}{\hat{T}(x)}\right)dx}, \tag{5}$$

where C is a constant.

In dividing (4) by y we have lost the solution $y(x)\equiv 0$, which is called trivial solution. However, it is included in (5) with $C=0$.

If $x_0\in J$, then the function

$$y(x)=y_0e^{\int_{x_0}^{x}\left(a(t)\frac{\hat{T}(t)-t\hat{T}'(t)}{\hat{T}(t)}+\frac{\hat{T}'(t)}{\hat{T}(t)}\right)dt}$$

clearly satisfies the equation (3) in J and passes through the point (x_0,y_0). Thus, if it is the solution of the initial value problem (4), (6), where

$$y(x_0)=y_0. \tag{6}$$

To find the solution of the iso-differential equation (2) we shall use the method of variation of parameters due to Lagrange. In (5) we assume that C is a function of x, i.e.,

$$y(x)=C(x)e^{\int\left(a(x)\frac{\hat{T}(x)-x\hat{T}'(x)}{\hat{T}(x)}+\frac{\hat{T}'(x)}{\hat{T}(x)}\right)dx} \tag{7}$$

and search for $C(x)$ so that (7) becomes a solution of the iso-differential equation (2). For this, substituting (7) in (3) we find

$$C'(x)e^{\int\left(a(x)\frac{\hat{T}(x)-x\hat{T}'(x)}{\hat{T}(x)}+\frac{\hat{T}'(x)}{\hat{T}(x)}\right)dx}+C(x)\left(a(x)\frac{\hat{T}(x)-x\hat{T}'(x)}{\hat{T}(x)}+\frac{\hat{T}'(x)}{\hat{T}(x)}\right)e^{\int\left(a(x)\frac{\hat{T}(x)-x\hat{T}'(x)}{\hat{T}(x)}+\frac{\hat{T}'(x)}{\hat{T}(x)}\right)dx}$$

$$=\left(a(x)\frac{\hat{T}(x)-x\hat{T}'(x)}{\hat{T}(x)}+\frac{\hat{T}'(x)}{\hat{T}(x)}\right)C(x)e^{\int\left(a(x)\frac{\hat{T}(x)-x\hat{T}'(x)}{\hat{T}(x)}+\frac{\hat{T}'(x)}{\hat{T}(x)}\right)dx}+b(x)\frac{\hat{T}(x)-x\hat{T}'(x)}{\hat{T}(x)},$$

whereupon

$$C'(x)=b(x)\frac{\hat{T}(x)-x\hat{T}'(x)}{\hat{T}(x)}e^{-\int\left(a(x)\frac{\hat{T}(x)-x\hat{T}'(x)}{\hat{T}(x)}+\frac{\hat{T}'(x)}{\hat{T}(x)}\right)dx}.$$

Then

$$C(x)=\int b(x)\frac{\hat{T}(x)-x\hat{T}'(x)}{\hat{T}(x)}e^{-\int\left(a(x)\frac{\hat{T}(x)-x\hat{T}'(x)}{\hat{T}(x)}+\frac{\hat{T}'(x)}{\hat{T}(x)}\right)dx}dx+C_1,$$

where C_1 is a constant.

Now substituting this $C(x)$ in (7), we find the solution (2) as

$$y(x)=e^{\int\left(a(x)\frac{\hat{T}(x)-x\hat{T}'(x)}{\hat{T}(x)}+\frac{\hat{T}'(x)}{\hat{T}(x)}\right)dx}\Bigg(C_1+\int b(x)\frac{\hat{T}(x)-x\hat{T}'(x)}{\hat{T}(x)}e^{-\int\left(a(x)\frac{\hat{T}(x)-x\hat{T}'(x)}{\hat{T}(x)}+\frac{\hat{T}'(x)}{\hat{T}(x)}\right)dx}dx\Bigg) \tag{8}$$

Definition 3.0.114. *The function* (8) *will be called the general solution of* (2).

From (8) the solution of the initial value problem (2), (6), where $x_0 \in J$, is easily obtained as

$$y(x) = e^{\int_{x_0}^x \left(a(t)\frac{\hat{T}(t)-t\hat{T}'(t)}{\hat{T}(t)}+\frac{\hat{T}'(t)}{\hat{T}(t)}\right)dt}\left(y_0+\int_{x_0}^x b(t)\frac{\hat{T}(t)-t\hat{T}'(t)}{\hat{T}(t)}e^{-\int_{x_0}^t\left(a(s)\frac{\hat{T}(s)-s\hat{T}'(s)}{\hat{T}(s)}+\frac{\hat{T}'(s)}{\hat{T}(s)}\right)ds}dt\right).$$

Example 3.0.115. *Let* $\hat{T}(x) = e^x$, $a(x) = x$, $b(x) = e^{x+\frac{x^2}{2}-\frac{x^3}{6}}$. *Then*

$$\frac{\hat{T}(x)-x\hat{T}'(x)}{\hat{T}(x)} = \frac{e^x-xe^x}{e^x} = 1-x,$$

$$\frac{\hat{T}'(x)}{\hat{T}(x)} = \frac{e^x}{e^x} = 1.$$

The equation (2) *takes the form*

$$y' = \left(x(1-x)+1\right)y+e^{x+\frac{x^2}{2}-\frac{x^3}{6}}(1-x)$$

$$y' = (1+x-x^2)y+(1-x)e^{x+\frac{x^2}{2}-\frac{x^3}{6}}.$$

Then

$$y(x) = e^{\int\left(1+x-\frac{x^2}{2}\right)dx}\left(C+\int(1-x)e^{x+\frac{x^2}{2}-\frac{x^3}{6}}e^{-\int\left(1+x-\frac{x^2}{2}\right)dx}dx\right)$$

$$= e^{x+\frac{x^2}{2}-\frac{x^3}{6}}\left(C+\int(1-x)e^{x+\frac{x^2}{2}-\frac{x^3}{3}}e^{-x-\frac{x^2}{2}+\frac{x^3}{6}}dx\right)$$

$$= e^{x+\frac{x^2}{2}-\frac{x^3}{6}}\left(C+x-\frac{x^2}{2}\right),$$

where C is a constant, is the solution of the considered equation.

Exercise 3.0.116. *Let* $\hat{T}(x) = e^x$,

$$a(x) = \frac{4-x}{x(1-x)}, \qquad b(x) = -\frac{2x^2+4}{x(1-x)}, \qquad x \neq 0,1.$$

Find the solution of (2) *for which*

$$\lim_{x\longrightarrow 1} y(x) = 1.$$

Answer. $y(x) = -x^4+x^2+1$.

Exercise 3.0.117. *Let* a, $b \in C(J)$. *Prove that*

$$y(x) = e^{\int\left(a(x)\frac{\hat{T}(x)-x\hat{T}'(x)}{\hat{T}^2(x)}+\frac{\hat{T}'(x)}{\hat{T}(x)}\right)dx}\Bigg(C$$

$$+\int b(x)\frac{\hat{T}(x)-x\hat{T}'(x)}{\hat{T}(x)}e^{-\int\left(a(x)\frac{\hat{T}(x)-x\hat{T}'(x)}{\hat{T}^2(x)}+\frac{\hat{T}'(x)}{\hat{T}(x)}\right)dx}dx\Bigg),$$

where C is a constant, is the general solution to the equation

$$\left(\hat{y}^{\wedge\wedge}\right)^{\circledast} = \hat{a}^{\wedge}(\hat{x})\hat{y}^{\wedge\wedge} + \hat{b}^{\wedge}(\hat{x}).$$

Exercise 3.0.118. *Let $a, b \in C(J)$. Prove that*

$$y(x) = e^{\int \left(a\left(\frac{x}{\hat{T}(x)}\right)\left(\hat{T}(x)-x\hat{T}'(x)\right)+1\right)dx}\Bigg(C+$$

$$\int b\left(\frac{x}{\hat{T}(x)}\right)\left(\hat{T}(x)-x\hat{T}'(x)\right)e^{-\int\left(a\left(\frac{x}{\hat{T}(x)}\right)\left(\hat{T}(x)-x\hat{T}'(x)\right)+1\right)dx}dx\Bigg)$$

is the general solution to the equation

$$\left(\hat{y}^{\wedge\wedge}\right)^{\circledast} = \hat{a}(\hat{x})\hat{\times}y(x) + \hat{b}(\hat{x}).$$

Here C is a constant.

Exercise 3.0.119. *Let $a, b \in C(J)$. Prove that*

$$y(x) = e^{\int\left(a(x)\left(\hat{T}(x)-x\hat{T}'(x)\right)+\frac{\hat{T}'(x)}{\hat{T}(x)}\right)dx}\Bigg(C$$

$$+\int b(x)\left(\hat{T}(x)-x\hat{T}'(x)\right)e^{-\int\left(a(x)\left(\hat{T}(x)-x\hat{T}'(x)\right)+\frac{\hat{T}'(x)}{\hat{T}(x)}\right)dx}dx\Bigg),$$

where C is a constant, is the general solution to the equation

$$\left(\hat{y}^{\wedge\wedge}\right)^{\circledast} = a(x)y(x) + b(x).$$

Certain nonlinear first-order iso-differential equations can be reduced to linear equations by an appropriate change of variables. For example, it is always possible for the iso-Bernoulli equations

$$\left(\hat{y}^{\wedge\wedge}\right)^{\circledast} = \hat{a}^{\wedge}(\hat{x})\hat{\times}\hat{y}^{\wedge\wedge} + \hat{b}^{\wedge}(\hat{x})\hat{\times}\left(\hat{y}^{\wedge\wedge}\right)^{\hat{n}}, \tag{9}$$

where $n \in \mathbb{N}, n \neq 1$. This equation we can rewrite in the following form

$$\frac{y'\hat{T}(x) - y\hat{T}'(x)}{\hat{T}(x)\left(\hat{T}(x) - x\hat{T}'(x)\right)} = a(x)\frac{y}{\hat{T}(x)} + \frac{b(x)}{\hat{T}(x)}\hat{T}(x)\frac{y^n}{\hat{T}(x)},$$

or

$$\frac{y'\hat{T}(x) - y\hat{T}'(x)}{\hat{T}(x) - x\hat{T}'(x)} = a(x)y + b(x)y^n,$$

or

$$y'\hat{T}(x) - y\hat{T}'(x) = a(x)\left(\hat{T}(x) - x\hat{T}'(x)\right)y + b(x)\left(\hat{T}(x) - x\hat{T}'(x)\right)y^n,$$

or

$$y'\hat{T}(x) = \Big(a(x)\Big(\hat{T}(x) - x\hat{T}'(x)\Big) + \hat{T}'(x)\Big)y + b(x)\Big(\hat{T}(x) - x\hat{T}'(x)\Big)y^n,$$

or

$$y' = \left(a(x)\frac{\hat{T}(x) - x\hat{T}'(x)}{\hat{T}(x)} + \frac{\hat{T}'(x)}{\hat{T}(x)}\right)y + b(x)\frac{\hat{T}(x) - x\hat{T}'(x)}{\hat{T}(x)}y^n. \tag{10}$$

In (9), the cases $n = 0, 1$ are excluded because in these cases this equation is obviously linear.

The equation (10) is equivalent to the equation

$$y^{-n}y' = \left(a(x)\frac{\hat{T}(x) - x\hat{T}'(x)}{\hat{T}(x)} + \frac{\hat{T}'(x)}{\hat{T}(x)}\right)y^{1-n} + b(x)\frac{\hat{T}(x) - x\hat{T}'(x)}{\hat{T}(x)}. \tag{11}$$

We put $z = y^{1-n}$. Then

$$z' = (1-n)y^{-n}y'$$

or

$$y^{-n}y' = \frac{z'}{1-n}.$$

From here, the equation (11) admits the following form

$$z' = (1-n)\left(a(x)\frac{\hat{T}(x) - x\hat{T}'(x)}{\hat{T}(x)} + \frac{\hat{T}'(x)}{\hat{T}(x)}\right)z + (1-n)b(x)\frac{\hat{T}(x) - x\hat{T}'(x)}{\hat{T}(x)}, \tag{12}$$

which is a linear iso-differential equation. For the general solution of the equation (12) we have the following representation

$$z = e^{(1-n)\int\left(a(x)\frac{\hat{T}(x) - x\hat{T}'(x)}{\hat{T}(x)} + \frac{\hat{T}'(x)}{\hat{T}(x)}\right)dx}\Bigg(C +$$

$$(1-n)\int b(x)\frac{\hat{T}(x) - x\hat{T}'(x)}{\hat{T}(x)}e^{(n-1)\int\left(a(x)\frac{\hat{T}(x) - x\hat{T}'(x)}{\hat{T}(x)} + \frac{\hat{T}'(x)}{\hat{T}(x)}\right)dx}dx\Bigg),$$

where C is a constant. Hence, for the general solution of the iso-differential equation (9) we have, $y \neq 0$,

$$y = e^{\int\left(a(x)\frac{\hat{T}(x) - x\hat{T}'(x)}{\hat{T}(x)} + \frac{\hat{T}'(x)}{\hat{T}(x)}\right)dx}\Bigg(C +$$

$$(1-n)\int b(x)\frac{\hat{T}(x) - x\hat{T}'(x)}{\hat{T}(x)}e^{(n-1)\int\left(a(x)\frac{\hat{T}(x) - x\hat{T}'(x)}{\hat{T}(x)} + \frac{\hat{T}'(x)}{\hat{T}(x)}\right)dx}dx\Bigg)^{\frac{1}{1-n}},$$

also $y = 0$ is a solution of (9).

Example 3.0.120. *Let* $\hat{T}(x) = e^x$, $a(x) = -\frac{x+1}{x(1-x)}$, $b(x) = \frac{x}{1-x}$, $x \neq 0, 1$. *Then*

$$\frac{\hat{T}(x) - x\hat{T}'(x)}{\hat{T}(x)} = \frac{e^x - xe^x}{e^x}$$

$$= 1 - x,$$

$$\frac{\hat{T}'(x)}{\hat{T}(x)} = 1,$$

$$a(x)\frac{\hat{T}(x) - x\hat{T}'(x)}{\hat{T}(x)} + \frac{\hat{T}'(x)}{\hat{T}(x)} = -\frac{x+1}{x(1-x)}(1-x) + 1$$

$$= -\frac{1}{x},$$

$$b(x)\frac{\hat{T}(x) - x\hat{T}'(x)}{\hat{T}(x)} = \frac{x}{1-x}(1-x)$$

$$= x.$$

The equation (9) *takes the form*

$$y' = -\frac{1}{x}y + xy^2. \tag{13}$$

We note that $y = 0$ *is an its solution.*

For $y \neq 0$, *using* (13), *we get*

$$y^{-2}y' = -\frac{1}{x}y^{-1} + x.$$

We set $z = y^{-1}$. *Then*

$$z' = \frac{1}{x}z - x$$

and

$$z = e^{\int \frac{1}{x}dx}\Big(C - \int xe^{-\int \frac{1}{x}dx}dx\Big) = x\Big(C - \int dx\Big) = Cx - x^2,$$

where C is a constant. Consequently

$$y = \frac{1}{Cx - x^2}, \qquad C \neq x.$$

Exercise 3.0.121. *Let* $\hat{T}(x) = e^x$, $a(x) = \frac{\cos x}{1-x}$, $b(x) = \frac{\cos x}{1-x}$, $x \neq 1$. *Determine the equation* (9) *and find its general solution.*

Answer. The equation (9) admits the following form

$$y' = \Big(\cos x\Big)y + \Big(\cos x\Big)y^4.$$

$y = 0$ is an its solution and its general solution is given by

$$y = \frac{1}{\Big(Ce^{-3\sin x} - 1\Big)^{\frac{1}{3}}}, \qquad C \neq e^{3\sin x},$$

where C is a constant.

Exercise 3.0.122. *Prove that*

$$y = e^{\int\left(a(x)\frac{\hat{T}(x)-x\hat{T}'(x)}{\hat{T}(x)}+\frac{\hat{T}'(x)}{\hat{T}(x)}\right)dx}\Bigg(C+$$

$$(1-n)\int b(x)\frac{\hat{T}(x)-x\hat{T}'(x)}{\hat{T}^{n+1}(x)}e^{(n-1)\int\left(a(x)\frac{\hat{T}(x)-x\hat{T}'(x)}{\hat{T}(x)}+\frac{\hat{T}'(x)}{\hat{T}(x)}\right)dx}dx\Bigg)^{\frac{1}{1-n}},$$

where C is a constant, is the general solution to the equation

$$\left(\hat{y}^{\wedge\wedge}\right)^{\circledast} = \hat{a}^{\wedge}(\hat{x})\hat{\times}\hat{y}^{\wedge}(\hat{x}) + \hat{b}^{\wedge}(\hat{x})\hat{\times}\left(\hat{y}^{\wedge\wedge}\right)^{n}.$$

Other nonlinear iso-differential equation is the iso-Riccati equation

$$\left(\hat{y}^{\wedge\wedge}\right)^{\circledast} = \hat{a}^{\wedge}(\hat{x})\hat{\times}\hat{y}^{\wedge\wedge} + \hat{b}^{\wedge}(\hat{x})\hat{\times}\left(\hat{y}^{\wedge\wedge}\right)^{\hat{2}} + \hat{c}^{\wedge}(\hat{x}), \tag{14}$$

where $c \in C(J)$.

The equation (14) we can represent into the form

$$\frac{y'\hat{T}(x)-y\hat{T}'(x)}{\hat{T}(x)}\left(\hat{T}(x)-x\hat{T}'(x)\right) = a(x)\frac{y}{\hat{T}(x)} + b(x)\frac{y^2}{\hat{T}(x)} + \frac{c(x)}{\hat{T}(x)}$$

or

$$\frac{y'\hat{T}(x)-y\hat{T}'(x)}{\hat{T}(x)-x\hat{T}'(x)} = a(x)y + b(x)y^2 + c(x),$$

or

$$y'\hat{T}(x)-y\hat{T}'(x) = a(x)\left(\hat{T}(x)-x\hat{T}'(x)\right)y + b(x)\left(\hat{T}(x)-x\hat{T}'(x)\right)y^2$$

$$+c(x)\left(\hat{T}(x)-x\hat{T}'(x)\right),$$

or

$$y'\hat{T}(x) = \left(a(x)\left(\hat{T}(x)-x\hat{T}'(x)\right)+\hat{T}'(x)\right)y + b(x)\left(\hat{T}(x)-x\hat{T}'(x)\right)y^2$$

$$+c(x)\left(\hat{T}(x)-x\hat{T}'(x)\right),$$

or

$$y' = \left(a(x)\frac{\hat{T}(x)-x\hat{T}'(x)}{\hat{T}(x)}+\frac{\hat{T}'(x)}{\hat{T}(x)}\right)y + b(x)\frac{\hat{T}(x)-x\hat{T}'(x)}{\hat{T}(x)}y^2 + c(x)\frac{\hat{T}(x)-x\hat{T}'(x)}{\hat{T}(x)}.$$

If one solution $y_1(x)$ of the iso-Riccati equation (14) is known, then the substitution

$$y = y_1(x) + z^{-1} \qquad \text{or} \qquad y_1(x) - y = -z^{-1}$$

converts it into a first-order linear iso-differential equation in z. Indeed, we have

$$y' = y_1'(x) - \frac{1}{z^2}z',$$

or

$$\frac{1}{z^2}z' = y_1'(x) - y'$$

$$= \left(a(x)\frac{\hat{T}(x)-x\hat{T}'(x)}{\hat{T}(x)} + \frac{\hat{T}'(x)}{\hat{T}(x)}\right)\left(y_1(x)-y\right)$$

$$+b(x)\frac{\hat{T}(x)-x\hat{T}'(x)}{\hat{T}(x)}\left(y_1(x)-y\right)\left(y_1(x)+y\right)$$

$$= \left(a(x)\frac{\hat{T}(x)-x\hat{T}'(x)}{\hat{T}(x)} + \frac{\hat{T}'(x)}{\hat{T}(x)}\right)\left(-\frac{1}{z}\right)$$

$$+b(x)\frac{\hat{T}(x)-x\hat{T}'(x)}{\hat{T}(x)}\left(-\frac{1}{z}\right)\left(2y_1(x)+\frac{1}{z}\right)$$

$$= \left(\left(a(x)+2b(x)y_1(x)\right)\frac{\hat{T}(x)-x\hat{T}'(x)}{\hat{T}(x)} + \frac{\hat{T}'(x)}{\hat{T}(x)}\right)\left(-\frac{1}{z}\right)$$

$$-b(x)\frac{\hat{T}(x)-x\hat{T}'(x)}{\hat{T}(x)}\frac{1}{z^2},$$

or

$$z' = -\left(\left(a(x)+2b(x)y_1(x)\right)\frac{\hat{T}(x)-x\hat{T}'(x)}{\hat{T}(x)} + \frac{\hat{T}'(x)}{\hat{T}(x)}\right)z$$

$$-b(x)\frac{\hat{T}(x)-x\hat{T}'(x)}{\hat{T}(x)}.$$

Example 3.0.123. *Let* $\hat{T}(x) = e^x$,

$$a(x) = \frac{2x+1}{x-1}, \qquad b(x) = \frac{1}{1-x}, \qquad c(x) = \frac{1+x^2}{1-x}, \qquad x \neq 1.$$

Then the equation (14) *takes the form*

$$y' = -2xy + y^2 + 1 + x^2.$$

We note that $y_1(x) = x$ *is its particular solution. Let* $y = x + \frac{1}{z}$. *Then*

$$y' = 1 - \frac{z'}{z^2},$$

$$1 - \frac{z'}{z^2} = -2x\left(x+\frac{1}{z}\right) + \left(x+\frac{1}{z}\right)^2 + 1 + x^2,$$

$$z' = -1,$$

$$z = -x + C,$$

where C *is a constant.*

Therefore the general solution of (14) *is given by*

$$y = x + \frac{1}{C-x}, \qquad x \neq C.$$

Exercise 3.0.124. *Let* $\hat{T}(x) = 1+x^2$,

$$a(x) = -\frac{1+3x^2}{x(1-x^2)}, \qquad b(x) = -\frac{1+x^2}{1-x^2}, \qquad c(x) = 4\frac{1+x^2}{x^2(1-x^2)}, \qquad x \neq 0, \pm 1.$$

Determine the equations (14) *and its general solution.*

Solution. We have

$$\hat{T}(x) - x\hat{T}'(x) = x^2 + 1 - 2x^2$$

$$= 1 - x^2,$$

$$a(x)\frac{\hat{T}(x)-x\hat{T}'(x)}{\hat{T}(x)} + \frac{\hat{T}'(x)}{\hat{T}(x)} = -\frac{1+3x^2}{x(1-x^2)}\frac{1-x^2}{1+x^2} + \frac{2x}{1+x^2}$$

$$= -\frac{1+3x^2}{x(1+x^2)} + \frac{2x}{1+x^2}$$

$$= \frac{-1-3x^2+2x^2}{x(1+x^2)}$$

$$= -\frac{1}{x},$$

$$b(x)\frac{\hat{T}(x)-x\hat{T}'(x)}{\hat{T}(x)} = -\frac{1+x^2}{1-x^2}\frac{1-x^2}{1+x^2}$$

$$= -1,$$

$$c(x)\frac{\hat{T}(x)-x\hat{T}'(x)}{\hat{T}(x)} = 4\frac{1+x^2}{x^2(1-x^2)}\frac{1-x^2}{1+x^2}$$

$$= \frac{4}{x^2}.$$

Then the equation (14) can be represented in the following form

$$y'(x) = -\frac{1}{x}y(x) - y^2(x) + \frac{4}{x^2}. \tag{15}$$

We will search a particular solution of the equation (15) in the form

$$y_1(x) = \frac{a}{x},$$

where a is a constant. We put $y_1(x)$ in (15) and we get

$$-\frac{a}{x^2} = -\frac{a}{x^2} - \frac{a^2}{x^2} + \frac{4}{x^2}$$

or

$$a^2 = 4.$$

Let

$$y_1(x) = \frac{2}{x}.$$

We set

$$y(x) = y_1(x) + \frac{1}{z(x)}$$

$$= \frac{2}{x} + \frac{1}{z(x)}.$$

Then

$$y'(x) = -\frac{2}{x^2} - \frac{z'(x)}{z^2(x)}.$$

From here, using (15), we get

$$-\frac{2}{x^2} - \frac{z'(x)}{z^2(x)} = -\frac{1}{x}\left(\frac{2}{x} + \frac{1}{z(x)}\right) - \left(\frac{2}{x} + \frac{1}{z(x)}\right)^2 + \frac{4}{x^2} \implies$$

$$-\frac{2}{x^2} - \frac{z'(x)}{z^2(x)} = -\frac{2}{x^2} - \frac{1}{xz(x)} - \frac{4}{x^2} - \frac{4}{xz(x)} + \frac{1}{z^2(x)} + \frac{4}{x^2} \implies$$

$$-\frac{z'(x)}{z^2(x)} - \frac{5}{xz(x)} + \frac{1}{z^2(x)} \implies$$

$$z'(x) = \frac{5}{x}z(x) - 1,$$

which is a linear equation. For its general solution we have

$$z(x) = e^{5\int \frac{dx}{x}}\left(C - \int e^{-5\int \frac{dx}{x}} dx\right)$$

$$= x^5\left(C - \int \frac{1}{x^5} dx\right)$$

$$= x^5\left(C + \frac{1}{4x^4}\right)$$

$$= Cx^5 + \frac{x}{4}.$$

From here,

$$y(x) = \frac{2}{x} + \frac{4}{4Cx^5 + x}$$

is the general solution to the equation (14). Here C is a constant.

Exercise 3.0.125. *Let* $\hat{T}(x) = 1 + x^4$,

$$a(x) = \frac{1 + 2x - 3x^4 + 2x^5}{1 - 3x^4}, \qquad b(x) = -\frac{1 + x^4}{x(1 - 3x^4)}, \qquad c(x) = -\frac{x(1 + x^4)}{1 - 3x^4}, \qquad x \neq 0, \pm\sqrt[4]{3}.$$

Determine the equation (14) *and find its general solution.*

Answer. The equation (14) can be represented in the following form

$$y'(x) = \frac{2x + 1}{x} y(x) - \frac{1}{x} y^2(x) - x.$$

Its general solution is given by

$$y(x) = x + \frac{x}{x + C}, \qquad C \neq -x,$$

C is a constant. Also,

$$y(x) = x$$

is an its solution.

Let

$$\hat{T} : [p,q] \longrightarrow [c,d], \quad \hat{T}(x) > 0, \quad x\hat{T}(x) \in [c,d] \quad \text{for} \quad \forall x \in [p,q], \quad \hat{T} \in C^1([p,q]), \tag{16}$$

$$g(x) = x\hat{T}(x), \quad x \in [p,q], \quad \text{is strictly increasing(decreasing) in} \quad [p,q]. \tag{17}$$

We have that $g \in C^1([p,q])$ and there exists $\phi_1 : [c,d] \longrightarrow [p,q]$ such that

$$\phi_1 \in C^1([c,d]), \qquad \phi_1\Big(g(x)\Big) = x \quad \text{for} \quad \forall x \in [p,q].$$

If we set

$$z := x\hat{T}(x), \qquad x \in [p,q],$$

then

$$z \in [c,d], \qquad x = \phi_1(z).$$

Now we consider the equation

$$\Big(\hat{y}^\wedge(\hat{x})\Big)^{\circledast} = \hat{a}^\wedge(\hat{x})\hat{\times}\hat{y}^\wedge(x) + \hat{b}^\wedge(\hat{x}), \tag{18}$$

where a, $b \in C^1([p,q])$.

Definition 3.0.126. *The equation* (18) *will be called first-order linear iso-differential equation.*

The equation (18) we can rewrite in the following form

$$\frac{y'\Big(x\hat{T}(x)\Big)\Big(\hat{T}(x)+x\hat{T}'(x)\Big)\hat{T}(x)-y\Big(x\hat{T}(x)\Big)\hat{T}'(x)}{\hat{T}(x)\Big(\hat{T}(x)-x\hat{T}'(x)\Big)} = \frac{a(x)}{\hat{T}(x)}\hat{T}(x)\frac{y\Big(x\hat{T}(x)\Big)}{\hat{T}(x)} + \frac{b(x)}{\hat{T}(x)}$$

or

$$\frac{y'\Big(x\hat{T}(x)\Big)\Big(\hat{T}(x)+x\hat{T}'(x)\Big)\hat{T}(x)-y\Big(x\hat{T}(x)\Big)\hat{T}'(x)}{\hat{T}(x)\Big(\hat{T}(x)-x\hat{T}'(x)\Big)} = a(x)\frac{y\Big(x\hat{T}(x)\Big)}{\hat{T}(x)} + b(x),$$

or

$$y'\Big(x\hat{T}(x)\Big)\Big(\hat{T}(x)+x\hat{T}'(x)\Big)\hat{T}(x) - y\Big(x\hat{T}(x)\Big)\hat{T}'(x) = a(x)\Big(\hat{T}(x)-x\hat{T}'(x)\Big)y\Big(x\hat{T}(x)\Big)$$

$$+b(x)\Big(\hat{T}(x)-x\hat{T}'(x)\Big),$$

or

$$y'\Big(x\hat{T}(x)\Big)\Big(\hat{T}(x)+x\hat{T}'(x)\Big)\hat{T}(x) = \Big(a(x)\Big(\hat{T}(x)-x\hat{T}'(x)\Big)+\hat{T}'(x)\Big)y\Big(x\hat{T}(x)\Big)$$

$$+b(x)\Big(\hat{T}(x)-x\hat{T}'(x)\Big),$$

or

$$y'\left(x\hat{T}(x)\right) = \frac{a(x)\left(\hat{T}(x)-x\hat{T}'(x)\right)+\hat{T}'(x)}{\left(\hat{T}(x)+x\hat{T}'(x)\right)\hat{T}(x)}y\left(x\hat{T}(x)\right) + \frac{b(x)\left(\hat{T}(x)-x\hat{T}'(x)\right)}{\left(\hat{T}(x)+x\hat{T}'(x)\right)\hat{T}(x)}. \tag{19}$$

Let

$$\psi_1(z) := \frac{a\left(\phi_1(z)\right)\left(\hat{T}\left(\phi_1(z)\right)-\phi_1(z)\hat{T}'\left(\phi_1(z)\right)\right)+\hat{T}'\left(\phi_1(z)\right)}{\left(\hat{T}\left(\phi_1(z)\right)+\phi_1(z)\hat{T}'\left(\phi_1(z)\right)\right)\hat{T}\left(\phi_1(z)\right)},$$

$$\psi_2(z) := \frac{b\left(\phi_1(z)\right)\left(\hat{T}\left(\phi_1(z)\right)-\phi_1(z)\hat{T}'\left(\phi_1(z)\right)\right)}{\left(\hat{T}\left(\phi_1(z)\right)+\phi_1(z)\hat{T}'\left(\phi_1(z)\right)\right)\hat{T}\left(\phi_1(z)\right)}.$$

Then the equation (19) we can rewrite as follows

$$y'(z) = \psi_1(z)y(z) + \psi_2(z).$$

Its general solution is given by

$$y(z) = e^{\int \psi_1(z)dz}\left(C + \int e^{-\int \psi_1(z)dz}\psi_2(z)dz\right),$$

where C is a constant.

Exercise 3.0.127. *Suppose* (16), (17), *and consider the equation*

$$\left(\hat{y}^{\wedge}(x)\right)^{\circledast} = \hat{a}^{\wedge}(x)\hat{\times}\hat{y}^{\wedge}(x) + \hat{b}^{\wedge}(\hat{x}),$$

where $a,\, b \in \mathcal{C}^1([p,q])$. *Deduct its general solution.*

Solution. The given equation we can rewrite in the following form

$$\frac{y'\left(x\hat{T}(x)\right)\left(\hat{T}(x)+x\hat{T}'(x)\right)-y\left(x\hat{T}(x)\right)\hat{T}'(x)}{\hat{T}(x)\left(\hat{T}(x)-x\hat{T}'(x)\right)} = a\left(x\hat{T}(x)\right)\hat{T}(x)\frac{y\left(x\hat{T}(x)\right)}{\hat{T}(x)} + \frac{b(x)}{\hat{T}(x)}$$

or

$$\frac{y'\left(x\hat{T}(x)\right)\left(\hat{T}(x)+x\hat{T}'(x)\right)-y\left(x\hat{T}(x)\right)\hat{T}'(x)}{\hat{T}(x)\left(\hat{T}(x)-x\hat{T}'(x)\right)} = a\left(x\hat{T}(x)\right)y\left(x\hat{T}(x)\right) + \frac{b(x)}{\hat{T}(x)},$$

or

$$y'\left(x\hat{T}(x)\right)\left(\hat{T}(x)+x\hat{T}'(x)\right)\hat{T}(x) - y\left(x\hat{T}(x)\right)\hat{T}'(x)$$

$$= a\left(x\hat{T}(x)\right)\hat{T}(x)\left(\hat{T}(x)-x\hat{T}'(x)\right)y\left(x\hat{T}(x)\right) + b(x)\left(\hat{T}(x)-x\hat{T}'(x)\right),$$

or

$$y'\left(x\hat{T}(x)\right)\left(\hat{T}(x)+x\hat{T}'(x)\right)\hat{T}(x)$$

$$= \left(a\left(x\hat{T}(x)\right)\hat{T}(x)\left(\hat{T}(x)-x\hat{T}'(x)\right)+\hat{T}'(x)\right)y\left(x\hat{T}(x)\right) + b(x)\left(\hat{T}(x)-x\hat{T}'(x)\right),$$

or

$$y'\left(x\hat{T}(x)\right) = \left(a\left(x\hat{T}(x)\right)\frac{\hat{T}(x)-x\hat{T}'(x)}{\hat{T}(x)+x\hat{T}'(x)} + \frac{\hat{T}'(x)}{\hat{T}(x)\left(\hat{T}(x)-x\hat{T}'(x)\right)}\right)y\left(x\hat{T}(x)\right)$$

$$+b(x)\frac{\hat{T}(x)-x\hat{T}'(x)}{\hat{T}(x)\left(\hat{T}(x)+x\hat{T}'(x)\right)}.$$

We set

$$\psi_3(z) = a(z)\frac{\hat{T}\left(\phi_1(z)\right)-\phi_1(z)\hat{T}'\left(\phi_1(z)\right)}{\hat{T}\left(\phi_1(z)\right)+\phi_1(z)\hat{T}'\left(\phi_1(z)\right)} + \frac{\hat{T}'\left(\phi_1(z)\right)}{\hat{T}\left(\phi_1(z)\right)\left(\hat{T}\left(\phi_1(z)\right)-\phi_1(z)\hat{T}'\left(\phi_1(z)\right)\right)}.$$

We obtain the equation

$$y'(z) = \psi_3(z)y(z) + \psi_2(z).$$

Its general solution is given by

$$y(z) = e^{\int \psi_3(z)dz}\left(C + \int \psi_2(z)e^{-\int \psi_3(z)dz}dz\right),$$

where C is a constant.

Exercise 3.0.128. *Suppose* (16), (17), *and consider the equation*

$$\left(y^\wedge(x)\right)^{\circledast} = \hat{a}^\wedge(\hat{x})\,\hat{\times}\,y^\wedge(x) + b^\wedge(\hat{x}),$$

where a, b $\in \mathcal{C}([p,q])$. *Deduct its general solution.*

Solution. The given equation we can rewrite in the form

$$\frac{y'\left(x\hat{T}(x)\right)\hat{T}(x)\left(\hat{T}(x)+x\hat{T}'(x)\right)}{\hat{T}(x)-x\hat{T}'(x)} = \frac{a(x)}{\hat{T}(x)}\hat{T}(x)y\left(x\hat{T}(x)\right)+b(x)$$

or

$$\frac{y'\left(x\hat{T}(x)\right)\hat{T}(x)\left(\hat{T}(x)+x\hat{T}'(x)\right)}{\hat{T}(x)-x\hat{T}'(x)} = a(x)y\left(x\hat{T}(x)\right)+b(x),$$

or

$$y'\left(x\hat{T}(x)\right) = a(x)\frac{\hat{T}(x)-x\hat{T}'(x)}{\hat{T}(x)\left(\hat{T}(x)+x\hat{T}'(x)\right)}y\left(x\hat{T}(x)\right)+b(x)\frac{\hat{T}(x)-x\hat{T}'(x)}{\hat{T}(x)\left(\hat{T}(x)+x\hat{T}'(x)\right)}.$$

We set

$$\psi_4(z) = a\left(\phi_1(z)\right)\frac{\hat{T}\left(\phi_1(z)\right)-\phi_1(z)\hat{T}'\left(\phi_1(z)\right)}{\hat{T}\left(\phi_1(z)\right)\left(\hat{T}\left(\phi_1(z)\right)+\phi_1(z)\hat{T}'\left(\phi_1(z)\right)\right)},$$

$$\psi_5(z) = b\left(\phi_1(z)\right)\frac{\hat{T}\left(\phi_1(z)\right)-\phi_1(z)\hat{T}'\left(\phi_1(z)\right)}{\hat{T}\left(\phi_1(z)\right)\left(\hat{T}\left(\phi_1(z)\right)+\phi_1(z)\hat{T}'\left(\phi_1(z)\right)\right)}.$$

Then we obtain the equation

$$y'(z) = \psi_4(z)y(z) + \psi_5(z).$$

Its general solution is given by

$$y(z) = e^{\int \psi_4(z)dz}\Big(C + \int e^{-\int \psi_4(z)dz}\psi_5(z)dz\Big),$$

where C is a solution.

Below we will suppose (16), (17).

Now we consider the equation

$$\Big(\hat{y}^{\wedge}(x)\Big)^{\circledast} = \hat{a}^{\wedge}(\hat{x})\hat{\times}y^{\wedge}(x) + \hat{b}^{\wedge}(\hat{x}), \tag{20}$$

where a, $b \in C([p,q])$.

The equation (20) we can rewrite in the following form.

$$\frac{y'\Big(x\hat{T}(x)\Big)\Big(\hat{T}(x)+x\hat{T}'(x)\Big) - y\Big(x\hat{T}(x)\Big)\hat{T}'(x)}{\hat{T}(x)\Big(\hat{T}(x)-x\hat{T}'(x)\Big)} = \frac{a(x)}{\hat{T}(x)}\hat{T}(x)y\Big(x\hat{T}(x)\Big) + \frac{b(x)}{\hat{T}(x)}$$

or

$$\frac{y'\Big(x\hat{T}(x)\Big)\Big(\hat{T}(x)+x\hat{T}'(x)\Big)\hat{T}(x) - y\Big(x\hat{T}(x)\Big)\hat{T}'(x)}{\hat{T}(x)-x\hat{T}'(x)} = a(x)\hat{T}(x)y\Big(x\hat{T}(x)\Big) + b(x),$$

or

$$y'\Big(x\hat{T}(x)\Big)\Big(\hat{T}(x)+x\hat{T}'(x)\Big)\hat{T}(x) - y\Big(x\hat{T}(x)\Big)\hat{T}'(x)$$

$$= a(x)\hat{T}(x)\Big(\hat{T}(x)-x\hat{T}'(x)\Big)y\Big(x\hat{T}(x)\Big) + b(x)\Big(\hat{T}(x)-x\hat{T}'(x)\Big),$$

or

$$y'\Big(x\hat{T}(x)\Big)\Big(\hat{T}(x)+x\hat{T}'(x)\Big)\hat{T}(x)$$

$$= \Big(a(x)\hat{T}(x)\Big(\hat{T}(x)-x\hat{T}'(x)\Big)+\hat{T}'(x)\Big)y\Big(x\hat{T}(x)\Big) + b(x)\Big(\hat{T}(x)-x\hat{T}'(x)\Big),$$

or

$$y'\Big(x\hat{T}(x)\Big) = \Big(a(x)\frac{\hat{T}(x)-x\hat{T}'(x)}{\hat{T}(x)+x\hat{T}'(x)} + \frac{\hat{T}'(x)}{\hat{T}(x)\Big(\hat{T}(x)+x\hat{T}'(x)\Big)}\Big)y\Big(x\hat{T}(x)\Big)$$

$$+b(x)\frac{\hat{T}(x)-x\hat{T}'(x)}{\hat{T}(x)\Big(\hat{T}(x)+x\hat{T}'(x)\Big)}.$$

We set

$$\psi_6(z) = a\Big(\phi_1(z)\Big)\frac{\hat{T}\Big(\phi_1(z)\Big)-\phi_1(z)\hat{T}'\Big(\phi_1(z)\Big)}{\hat{T}\Big(\phi_1(z)\Big)\Big(\hat{T}\Big(\phi_1(z)\Big)+\phi_1(z)\hat{T}'\Big(\phi_1(z)\Big)\Big)}$$

$$+\frac{\hat{T}'\Big(\phi_1(z)\Big)}{\hat{T}\Big(\phi_1(z)\Big)\Big(\hat{T}\Big(\phi_1(z)\Big)+\phi_1(z)\hat{T}'\Big(\phi_1(z)\Big)\Big)}.$$

Then the equation (20) admits the following representation

$$y'(z) = \psi_6(z)y(z) + \psi_2(z).$$

Its general solution is

$$y(z) = e^{\int \psi_6(z)dz}\Big(C + \int \psi_2(z)e^{-\int \psi_6(z)dz}dz\Big),$$

where C is a constant.

Definition 3.0.129. *The equation* (20) *will be called first-order linear iso-differential equation.*

Exercise 3.0.130. *Let* $a,\ b \in C([p,q])$. *Determine the equation*

$$\Big(\hat{y}^{\wedge}(x)\Big)^{\circledast} = a^{\wedge}(\hat{x})y^{\wedge}(x) + \hat{b}(\hat{x}) \tag{21}$$

and find its general solution.

Definition 3.0.131. *The equation* (21) *will be called first-order linear iso-differential equation.*

Answer. The equation (21) can be represented in the following form.

$$y'\Big(x\hat{T}(x)\Big) = \Big(a(x)\frac{\hat{T}(x) - x\hat{T}'(x)}{\hat{T}(x) + x\hat{T}'(x)} + \frac{\hat{T}'(x)}{\hat{T}(x)\Big(\hat{T}(x) + x\hat{T}'(x)\Big)}\Big)y\Big(x\hat{T}(x)\Big)$$

$$+b\Big(\frac{x}{\hat{T}(x)}\Big)\frac{\hat{T}(x) - x\hat{T}'(x)}{\hat{T}(x)\Big(\hat{T}(x) + x\hat{T}'(x)\Big)}.$$

We set

$$\psi_7(z) = b\Big(\frac{\phi_1(z)}{\hat{T}\big(\phi_1(z)\big)}\Big)\frac{\hat{T}\big(\phi_1(z)\big) - \phi_1(z)\hat{T}'\big(\phi_1(z)\big)}{\hat{T}\big(\phi_1(z)\big)\Big(\hat{T}\big(\phi_1(z)\big) + \phi_1(z)\hat{T}'\big(\phi_1(z)\big)\Big)}.$$

Then the equation (21) we can rewrite in the following form.

$$y'(z) = \psi_6(z)y(z) + \psi_7(z).$$

Its general solution is given by

$$y(z) = e^{\int \psi_6(z)dz}\Big(C + \int \psi_7(z)e^{-\int \psi_6(z)dz}dz\Big).$$

Exercise 3.0.132. *Let* $a,\ b \in C([p,q])$. *Determine the equation*

$$\Big(\hat{y}^{\wedge}(x)\Big)^{\circledast} = a(\hat{x})\hat{\times}y^{\wedge}(x) + b(x) \tag{22}$$

and find its general solution.

Definition 3.0.133. *The equation* (22) *will be called first-order linear iso-differential equation.*

Answer. The equation (22) we can represent in the following form.

$$y'\left(x\hat{T}(x)\right) = \left(a\left(\frac{x}{\hat{T}(x)}\right)\frac{\hat{T}(x)\left(\hat{T}(x)-x\hat{T}'(x)\right)}{\hat{T}(x)+x\hat{T}'(x)} + \frac{\hat{T}'(x)}{\hat{T}(x)\left(\hat{T}(x)+x\hat{T}'(x)\right)}\right)y\left(x\hat{T}(x)\right)$$

$$+b(x)\frac{\hat{T}(x)-x\hat{T}'(x)}{\hat{T}(x)+x\hat{T}'(x)}.$$

Let

$$\psi_8(z) = a\left(\frac{\phi_1(z)}{\hat{T}\left(\phi_1(z)\right)}\right)\hat{T}\left(\phi_1(z)\right)\frac{\hat{T}\left(\phi_1(z)\right)-\phi_1(z)\hat{T}'\left(\phi_1(z)\right)}{\hat{T}\left(\phi_1(z)\right)+\phi_1(z)\hat{T}'\left(\phi_1(z)\right)}$$

$$+\frac{\hat{T}'\left(\phi_1(z)\right)}{\hat{T}\left(\phi_1(z)\right)\left(\hat{T}\left(\phi_1(z)\right)+\phi_1(z)\hat{T}'\left(\phi_1(z)\right)\right)},$$

$$\psi_9(z) = b\left(\phi_1(z)\right)\frac{\hat{T}\left(\phi_1(z)\right)-\phi_1(z)\hat{T}'\left(\phi_1(z)\right)}{\hat{T}\left(\phi_1(z)\right)+\phi_1(z)\hat{T}'\left(\phi_1(z)\right)}.$$

Then the equation (22) we can rewrite in the form.

$$y'(z) = \psi_8(z)y(z) + \psi_9(z).$$

Its general solution is given by

$$y(z) = e^{\int \psi_8(z)dz}\left(C + \int e^{-\int \psi_8(z)dz}\psi_9(z)dz\right),$$

where C is a constant.

Now we consider the equation

$$\left(y^{\wedge}(x)\right)^{\circledast} = \hat{a}^{\wedge}(\hat{x})\hat{\times}\hat{y}^{\wedge}(x) + \hat{b}^{\wedge}(\hat{x}), \tag{23}$$

where $a, b \in C([p,q])$.

Definition 3.0.134. *The equation* (23) *will be called first-order linear iso-differential equation.*

The equation (23) we rewrite as follows.

$$\frac{y'\left(x\hat{T}(x)\right)\hat{T}(x)\left(\hat{T}(x)+x\hat{T}'(x)\right)}{\hat{T}(x)-x\hat{T}'(x)} = \frac{a(x)}{\hat{T}(x)}\hat{T}(x)\frac{y\left(x\hat{T}(x)\right)}{\hat{T}(x)} + \frac{b(x)}{\hat{T}(x)}$$

or

$$\frac{y'\left(x\hat{T}(x)\right)\hat{T}(x)\left(\hat{T}(x)+x\hat{T}'(x)\right)}{\hat{T}(x)-x\hat{T}'(x)} = a(x)\frac{y\left(x\hat{T}(x)\right)}{\hat{T}(x)} + \frac{b(x)}{\hat{T}(x)},$$

or

$$y'\left(x\hat{T}(x)\right)\hat{T}(x)\left(\hat{T}(x)+x\hat{T}'(x)\right) = a(x)\tfrac{\hat{T}(x)-x\hat{T}'(x)}{\hat{T}(x)}y\left(x\hat{T}(x)\right)$$

$$+b(x)\tfrac{\hat{T}(x)-x\hat{T}'(x)}{\hat{T}(x)},$$

or

$$y'\left(x\hat{T}(x)\right) = a(x)\frac{\hat{T}(x)-x\hat{T}'(x)}{\hat{T}^2(x)\left(\hat{T}(x)+x\hat{T}'(x)\right)}y\left(x\hat{T}(x)\right) + b(x)\frac{\hat{T}(x)-x\hat{T}'(x)}{\hat{T}^2(x)\left(\hat{T}(x)+x\hat{T}'(x)\right)}.$$

Let

$$\psi_{10}(z) = a\left(\phi_1(z)\right)\frac{\hat{T}\left(\phi_1(z)\right)-\phi_1(z)\hat{T}'\left(\phi_1(z)\right)}{\hat{T}^2\left(\phi_1(z)\right)\left(\hat{T}\left(\phi_1(z)\right)\right)+\phi_1(z)\hat{T}'\left(\phi_1(z)\right)},$$

$$\psi_{11}(z) = b\left(\phi_1(z)\right)\frac{\hat{T}\left(\phi_1(z)\right)-\phi_1(z)\hat{T}'\left(\phi_1(z)\right)}{\hat{T}^2\left(\phi_1(z)\right)\left(\hat{T}\left(\phi_1(z)\right)\right)+\phi_1(z)\hat{T}'\left(\phi_1(z)\right)}.$$

Then we obtain the equation

$$y'(z) = \psi_{10}(z)y(z)+\psi_{11}(z).$$

Its general solution is given by

$$y(z) = e^{\int\psi_{10}(z)dz}\left(C+\int\psi_{11}(z)e^{-\int\psi_{10}(z)dz}dz\right),$$

where C is a constant.

Exercise 3.0.135. *Let a, $b\in C([p,q])$. Find the general solution of the following iso-Bernoulli equation*

$$\left(\hat{y}^{\wedge}(x)\right)^{\circledast} = \hat{a}^{\wedge}(\hat{x})\hat{\times}\left(\hat{y}^{\wedge}(x)\right)^{\hat{2}} + b(x)\hat{\times}y^{\wedge}(x).$$

Exercise 3.0.136. *Let a, $b\in C([p,q])$. Find the general solution of the following iso-Bernoulli equation*

$$\left(y^{\wedge}(x)\right)^{\circledast} = a^{\wedge}(x)\left(y^{\wedge}(x)\right)^{\hat{4}} + \hat{b}(\hat{x})\hat{\times}y^{\wedge}(x).$$

Exercise 3.0.137. *Let a, b, $c\in C([p,q])$. Let also, $y_1(x)$ be a particular solution to the iso-Riccati equation*

$$\left(\hat{y}^{\wedge}(x)\right) = a(x)\hat{\times}\left(\hat{y}^{\wedge}(x)\right)^{\hat{2}} + \hat{b}(x)\hat{\times}y^{\wedge}(x)+\hat{c}(\hat{x}).$$

Prove that

$$y(x) = y_1(x) + \frac{1}{y_2(x)}$$

transforms it to an iso-Bernoulli equation with respect to $y_2(x)$.

Exercise 3.0.138. *Let a, b,* $c \in C([p,q])$. *Let also,* $y_1(x)$ *be a particular solution to the iso-Riccati equation*

$$\left(y^\wedge(x)\right) = \hat{a}(\hat{x})\hat{\times}\left(y^\wedge(x)\right)^{\hat{2}} + \hat{b}(x)\hat{\times}y^\wedge(x) + \hat{c}(\hat{x}).$$

Prove that

$$y(x) = y_1(x) + \frac{1}{y_2(x)}$$

transforms it to an iso-Bernoulli equation with respect to $y_2(x)$.

Let

$$\hat{T} : [p,q] \longrightarrow [c,d], \qquad \hat{T}(x) > 0, \qquad \tfrac{x}{\hat{T}(x)} \in [c,d] \quad \text{for} \quad \forall x \in [p,q], \qquad \hat{T} \in C^1([p,q]), \tag{24}$$

$$h(x) = \frac{x}{\hat{T}(x)}, \qquad x \in [p,q] \quad \text{is strictly increasing(decreasing) in} \quad [p,q]. \tag{25}$$

We have that $h \in C^1([p,q])$ and there exists $\phi_2 : [c,d] \longrightarrow [p,q]$ such that

$$\phi_2 \in C^1([c,d]), \qquad \phi_2\left(h(x)\right) = x \quad \text{for} \quad \forall x \in [p,q].$$

If we set

$$z_1 := \frac{x}{\hat{T}(x)}, \qquad x \in [p,q],$$

then

$$z_1 \in [c,d], \qquad x = \phi_2(z_1).$$

Below we will suppose (24), (25).
Now we consider the equation

$$\left(\hat{y}(\hat{x})\right)^{\circledast} = \hat{a}^\wedge(\hat{x})\hat{\times}\hat{y}(\hat{x}) + \hat{b}^\wedge(\hat{x}), \tag{26}$$

where $a,\ b \in C([p,q])$.

Definition 3.0.139. *The equation* (26) *will be called first-order linear iso-differential equation.*

The equation (26) we can represent as follows.

$$\frac{y'\left(\frac{x}{\hat{T}(x)}\right)\left(\hat{T}(x) - x\hat{T}'(x)\right) - y\left(\frac{x}{\hat{T}(x)}\right)\hat{T}(x)\hat{T}'(x)}{\hat{T}^2(x)\left(\hat{T}(x) - x\hat{T}'(x)\right)} = \frac{a(x)}{\hat{T}(x)}\hat{T}(x)\frac{y\left(\frac{x}{\hat{T}(x)}\right)}{\hat{T}(x)} + \frac{b(x)}{\hat{T}(x)}$$

or

$$\frac{y'\left(\frac{x}{\hat{T}(x)}\right)\left(\hat{T}(x)-x\hat{T}'(x)\right)-y\left(\frac{x}{\hat{T}(x)}\right)\hat{T}(x)\hat{T}'(x)}{\hat{T}^2(x)\left(\hat{T}(x)-x\hat{T}'(x)\right)} = a(x)\frac{y\left(\frac{x}{\hat{T}(x)}\right)}{\hat{T}(x)}+\frac{b(x)}{\hat{T}(x)},$$

or

$$y'\left(\tfrac{x}{\hat{T}(x)}\right)\left(\hat{T}(x)-x\hat{T}'(x)\right)-y\left(\tfrac{x}{\hat{T}(x)}\right)\hat{T}(x)\hat{T}'(x)$$

$$= a(x)\hat{T}(x)\left(\hat{T}(x)-x\hat{T}'(x)\right)y\left(\tfrac{x}{\hat{T}(x)}\right)+b(x)\hat{T}(x)\left(\hat{T}(x)-x\hat{T}'(x)\right),$$

or

$$y'\left(\tfrac{x}{\hat{T}(x)}\right)\left(\hat{T}(x)-x\hat{T}'(x)\right)$$

$$= \left(a(x)\hat{T}(x)\left(\hat{T}(x)-x\hat{T}'(x)\right)+\hat{T}(x)\hat{T}'(x)\right)y\left(\tfrac{x}{\hat{T}(x)}\right)+b(x)\hat{T}(x)\left(\hat{T}(x)-x\hat{T}'(x)\right),$$

or

$$y'\left(\frac{x}{\hat{T}(x)}\right)=\left(a(x)\hat{T}(x)+\frac{\hat{T}(x)\hat{T}'(x)}{\hat{T}(x)-x\hat{T}'(x)}\right)y\left(\frac{x}{\hat{T}(x)}\right)+b(x)\hat{T}(x).$$

Let

$$\tau_1(z_1)=a\left(\phi_2(z_1)\right)\hat{T}\left(\phi_2(z_1)\right)+\frac{\hat{T}\left(\phi_2(z_1)\right)\hat{T}'\left(\phi_2(z_1)\right)}{\hat{T}\left(\phi_2(z_1)\right)-\phi_2(z_1)\hat{T}'\left(\phi_2(z_1)\right)},$$

$$\tau_2(z_1)=b\left(\phi_2(z_1)\right)\hat{T}\left(\phi_2(z_1)\right).$$

Then the equation (26) admits the form

$$y'(z_1)=\tau_1(z_1)y(z_1)+\tau_2(z_1).$$

Its general solution is given by

$$y(z_1)=e^{\int\tau_1(z_1)dz_1}\left(C+\int\tau_2(z_1)e^{-\int\tau_1(z_1)dz_1}dz_1\right),$$

where C is a constant.

Exercise 3.0.140. *Let $a,\ b\in C([p,q])$. Consider the equation*

$$\left(\hat{y}(\hat{x})\right)^{\circledast}=\hat{a}(\hat{x})\hat{\times}\hat{y}(\hat{x})\hat{+}\hat{b}(\hat{x}).$$

Find its general solution.

Solution. The given equation admits the following representation

$$\frac{y'\left(\frac{x}{\hat{T}(x)}\right)\left(\hat{T}(x)-x\hat{T}'(x)\right)-y\left(\frac{x}{\hat{T}(x)}\right)\hat{T}(x)\hat{T}'(x)}{\hat{T}^2(x)\left(\hat{T}(x)-x\hat{T}'(x)\right)}=\frac{a\left(\frac{x}{\hat{T}(x)}\right)}{\hat{T}(x)}\hat{T}(x)\frac{y\left(\frac{x}{\hat{T}(x)}\right)}{\hat{T}(x)}+\frac{b\left(\frac{x}{\hat{T}(x)}\right)}{\hat{T}(x)}$$

or

$$\frac{y'\left(\frac{x}{\hat{T}(x)}\right)\left(\hat{T}(x)-x\hat{T}'(x)\right)-y\left(\frac{x}{\hat{T}(x)}\right)\hat{T}(x)\hat{T}'(x)}{\hat{T}^2(x)\left(\hat{T}(x)-x\hat{T}'(x)\right)}=a\left(\frac{x}{\hat{T}(x)}\right)\frac{y\left(\frac{x}{\hat{T}(x)}\right)}{\hat{T}(x)}+\frac{b\left(\frac{x}{\hat{T}(x)}\right)}{\hat{T}(x)},$$

or

$$y'\left(\tfrac{x}{\hat{T}(x)}\right)\left(\hat{T}(x)-x\hat{T}'(x)\right)-y\left(\tfrac{x}{\hat{T}(x)}\right)\hat{T}(x)\hat{T}'(x)$$
$$=a\left(\tfrac{x}{\hat{T}(x)}\right)\hat{T}(x)\left(\hat{T}(x)-x\hat{T}'(x)\right)y\left(\tfrac{x}{\hat{T}(x)}\right)+b\left(\tfrac{x}{\hat{T}(x)}\right)\hat{T}(x)\left(\hat{T}(x)-x\hat{T}'(x)\right),$$

or

$$y'\left(\tfrac{x}{\hat{T}(x)}\right)\left(\hat{T}(x)-x\hat{T}'(x)\right)=\left(a\left(\tfrac{x}{\hat{T}(x)}\right)\hat{T}(x)\left(\hat{T}(x)-x\hat{T}'(x)\right)+\hat{T}(x)\hat{T}'(x)\right)y\left(\tfrac{x}{\hat{T}(x)}\right)$$
$$+b\left(\tfrac{x}{\hat{T}(x)}\right)\hat{T}(x)\left(\hat{T}(x)-x\hat{T}'(x)\right),$$

or

$$y'\left(\tfrac{x}{\hat{T}(x)}\right)=\left(a\left(\tfrac{x}{\hat{T}(x)}\right)\hat{T}(x)+\tfrac{\hat{T}(x)\hat{T}'(x)}{\hat{T}(x)-x\hat{T}'(x)}\right)y\left(\tfrac{x}{\hat{T}(x)}\right)+b\left(\tfrac{x}{\hat{T}(x)}\right)\hat{T}(x).$$

Let

$$\tau_3(z_1)=a(z_1)\hat{T}\left(\phi_2(z_1)\right)+\frac{\hat{T}\left(\phi_2(z_1)\right)\hat{T}'\left(\phi_2(z_1)\right)}{\hat{T}\left(\phi_2(z_1)\right)-\phi_2(z_1)\hat{T}'\left(\phi_2(z_1)\right)},$$

$$\tau_4(z_1)=b(z_1)\hat{T}\left(\phi_2(z_1)\right).$$

Thus, we obtain the equation

$$y'(z_1)=\tau_3(z_1)y(z_1)+\tau_4(z_1).$$

Its general solution is given by

$$y(z_1)=e^{\int\tau_3(z_1)dz_1}\left(C+\int\tau_4(z_1)e^{-\int\tau_3(z_1)dz_1}dz_1\right),$$

where C is a constant.

Exercise 3.0.141. *Let a, $b\in\mathcal{C}([p,q])$. Consider the equation*

$$\left(\hat{y}(\hat{x})\right)^{\circledast}=a(x)\hat{y}(\hat{x})+b(x).$$

Find its general solution.

Solution. We can rewrite the given equation as follows.

$$\frac{y'\left(\frac{x}{\hat{T}(x)}\right)\left(\hat{T}(x)-x\hat{T}'(x)\right)-y\left(\frac{x}{\hat{T}(x)}\right)\hat{T}(x)\hat{T}'(x)}{\hat{T}^2(x)\left(\hat{T}(x)-x\hat{T}'(x)\right)}=a(x)\frac{y\left(\frac{x}{\hat{T}(x)}\right)}{\hat{T}(x)}+b(x)$$

or

$$y'\left(\frac{x}{\hat{T}(x)}\right)\left(\hat{T}(x)-x\hat{T}'(x)\right)-y\left(\frac{x}{\hat{T}(x)}\right)\hat{T}(x)\hat{T}'(x)$$

$$=a(x)\hat{T}(x)\left(\hat{T}(x)-x\hat{T}'(x)\right)y\left(\frac{x}{\hat{T}(x)}\right)+b(x)\hat{T}^2(x)\left(\hat{T}(x)-x\hat{T}'(x)\right),$$

or

$$y'\left(\frac{x}{\hat{T}(x)}\right)\Big)\left(\hat{T}(x)-x\hat{T}'(x)\right)$$

$$=\left(a(x)\hat{T}(x)\left(\hat{T}(x)-x\hat{T}'(x)\right)+\hat{T}(x)\hat{T}'(x)\right)y\left(\frac{x}{\hat{T}(x)}\right)+b(x)\hat{T}^2(x)\left(\hat{T}(x)-x\hat{T}'(x)\right),$$

or

$$y'\left(\frac{x}{\hat{T}(x)}\right)=\left(a(x)\hat{T}(x)+\frac{\hat{T}(x)\hat{T}'(x)}{\hat{T}(x)-x\hat{T}'(x)}\right)y\left(\frac{x}{\hat{T}(x)}\right)+b(x)\hat{T}^2(x).$$

Let

$$\tau_5(z_1)=b\left(\phi_2(z_1)\right)\hat{T}^2\left(\phi_2(z_1)\right).$$

Then we obtain the equation

$$y'(z_1)=\tau_1(z_1)y(z_1)+\tau_5(z_1).$$

Its general solution is given by

$$y(z_1)=e^{\int\tau_1(z_1)dz_1}\left(C+\int e^{-\int\tau_1(z_1)dz_1}\tau_5(z_1)dz_1\right),$$

where C is a constant.

Now we consider the equation

$$\left(y^{\vee}(x)\right)^{\circledast}=a^{\vee}(x)\hat{\times}y^{\vee}(x)+b^{\vee}(x), \tag{27}$$

where $a,\, b\in C([p,q])$.

Definition 3.0.142. *The equation* (27) *will be called first-order linear iso-differential equation.*

The equation (27) we can rewrite in the following form.

$$\frac{1}{\hat{T}(x)}y'\left(\frac{x}{\hat{T}(x)}\right)=a\left(\frac{x}{\hat{T}(x)}\right)\hat{T}(x)y\left(\frac{x}{\hat{T}(x)}\right)+b\left(\frac{x}{\hat{T}(x)}\right)$$

or

$$y'\left(\frac{x}{\hat{T}(x)}\right)=a\left(\frac{x}{\hat{T}(x)}\right)\hat{T}^2(x)y\left(\frac{x}{\hat{T}(x)}\right)+\hat{T}(x)b\left(\frac{x}{\hat{T}(x)}\right).$$

Let

$$\tau_6(z_1)=a(z_1)\hat{T}^2\left(\phi_2(z_1)\right).$$

Thus, we get the equation

$$y'(z_1)=\tau_6(z_1)y(z_1)+\tau_4(z_1).$$

Its general solution is given by

$$y(z_1) = e^{\int \tau_6(z_1)dz_1}\Big(C + \int \tau_4(z_1)e^{-\int \tau_6(z_1)dz_1}dz_1\Big),$$

where C is a constant.

Now we consider the equation

$$\Big(\hat{y}(\hat{x})\Big)^{\circledast} = a(x)y^{\vee}(x) + b(x), \tag{28}$$

where $a,\, b \in C([p,q])$.

Definition 3.0.143. *The equation* (28) *will be called first-order linear iso-differential equation.*

The equation (28) we can rewrite in the following form

$$\frac{y\Big(\frac{x}{\hat{T}(x)}\Big)'\Big(\hat{T}(x) - x\hat{T}'(x)\Big) - y\Big(\frac{x}{\hat{T}(x)}\Big)\hat{T}(x)\hat{T}'(x)}{\hat{T}^2(x)\Big(\hat{T}(x) - x\hat{T}'(x)\Big)} = a(x)y\Big(\frac{x}{\hat{T}(x)}\Big) + b(x)$$

or

$$y\Big(\frac{x}{\hat{T}(x)}\Big)'\Big(\hat{T}(x) - x\hat{T}'(x)\Big) - y\Big(\frac{x}{\hat{T}(x)}\Big)\hat{T}(x)\hat{T}'(x)$$
$$= a(x)\hat{T}^2(x)\Big(\hat{T}(x) - x\hat{T}'(x)\Big)y\Big(\frac{x}{\hat{T}(x)}\Big) + b(x)\hat{T}^2(x)\Big(\hat{T}(x) - x\hat{T}'(x)\Big),$$

or

$$y\Big(\frac{x}{\hat{T}(x)}\Big)'\Big(\hat{T}(x) - x\hat{T}'(x)\Big)$$
$$= \Big(a(x)\hat{T}^2(x)\Big(\hat{T}(x) - x\hat{T}'(x)\Big) + \hat{T}(x)\hat{T}'(x)\Big)y\Big(\frac{x}{\hat{T}(x)}\Big) + b(x)\hat{T}^2(x)\Big(\hat{T}(x) - x\hat{T}'(x)\Big),$$

or

$$y\Big(\frac{x}{\hat{T}(x)}\Big)' = \Big(a(x)\hat{T}^2(x) + \frac{\hat{T}(x)\hat{T}'(x)}{\hat{T}(x) - x\hat{T}'(x)}\Big)y\Big(\frac{x}{\hat{T}(x)}\Big) + b(x)\hat{T}^2(x).$$

Let

$$\tau_7(z_1) = a\Big(\phi_2(z_1)\Big)\hat{T}^2(\phi_2(z_1)) + \frac{\hat{T}\Big(\phi_2(z_1)\Big)\hat{T}'\Big(\phi_2(z_1)\Big)}{\hat{T}\Big(\phi_2(z_1)\Big) - \phi_2(z_1)\hat{T}'\Big(\phi_2(z_1)\Big)}.$$

Then the equation (28) admits the following form

$$y'(z_1) = \tau_7(z_1)y(z_1) + \tau_5(z_1).$$

Its general solution can be represented in the form

$$y(z_1) = e^{\int \tau_7(z_1)dz_1}\Big(C + \int \tau_5(z_1)e^{-\int \tau_7(z_1)dz_1}dz_1\Big),$$

where C is a constant.

Now we consider the equation

$$\left(y^{\vee}(x)\right)^{\circledast} = \hat{a}^{\wedge}(\hat{x})\hat{\times}\hat{y}(\hat{x}) + \hat{b}^{\wedge}(\hat{x}), \tag{29}$$

where $a, b \in \mathcal{C}([p,q])$.

Definition 3.0.144. *The equation* (29) *will be called first-order linear iso-differential equation.*

The equation (29) we can rewrite as follows.

$$\frac{1}{\hat{T}(x)}y'\left(\frac{x}{\hat{T}(x)}\right) = \frac{a(x)}{\hat{T}(x)}\hat{T}(x)\frac{y\left(\frac{x}{\hat{T}(x)}\right)}{\hat{T}(x)} + \frac{b(x)}{\hat{T}(x)}$$

or

$$\frac{1}{\hat{T}(x)}y\left(\frac{x}{\hat{T}(x)}\right)' = a(x)\frac{y\left(\frac{x}{\hat{T}(x)}\right)}{\hat{T}(x)} + \frac{b(x)}{\hat{T}(x)},$$

or

$$y\left(\frac{x}{\hat{T}(x)}\right)' = a(x)y\left(\frac{x}{\hat{T}(x)}\right) + b(x).$$

Let

$$\tau_8(z_1) = a\Big(\phi_2(z_1)\Big), \qquad \tau_9(z_1) = b\Big(\phi_2(z_1)\Big).$$

Then we get the equation

$$y'(z_1) = \tau_8(z_1)y(z_1) + \tau_9(z_1).$$

Its general solution is given by

$$y(z_1) = e^{\int \tau_8(z_1)dz_1}\left(C + \int \tau_9(z_1)e^{-\int \tau_8(z_1)dz_1}dz_1\right),$$

where C is a constant.

Exercise 3.0.145. *Let $a, b \in \mathcal{C}(\mathbb{R})$, $\hat{T} \in \mathcal{C}^1(\mathbb{R})$, $\hat{T}(x) > 0$ for every $x \in \mathbb{R}$ and*

$$A(x) := a(x)\frac{\hat{T}(x) - x\hat{T}'(x)}{\hat{T}(x)} + \frac{\hat{T}'(x)}{\hat{T}(x)} \geq c > 0,$$

$$\left|B(x) := b(x)\frac{\hat{T}(x) - x\hat{T}'(x)}{\hat{T}(x)}\right| \leq M < \infty,$$

where c and M are constants. Prove that there exists unique bounded solution $y(x)$ of the equation

$$\left(\hat{y}^{\wedge}(\hat{x})\right)^{\circledast} = \hat{a}^{\wedge}(\hat{x})\hat{\times}\hat{y}^{\wedge}(\hat{x}) + \hat{b}^{\wedge}(\hat{x}).$$

Solution. The given equation we can rewrite in the form.

$$y'(x) = \left(a(x)\frac{\hat{T}(x) - x\hat{T}'(x)}{\hat{T}(x)} + \frac{\hat{T}'(x)}{\hat{T}(x)}\right)y(x) + b(x)\frac{\hat{T}(x) - x\hat{T}'(x)}{\hat{T}(x)}$$

or

$$y'(x) = A(x)y(x) + B(x). \tag{30}$$

For its solution we have the representation

$$y(x) = e^{\int_{x_0}^{x} A(t)dt}\left(y_0 + \int_{x_0}^{x} B(t)e^{-\int_{x_0}^{t} A(s)ds}dt\right), \tag{31}$$

where $x_0 \in \mathbb{R}$ is arbitrarily chosen and fixed. We will search y_0 so that $\lim_{x\longrightarrow\infty} y(x) = 0$.

Since, for $x > x_0$,

$$e^{\int_{x_0}^{x} A(t)dt} \geq e^{c(x-x_0)} \longrightarrow_{x\longrightarrow\infty} \infty,$$

then we put

$$y_0 + \int_{x_0}^{\infty} B(t)e^{-\int_{x_0}^{t} A(s)ds}dt = 0$$

or

$$y_0 = -\int_{x_0}^{\infty} B(t)e^{-\int_{x_0}^{t} A(s)ds}dt,$$

and from here

$$\begin{aligned} y(x) &= -e^{\int_{x_0}^{x} A(t)dt} \int_x^{\infty} B(t)e^{-\int_{x_0}^{t} A(s)ds}dt \\ &= -\int_x^{\infty} B(t)e^{-\int_x^t A(s)ds}dt. \end{aligned}$$

For $x > x_0$ we have

$$\begin{aligned} |y(x)| &= \left|-\int_x^{\infty} B(t)e^{-\int_x^t A(s)ds}dt\right| \\ &\leq \int_x^{\infty} |B(t)|e^{-\int_x^t A(s)ds}dt \\ &\leq \int_x^{\infty} |B(t)|e^{-c(t-x)}dt \\ &\leq M\int_x^{\infty} e^{-c(t-x)}dt \\ &= \frac{M}{c} < \infty. \end{aligned}$$

Consequently, $y(x)$ is a bounded solution to the considered equation.

Now we suppose that the considered equation has two bounded solutions $y_1(x)$ and $y_2(x)$. Let

$$l(x) = y_1(x) - y_2(x).$$

Then $l(x)$ is a bounded solution of the equation

$$y'(x) = A(x)y(x).$$

We have

$$l(x) = c_1 e^{\int_{x_0}^{x} A(t)dt}.$$

Then, for $x > x_0$, we have

$$|l(x)| \geq |c_1| e^{c(x-x_0)} \longrightarrow_{x \longrightarrow \infty} \infty,$$

which is a contradiction.

Advanced Practical Exercises

Problem 3.0.146. *Let* $\hat{T}(x) = e^x$,

$$a(x) = \frac{2}{x-1}, \qquad b(x) = \frac{x+x^2+x^3}{x-1}, \qquad x \neq 1.$$

Find the general solution of the equation (2).

Answer. $y(x) = Ce^{-x} - x^3 + 2x^2 - 5x + 5$, where C is a constant.

Problem 3.0.147. *Let* $\hat{T}(x) = e^x$,

$$a(x) = \frac{1-2x}{(1-x)(1+2x)}, \qquad b(x) = \frac{4x}{(1-x)(1+2x)}, \qquad x \neq -\frac{1}{2}, 1.$$

Find the solutions of the equation (2).

Answer.

$$y(x) = (2x+1)\Big(C + \log|2x+1|\Big) + 1,$$

where C is a constant.

Problem 3.0.148. *Let* $a, b \in \mathcal{C}(J)$. *Prove that*

$$y(x) = e^{\int \left(a\left(x\hat{T}(x)\right)\left(\hat{T}(x) - x\hat{T}'(x)\right) + \frac{\hat{T}'(x)}{\hat{T}(x)} \right) dx} \Bigg(C$$

$$+ \int b(x)\left(\hat{T}(x) - x\hat{T}'(x)\right) e^{-\int \left(a\left(x\hat{T}(x)\right)\left(\hat{T}(x) - x\hat{T}'(x)\right) + \frac{\hat{T}'(x)}{\hat{T}(x)} \right) dx} dx \Bigg),$$

where C *is a constant, is the general solution to the equation*

$$\left(y^{\wedge\wedge}\right)^{\circledast} = a^{\wedge}(x) y(x) + b(x).$$

Problem 3.0.149. *Let* $a, b \in \mathcal{C}(J)$. *Prove that*

$$y(x) = e^{\int \left(a\left(x\hat{T}(x)\right)\left(\hat{T}(x) - x\hat{T}'(x)\right) + \frac{\hat{T}'(x)}{\hat{T}(x)} \right) dx} \Bigg(C$$

$$+ \int b\left(x\hat{T}(x)\right)\left(\hat{T}(x) - x\hat{T}'(x)\right) e^{-\int \left(a\left(x\hat{T}(x)\right)\left(\hat{T}(x) - x\hat{T}'(x)\right) + \frac{\hat{T}'(x)}{\hat{T}(x)} \right) dx} dx \Bigg),$$

where C *is a constant, is the general solution to the equation*

$$\left(y^{\wedge\wedge}\right)^{\circledast} = a^{\wedge}(x) y(x) + b^{\wedge}(x).$$

Problem 3.0.150. *Let $a, b \in \mathcal{C}(J)$. Prove that*

$$y(x) = e^{\int \left(a\left(x\hat{T}(x)\right)\hat{T}(x)\left(\hat{T}(x)-x\hat{T}'(x)\right)+\frac{\hat{T}'(x)}{\hat{T}(x)}\right)dx}\Big(C$$

$$+\int b(x)\left(\hat{T}(x)-x\hat{T}'(x)\right)e^{-\int \left(a\left(x\hat{T}(x)\right)\hat{T}(x)\left(\hat{T}(x)-x\hat{T}'(x)\right)+\frac{\hat{T}'(x)}{\hat{T}(x)}\right)dx}dx\Big),$$

where C is a constant, is the general solution to the equation

$$\left(\hat{y}^{\wedge\wedge}\right)^{\circledast} = a^{\wedge}(x)\hat{\times}y(x)+b(x).$$

Problem 3.0.151. *Let $a, b \in \mathcal{C}(J)$. Prove that*

$$y(x) = e^{\int \left(a\left(x\hat{T}(x)\right)\hat{T}(x)\left(\hat{T}(x)-x\hat{T}'(x)\right)+\frac{\hat{T}'(x)}{\hat{T}(x)}\right)dx}\Big(C$$

$$+\int b\left(x\hat{T}(x)\right)\left(\hat{T}(x)-x\hat{T}'(x)\right)e^{-\int \left(a\left(x\hat{T}(x)\right)\hat{T}(x)\left(\hat{T}(x)-x\hat{T}'(x)\right)+\frac{\hat{T}'(x)}{\hat{T}(x)}\right)dx}dx\Big),$$

where C is a constant, is the general solution to the equation

$$\left(\hat{y}^{\wedge\wedge}\right)^{\circledast} = a^{\wedge}(x)\hat{\times}y(x)+b^{\wedge}(x).$$

Problem 3.0.152. *Let $a, b \in \mathcal{C}(J)$. Prove that*

$$y(x) = e^{\int \left(a\left(x\hat{T}(x)\right)\left(\hat{T}(x)-x\hat{T}'(x)\right)+\frac{\hat{T}'(x)}{\hat{T}(x)}\right)dx}\Big(C$$

$$+\int b\left(x\hat{T}(x)\right)\frac{1}{\hat{T}(x)}\left(\hat{T}(x)-x\hat{T}'(x)\right)e^{-\int \left(a\left(x\hat{T}(x)\right)\left(\hat{T}(x)-x\hat{T}'(x)\right)+\frac{\hat{T}'(x)}{\hat{T}(x)}\right)dx}dx\Big),$$

where C is a constant, is the general solution to the equation

$$\left(\hat{y}^{\wedge\wedge}\right)^{\circledast} = a^{\wedge}(x)y(x)+\hat{b}^{\wedge}(x).$$

Problem 3.0.153. *Let $a, b \in \mathcal{C}(J)$. Prove that*

$$y(x) = e^{\int \left(a\left(x\hat{T}(x)\right)\hat{T}(x)\left(\hat{T}(x)-x\hat{T}'(x)\right)+\frac{\hat{T}'(x)}{\hat{T}(x)}\right)dx}\Big(C$$

$$+\int b\left(x\hat{T}(x)\right)\frac{1}{\hat{T}(x)}\left(\hat{T}(x)-x\hat{T}'(x)\right)e^{-\int \left(a\left(x\hat{T}(x)\right)\hat{T}(x)\left(\hat{T}(x)-x\hat{T}'(x)\right)+\frac{\hat{T}'(x)}{\hat{T}(x)}\right)dx}dx\Big),$$

where C is a constant, is the general solution to the equation

$$\left(\hat{y}^{\wedge\wedge}\right)^{\circledast} = a^{\wedge}(x)\hat{\times}y(x)+\hat{b}^{\wedge}(x).$$

Problem 3.0.154. *Let $a \in \mathcal{C}(I)$. Prove that*

$$y(x) = \hat{T}(x)\left(C + \int a(x)\frac{\hat{T}(x) - x\hat{T}'(x)}{\hat{T}(x)}dx\right),$$

where C is a constant, is the general solution to the equation

$$\left(\hat{y}^{\wedge\wedge}\right)^{\circledast} = a(x).$$

Problem 3.0.155. *Let $a,\ b \in \mathcal{C}(J)$. Prove that*

$$y = e^{\int\left(a(x)\left(\hat{T}(x)-x\hat{T}'(x)\right)+\frac{\hat{T}'(x)}{\hat{T}(x)}\right)dx}\Bigg(C+$$

$$(1-n)\int b(x)\frac{\hat{T}(x)-x\hat{T}'(x)}{\hat{T}(x)}e^{(n-1)\int\left(a(x)\left(\hat{T}(x)-x\hat{T}'(x)\right)+\frac{\hat{T}'(x)}{\hat{T}(x)}\right)dx}dx\Bigg)^{\frac{1}{1-n}},$$

where C is a constant, is the general solution to the equation

$$\left(\hat{y}^{\wedge\wedge}\right)^{\circledast} = a^{\wedge}(\hat{x})\hat{\times}\hat{y}^{\wedge\wedge} + \hat{b}^{\wedge}(\hat{x})\hat{\times}\left(\hat{y}^{\wedge\wedge}\right)^{\hat{n}}.$$

Problem 3.0.156. *Let $\hat{T}(x) = e^x$,*

$$a(x) = \frac{2x-1}{1-x}, \qquad b(x) = \frac{1}{x-1}, \qquad c(x) = \frac{5-x^2}{1-x}, \qquad x \neq 1.$$

Determine the equation (14) *and find its general solution.*

Answer. The equation (14) can be represented in the following form

$$y'(x) = x + 2 + \frac{4}{Ce^{4x}-1}, \qquad C \neq e^{-4x},$$

C is a constant. Also,

$$y(x) = x + 2$$

is an its solution.

Problem 3.0.157. *Let $\hat{T}(x) = e^x$,*

$$a(x) = \frac{1+2e^x}{x-1}, \qquad b(x) = \frac{1}{1-x}, \qquad c(x) = \frac{e^x + e^{2x}}{1-x}, \qquad x \neq 1.$$

Determine the equation (14) *and find its general solution.*

Answer. The equation (14) can be represented in the following form

$$y'(x) = -2e^x y(x) + y^2(x) + e^x + e^{2x}.$$

Its general solution is given by

$$y(x) = e^x - \frac{1}{C+x}, \qquad C \neq -x,$$

C is a constant. Also,

$$y(x) = e^x$$

is an its solution.

Chapter 4

Iso-Integral Inequalities

In this chapter we suppose that a is a positive real number, $x_0 \in \mathbb{R}$, $\hat{T} \in C^1\Big(|x-x_0| \leq a\Big)$, $\hat{T}(x) > 0$ for every $x : |x-x_0| \leq a$.

Theorem 4.0.158. *Let $u(x)$, $p(x)$ and $q(x)$ be nonnegative continuous functions in the interval $|x-x_0| \leq a$. Let also*

$$\hat{T}(x) - x\hat{T}'(x) \leq 0 \qquad \text{for} \qquad |x-x_0| \leq a, \tag{1}$$

$$\hat{u}^\wedge(\hat{x}) \leq \hat{p}^\wedge(\hat{x}) + \left| \hat{\int}_{x_0}^{x} \hat{q}^\wedge(\hat{t}) \hat{\times} \hat{u}^\wedge(\hat{t}) \hat{\times} \hat{d}\hat{t} \right| \tag{2}$$

for $|x-x_0| \leq a$. Then the following inequality holds.

$$\hat{u}^\wedge(\hat{x}) \leq \hat{p}^\wedge(\hat{x}) + \left| \hat{\int}_{x_0}^{x} \hat{p}^\wedge(\hat{t}) \hat{\times} \hat{q}^\wedge(\hat{t}) \hat{\times} \hat{e}^{\left| \hat{\int}_t^x \hat{q}^\wedge(\hat{s}) \hat{\times} \hat{d}\hat{s} \right|} \hat{\times} \hat{d}\hat{t} \right| \tag{3}$$

for $|x-x_0| \leq a$.

Remark 4.0.159. *Sometimes we will use the following representations of the inequalities* (2) *and* (3)

$$u(x) \leq p(x) + \hat{T}(x) \left| \int_{x_0}^{x} q(t) \frac{\hat{T}(t) - t\hat{T}'(t)}{\hat{T}^2(t)} u(t) dt \right|,$$

$$u(x) \leq p(x) + \hat{T}(x) \left| \int_{x_0}^{x} p(t) q(t) \frac{\hat{T}(t) - t\hat{T}'(t)}{\hat{T}^2(t)} e^{\left| \int_t^x q(s) \frac{\hat{T}(s) - s\hat{T}'(s)}{\hat{T}(s)} ds \right|} dt \right|$$

for $|x-x_0| \leq a$, respectively.

Remark 4.0.160. *We note that in the exponential function in* (3) *we have a Riemann integral.*

Remark 4.0.161. *To prove* (3) *we can not apply directly the classical Gronwall's-type inequality. Conversely, from* (3) *we can obtain the classical Gronwall's type inequality when $\hat{T} \equiv 1$.*

Proof. Let $x \in [x_0, x_0+a]$. Then the given inequality is

$$u(x) \le p(x) - \hat{T}(x) \int_{x_0}^{x} q(t) \frac{\hat{T}(t) - t\hat{T}'(t)}{\hat{T}^2(t)} u(t)dt. \tag{4}$$

We define

$$r(x) = \hat{T}(x) \int_{x_0}^{x} q(t) \frac{\hat{T}(t) - t\hat{T}'(t)}{\hat{T}^2(t)} u(t)dt.$$

Therefore, using (4),

$$u(x) \le p(x) - r(x). \tag{5}$$

Also,

$$r'(x) = \hat{T}'(x) \int_{x_0}^{x} q(t) \frac{\hat{T}(t) - t\hat{T}'(t)}{\hat{T}^2(t)} u(t)dt + \hat{T}(x) q(x) \frac{\hat{T}(x) - x\hat{T}'(x)}{\hat{T}^2(x)} u(x)$$

$$= \hat{T}'(x) \int_{x_0}^{x} q(t) \frac{\hat{T}(t) - t\hat{T}'(t)}{\hat{T}^2(t)} u(t)dt + q(x) \frac{\hat{T}(x) - x\hat{T}'(x)}{\hat{T}(x)} u(x).$$

Using (1), (4) and (5), from the last inequality we get

$$r'(x) \ge \frac{\hat{T}'(x)}{\hat{T}(x)} r(x) + q(x) \frac{\hat{T}(x) - x\hat{T}'(x)}{\hat{T}(x)} \Big(p(x) - r(x) \Big)$$

$$= \frac{\hat{T}'(x)}{\hat{T}(x)} r(x) + p(x) q(x) \frac{\hat{T}(x) - x\hat{T}'(x)}{\hat{T}(x)} - q(x) \frac{\hat{T}(x) - x\hat{T}'(x)}{\hat{T}(x)} r(x),$$

from where

$$r'(x) - \left(\frac{\hat{T}'(x)}{\hat{T}(x)} - q(x) \frac{\hat{T}(x) - x\hat{T}'(x)}{\hat{T}(x)} \right) r(x) \ge p(x) q(x) \frac{\hat{T}(x) - x\hat{T}'(x)}{\hat{T}(x)}.$$

We multiply the last inequality with

$$e^{-\int_{x_0}^{x} \left(\frac{\hat{T}'(t)}{\hat{T}(t)} - q(t) \frac{\hat{T}(t) - t\hat{T}'(t)}{\hat{T}(t)} \right) dt} = \frac{\hat{T}(x_0)}{\hat{T}(x)} e^{\int_{x_0}^{x} q(t) \frac{\hat{T}(t) - t\hat{T}'(t)}{\hat{T}(t)} dt}$$

and we get

$$\left(r(x) \frac{\hat{T}(x_0)}{\hat{T}(x)} e^{\int_{x_0}^{x} q(t) \frac{\hat{T}(t) - t\hat{T}'(t)}{\hat{T}(t)} dt} \right)'$$

$$\ge p(x) q(x) \frac{\hat{T}(x) - x\hat{T}'(x)}{\hat{T}^2(x)} \hat{T}(x_0) e^{\int_{x_0}^{x} q(t) \frac{\hat{T}(t) - t\hat{T}'(t)}{\hat{T}(t)} dt}.$$

The last inequality we integrate from x_0 to x and we obtain, using $r(x_0) = 0$,

$$r(x) \frac{\hat{T}(x_0)}{\hat{T}(x)} e^{\int_{x_0}^{x} q(t) \frac{\hat{T}(t) - t\hat{T}'(t)}{\hat{T}(t)} dt}$$

$$\ge \hat{T}(x_0) \int_{x_0}^{x} p(t) q(t) \frac{\hat{T}(t) - t\hat{T}'(t)}{\hat{T}^2(t)} e^{\int_{x_0}^{t} q(s) \frac{\hat{T}(s) - s\hat{T}'(s)}{\hat{T}(s)} ds} dt$$

or

$$r(x) \ge \hat{T}(x) \int_{x_0}^{x} p(t) q(t) \frac{\hat{T}(t) - t\hat{T}'(t)}{\hat{T}^2(t)} e^{-\int_{t}^{x} q(s) \frac{\hat{T}(s) - s\hat{T}'(s)}{\hat{T}(s)} ds} dt.$$

Now we apply (5) and we find

$$u(x) \leq p(x) - \hat{T}(x) \int_{x_0}^{x} p(t)q(t) \frac{\hat{T}(t) - t\hat{T}'(t)}{\hat{T}^2(t)} e^{-\int_t^x q(s) \frac{\hat{T}(s) - s\hat{T}'(s)}{\hat{T}(s)} ds} dt.$$

Now we will prove the assertion in the case when $x \in [x_0 - a, x_0]$. In this case the given inequality (2) has the following representation

$$u(x) \leq p(x) + \hat{T}(x) \int_{x_0}^{x} q(t)u(t) \frac{\hat{T}(t) - t\hat{T}'(t)}{\hat{T}^2(t)} dt. \tag{6}$$

We define

$$r_1(x) = \hat{T}(x) \int_{x_0}^{x} q(t)u(t) \frac{\hat{T}(t) - t\hat{T}'(t)}{\hat{T}^2(t)} dt.$$

Then, using (6),

$$u(x) \leq p(x) + r_1(x) \tag{7}$$

and, using (1) and (7),

$$r_1'(x) = \hat{T}'(x) \int_{x_0}^{x} q(t)u(t) \frac{\hat{T}(t) - t\hat{T}'(t)}{\hat{T}^2(t)} dt + \hat{T}(x)q(x)u(x) \frac{\hat{T}(x) - x\hat{T}'(x)}{\hat{T}^2(x)}$$

$$= \frac{\hat{T}'(x)}{\hat{T}(x)} r(x) + q(x)u(x) \frac{\hat{T}(x) - x\hat{T}'(x)}{\hat{T}(x)}$$

$$\geq \frac{\hat{T}'(x)}{\hat{T}(x)} r_1(x) + q(x) \frac{\hat{T}(x) - x\hat{T}'(x)}{\hat{T}(x)} \Big(p(x) + r_1(x) \Big)$$

$$= \left(\frac{\hat{T}'(x)}{\hat{T}(x)} + q(x) \frac{\hat{T}(x) - x\hat{T}'(x)}{\hat{T}(x)} \right) r_1(x) + p(x)q(x) \frac{\hat{T}(x) - x\hat{T}'(x)}{\hat{T}(x)},$$

from where

$$r_1'(x) - \left(\frac{\hat{T}'(x)}{\hat{T}(x)} + q(x) \frac{\hat{T}(x) - x\hat{T}'(x)}{\hat{T}(x)} \right) r_1(x) \geq p(x)q(x) \frac{\hat{T}(x) - x\hat{T}'(x)}{\hat{T}(x)} \implies$$

$$\left(r_1'(x) - \left(\frac{\hat{T}'(x)}{\hat{T}(x)} + q(x) \frac{\hat{T}(x) - x\hat{T}'(x)}{\hat{T}(x)} \right) r_1(x) \right) \frac{\hat{T}(x_0)}{\hat{T}(x)} e^{-\int_{x_0}^{x} q(t) \frac{\hat{T}(t) - t\hat{T}'(t)}{\hat{T}(t)} dt}$$

$$\geq p(x)q(x) \frac{\hat{T}(x) - x\hat{T}'(x)}{\hat{T}(x)} \frac{\hat{T}(x_0)}{\hat{T}(x)} e^{-\int_{x_0}^{x} q(t) \frac{\hat{T}(t) - t\hat{T}'(t)}{\hat{T}(t)} dt}$$

or

$$\left(r_1(x) \frac{\hat{T}(x_0)}{\hat{T}(x)} e^{-\int_{x_0}^{x} q(t) \frac{\hat{T}(t) - t\hat{T}'(t)}{\hat{T}(t)} dt} \right)'$$

$$\geq p(x)q(x) \frac{\hat{T}(x) - x\hat{T}'(x)}{\hat{T}(x)} \frac{\hat{T}(x_0)}{\hat{T}(x)} e^{-\int_{x_0}^{x} q(t) \frac{\hat{T}(t) - t\hat{T}'(t)}{\hat{T}(t)} dt}.$$

Now we integrate the last inequality from x_0 to x and since $x \leq x_0$ we get, using $r_1(x_0) = 0$,

$$r_1(x) \frac{\hat{T}(x_0)}{\hat{T}(x)} e^{-\int_{x_0}^{x} q(t) \frac{\hat{T}(t) - t\hat{T}'(t)}{\hat{T}(t)} dt}$$

$$\leq \hat{T}(x_0) \int_{x_0}^{x} p(t)q(t) \frac{\hat{T}(t) - t\hat{T}'(t)}{\hat{T}^2(t)} e^{-\int_{x_0}^{t} q(s) \frac{\hat{T}(s) - s\hat{T}'(s)}{\hat{T}(s)} ds} dt,$$

therefore

$$r_1(x) \le \hat{T}(x) \int_{x)}^{x} p(t)q(t) \frac{\hat{T}(t) - t\hat{T}'(t)}{\hat{T}^2(t)} e^{-\int_x^t q(s) \frac{\hat{T}(s) - s\hat{T}'(s)}{\hat{T}(s)} ds} dt.$$

From the last inequality and (7) we obtain

$$u(x) \le p(x) + \hat{T}(x) \int_{x_0}^{x} p(t)q(t) \frac{\hat{T}(t) - t\hat{T}'(t)}{\hat{T}^2(t)} e^{-\int_x^t q(s) \frac{\hat{T}(s) - s\hat{T}'(s)}{\hat{T}(s)} ds} dt.$$

□

Corollary 4.0.162. *If in the last theorem the function* $p(x) \equiv 0$*, then* $u(x) \equiv 0$.

Corollary 4.0.163. *If in the last theorem the function* $p(x)$ *is nondecreasing in* $[x_0, x_0 + a]$ *and the function* $\hat{T}(x)$ *is nonincreasing in* $[x_0, x_0 + a]$*, then*

$$u(x) \le p(x) e^{-\int_{x_0}^{x} q(s) \frac{\hat{T}(s) - s\hat{T}'(s)}{\hat{T}(s)} ds}$$

for $x \in [x_0, x_0 + a]$.

Proof. Since p is nondecreasing function in $[x_0, x_0 + a]$, then

$$p(t) \le p(x) \qquad \text{for} \qquad \forall t, x \in [x_0, x_0 + a], \qquad t \le x.$$

Because $\hat{T}(x)$ is nonincreasing function in $[x_0, x_0 + a]$, then

$$\hat{T}(t) \ge \hat{T}(x) \qquad \text{for} \qquad \forall t, x \in [x_0, x_0 + a], \qquad t \le x$$

or

$$\frac{1}{\hat{T}(t)} \le \frac{1}{\hat{T}(x)} \qquad \text{for} \qquad \forall t, x \in [x_0, x_0 + a], \qquad t \le x.$$

Therefore, from (3), we get for $x \in [x_0, x_0 + a]$

$$u(x) \le p(x) - \hat{T}(x) \int_{x_0}^{x} p(t)q(t) \frac{\hat{T}(t) - t\hat{T}'(t)}{\hat{T}^2(t)} e^{-\int_t^x q(s) \frac{\hat{T}(s) - s\hat{T}'(s)}{\hat{T}(s)} ds} dt$$

$$\le p(x) - \hat{T}(x) \frac{p(x)}{\hat{T}(x)} \int_{x_0}^{x} e^{-\int_t^x q(s) \frac{\hat{T}(s) - s\hat{T}'(s)}{\hat{T}(s)} ds} d e^{\int_x^t q(s) \frac{\hat{T}(s) - s\hat{T}'(s)}{\hat{T}(s)} ds}$$

$$= p(x) - p(x) e^{-\int_t^x q(s) \frac{\hat{T}(s) - s\hat{T}'(s)}{\hat{T}(s)} ds} \Big|_{t=x_0}^{t=x}$$

$$= p(x) - p(x) + p(x) e^{-\int_{x_0}^{x} q(s) \frac{\hat{T}(s) - s\hat{T}'(s)}{\hat{T}(s)} ds}$$

$$= p(x) e^{-\int_{x_0}^{x} q(s) \frac{\hat{T}(s) - s\hat{T}'(s)}{\hat{T}(s)} ds}.$$

□

Corollary 4.0.164. *If in the last Theorem the function $p(x)$ is nonincreasing function for $x \in [x_0 - a, x_0]$ and the function $\hat{T}(x)$ is nondecreasing in $[x_0 - a, x_0]$, then*

$$u(x) \leq p(x) e^{\int_{x_0}^{x} q(t) \frac{\hat{T}(t) - t\hat{T}'(t)}{\hat{T}(t)} dt}$$

for every $x \in [x_0 - a, x_0]$.

Proof. Since the function $p(x)$ is nonincreasing function in $[x_0 - a, x_0]$, then

$$p(t) \leq p(x) \qquad \text{for} \qquad t, x \in [x_0 - a, x_0], \qquad t \geq x.$$

Because the function $\hat{T}(x)$ is nondecreasing function in $[x_0 - a, x_0]$, then

$$\hat{T}(t) \geq \hat{T}(x) \qquad \text{for} \qquad \forall t, x \in [x_0 - a, x_0], \qquad t \geq x,$$

or

$$\frac{1}{\hat{T}(t)} \leq \frac{1}{\hat{T}(x)} \qquad \text{for} \qquad \forall t, x \in [x_0 - a, x_0], t \geq x.$$

Consequently

$$\frac{p(t)}{\hat{T}(t)} \leq \frac{p(x)}{\hat{T}(x)} \qquad \text{for} \qquad \forall t, x \in [x_0 - a, x_0], \qquad t \geq x.$$

From here and (3), for $x \in [x_0 - a, x_0]$, we obtain

$$u(x) \leq p(x) - \hat{T}(x) \int_x^{x_0} p(t) q(t) \frac{\hat{T}(t) - t\hat{T}'(t)}{\hat{T}(t)} e^{-\int_x^t q(s) \frac{\hat{T}(s) - s\hat{T}'(s)}{\hat{T}(s)} ds} dt$$

$$\leq p(x) - \hat{T}(x) \frac{p(x)}{\hat{T}(x)} \int_x^{x_0} q(t) \frac{\hat{T}(t) - t\hat{T}'(t)}{\hat{T}(t)} e^{-\int_x^t q(s) \frac{\hat{T}(s) - s\hat{T}'(s)}{\hat{T}(s)} ds} dt$$

$$= p(x) + p(x) \int_x^{x_0} e^{-\int_x^t q(s) \frac{\hat{T}(s) - s\hat{T}'(s)}{\hat{T}(s)} ds} de^{-\int_x^t q(s) \frac{\hat{T}(s) - s\hat{T}'(s)}{\hat{T}(s)} ds}$$

$$= p(x) + p(x) e^{-\int_x^t q(s) \frac{\hat{T}(s) - s\hat{T}'(s)}{\hat{T}(s)} ds} \Big|_{t=x}^{t=x_0}$$

$$= p(x) e^{\int_{x_0}^{x} q(s) \frac{\hat{T}(s) - s\hat{T}'(s)}{\hat{T}(s)} ds}.$$

□

In fact we can reformulate the last Theorem as follows.

Theorem 4.0.165. *Let $u(x)$, $p(x)$ and $q(x)$ be nonnegative continuous functions in the interval $|x - x_0| \leq a$, and $\hat{T}(x) - x\hat{T}'(x) \leq 0$ for every $x \in [x_0 - a, x_0 + a]$. If*

$$u(x) \leq p(x) - \hat{T}(x) \int_{x_0}^{x} q(t) u(t) \frac{\hat{T}(t) - t\hat{T}'(t)}{\hat{T}^2(t)} dt$$

for every $x \in [x_0, x_0+a]$, *then*

$$u(x) \le p(x) - \hat{T}(x) \int_{x_0}^{x} p(t)q(t) \frac{\hat{T}(t) - t\hat{T}'(t)}{\hat{T}^2(t)} e^{-\int_t^x q(s)\frac{\hat{T}(s)-s\hat{T}'(s)}{\hat{T}(s)}ds} dt$$

for every $x \in [x_0, x_0+a]$. *If*

$$u(x) \le p(x) + \hat{T}(x) \int_{x_0}^{x} q(t)u(t) \frac{\hat{T}(t) - t\hat{T}'(t)}{\hat{T}^2(t)} dt$$

for every $x \in [x_0 - a, x_0]$, *then*

$$u(x) \le p(x) + \hat{T}(x) \int_{x_0}^{x} p(t)q(t) \frac{\hat{T}(t) - t\hat{T}'(t)}{\hat{T}^2(t)} e^{-\int_x^t q(s)\frac{\hat{T}(s)-s\hat{T}'(s)}{\hat{T}(s)}ds} dt$$

for every $x \in [x_0 - a, x_0]$.

Exercise 4.0.166. *Let* $u(x)$ *and* $q(x)$ *be nonnegative continuous functions in the interval* $[x_0, x_0+a]$, $\hat{T}(x) - x\hat{T}'(x) \le 0$ *for every* $x \in [x_0, x_0+a]$. *Let also,*

$$u(x) \le \hat{T}(x) - \hat{T}(x) \int_{x_0}^{x} q(t)u(t) \frac{\hat{T}(t) - t\hat{T}'(t)}{\hat{T}(t)} dt$$

for every $x \in [x_0, x_0+a]$. *Prove that*

$$u(x) \le \hat{T}(x) e^{-\int_{x_0}^x q(s)\frac{\hat{T}(s)-s\hat{T}'(s)}{\hat{T}(s)}ds}$$

for every $x \in [x_0, x_0+a]$.

Exercise 4.0.167. *Let* $u(x)$ *and* $q(x)$ *be nonnegative continuous functions in the interval* $[x_0 - a, x_0]$, $\hat{T}(x) - x\hat{T}'(x) \le 0$ *for every* $x \in [x_0 - a, x_0]$. *Let also,*

$$u(x) \le \hat{T}(x) + \hat{T}(x) \int_{x_0}^{x} q(t)u(t) \frac{\hat{T}(t) - t\hat{T}'(t)}{\hat{T}(t)} dt$$

for every $x \in [x_0 - a, x_0]$. *Prove that*

$$u(x) \le \hat{T}(x) e^{\int_{x_0}^x q(s)\frac{\hat{T}(s)-s\hat{T}'(s)}{\hat{T}(s)}ds}$$

for every $x \in [x_0 - a, x_0]$.

Theorem 4.0.168. *Let* $u(x)$, $p(x)$ *and* $q(x)$ *be nonnegative continuous functions in the interval* $|x - x_0| \le a$, $\hat{T}(x) - x\hat{T}'(x) \ge 0$ *for every* $x \in [x_0 - a, x_0 + a]$. *If*

$$u(x) \le p(x) + \hat{T}(x) \int_{x_0}^{x} q(t) \frac{\hat{T}(t) - t\hat{T}'(t)}{\hat{T}^2(t)} u(t) dt$$

for every $x \in [x_0, x_0+a]$, *then*

$$u(x) \le p(x) + \hat{T}(x) \int_{x_0}^{x} p(t)q(t) \frac{\hat{T}(t) - t\hat{T}'(t)}{\hat{T}^2(t)} e^{\int_t^x q(s)\frac{\hat{T}(s)-s\hat{T}'(s)}{\hat{T}(s)}ds} dt \qquad (3')$$

for every $x \in [x_0, x_0+a]$. *If*

$$u(x) \le p(x) + \hat{T}(x)\int_x^{x_0} q(t)\frac{\hat{T}(t)-t\hat{T}'(t)}{\hat{T}^2(t)}u(t)dt$$

for $x \in [x_0-a, x_0]$, *then*

$$u(x) \le p(x) + \hat{T}(x)\int_x^{x_0} q(t)p(t)\frac{\hat{T}(t)-t\hat{T}'(t)}{\hat{T}^2(t)}e^{\int_x^t q(s)\frac{\hat{T}(s)-s\hat{T}'(s)}{\hat{T}(s)}ds}dt \qquad (3'')$$

for every $x \in [x_0-a, x_0]$.

Proof. We will prove the assertion in the case when $x \in [x_0, x_0+a]$. The proof in the case $x \in [x_0-a, x_0]$ is similar.

Let

$$r(x) = \int_{x_0}^x q(t)\frac{\hat{T}(t)-t\hat{T}'(t)}{\hat{T}^2(t)}u(t)dt.$$

Then

$$u(x) \le p(x) + \hat{T}(x)r(x) \qquad (8)$$

and, using (8),

$$r'(x) = u(x)q(x)\frac{\hat{T}(x)-x\hat{T}'(x)}{\hat{T}^2(x)}$$

$$\le q(x)\frac{\hat{T}(x)-x\hat{T}'(x)}{\hat{T}^2(x)}\Big(p(x)+\hat{T}(x)r(x)\Big)$$

$$= p(x)q(x)\frac{\hat{T}(x)-x\hat{T}'(x)}{\hat{T}^2(x)} + q(x)\frac{\hat{T}(x)-x\hat{T}'(x)}{\hat{T}(x)}r(x),$$

or

$$r'(x) - q(x)\frac{\hat{T}(x)-0x\hat{T}'(x)}{\hat{T}(x)}r(x) \le p(x)q(x)\frac{\hat{T}(x)-x\hat{T}'(x)}{\hat{T}^2(x)},$$

from where

$$\left(r'(x) - q(x)\frac{\hat{T}(x)-x\hat{T}'(x)}{\hat{T}(x)}r(x)\right)e^{-\int_{x_0}^x q(t)\frac{\hat{T}(t)-t\hat{T}'(t)}{\hat{T}(t)}dt}$$

$$\le p(x)q(x)\frac{\hat{T}(x)-x\hat{T}'(x)}{\hat{T}^2(x)}e^{-\int_{x_0}^x q(t)\frac{\hat{T}(t)-t\hat{T}'(t)}{\hat{T}(t)}dt},$$

or

$$\left(r(x)e^{-\int_{x_0}^x q(t)\frac{\hat{T}(t)-t\hat{T}'(t)}{\hat{T}(t)}dt}\right)' \le p(x)q(x)\frac{\hat{T}(x)-x\hat{T}'(x)}{\hat{T}^2(x)}e^{-\int_{x_0}^x q(t)\frac{\hat{T}(t)-t\hat{T}'(t)}{\hat{T}(t)}dt},$$

which we integrate from x_0 to x and we get

$$r(x)e^{-\int_{x_0}^x q(t)\frac{\hat{T}(t)-t\hat{T}'(t)}{\hat{T}(t)}dt} \le \int_{x_0}^x p(t)q(t)\frac{\hat{T}(t)-t\hat{T}'(t)}{\hat{T}^2(t)}e^{-\int_{x_0}^t q(s)\frac{\hat{T}(s)-s\hat{T}'(s)}{\hat{T}(s)}ds}dt,$$

or

$$r(x) \le \int_{x_0}^x p(t)q(t)\frac{\hat{T}(t)-t\hat{T}'(t)}{\hat{T}^2(t)}e^{\int_t^x q(s)\frac{\hat{T}(s)-s\hat{T}'(s)}{\hat{T}(s)}ds}dt.$$

From the last inequality and (8) we get

$$u(x) \le p(x) + \hat{T}(x) \int_{x_0}^{x} p(t)q(t) \frac{\hat{T}(t) - t\hat{T}'(t)}{\hat{T}^2(t)} e^{\int_t^x q(s) \frac{\hat{T}(s) - s\hat{T}'(s)}{\hat{T}(s)} ds} dt.$$

□

Corollary 4.0.169. *If in the last theorem* $p(x) \equiv 0$ *in* $[x_0 - a, x_0 + a]$, *then* $u(x) \equiv 0$ *in* $[x_0 - a, x_0 + a]$.

Corollary 4.0.170. *If in the last theorem* $p(x)$ *is increasing in* $[x_0, x_0 + a]$ *and* $\hat{T}(x)$ *is decreasing in* $[x_0, x_0 + a]$, *then*

$$u(x) \le p(x) e^{\int_{x_0}^{x} q(s) \frac{\hat{T}(s) - s\hat{T}'(s)}{\hat{T}(s)} ds} \quad \text{for} \quad \forall x \in [x_0, x_0 + a].$$

Proof. Since $p(x)$ is an increasing function in $[x_0, x_0 + a]$, then

$$p(t) \le p(x) \quad \text{for} \quad \forall t, x \in [x_0, x_0 + a], \quad t \le x.$$

Because $\hat{T}(x)$ is a decreasing function in $[x_0, x_0 + a]$, we have

$$\hat{T}(t) \ge \hat{T}(x) \quad \text{for} \quad \forall t, x \in [x_0, x_0 + a], \quad t \le x,$$

from where

$$\frac{1}{\hat{T}(t)} \le \frac{1}{\hat{T}(x)} \quad \text{for} \quad \forall t, x \in [x_0, x_0 + a], \quad t \le x.$$

Consequently

$$\frac{p(t)}{\hat{T}(t)} \le \frac{p(x)}{\hat{T}(x)} \quad \text{for} \quad t, x \in [x_0, x_0 + a], \quad t \le x.$$

From here and $(3')$ we conclude

$$u(x) \le p(x) + \hat{T}(x) \frac{p(x)}{\hat{T}(x)} \int_{x_0}^{x} q(t) \frac{\hat{T}(t) - t\hat{T}'(t)}{\hat{T}(t)} e^{\int_t^x q(s) \frac{\hat{T}(s) - s\hat{T}'(s)}{\hat{T}(s)} ds} dt$$

$$= p(x) - p(x) e^{\int_t^x q(s) \frac{\hat{T}(s) - s\hat{T}'(s)}{\hat{T}(s)} ds} \Big|_{t=x_0}^{t=x}$$

$$= p(x) - p(x) + p(x) e^{\int_{x_0}^{x} q(s) \frac{\hat{T}(s) - s\hat{T}'(s)}{\hat{T}(s)} ds}$$

$$= p(x) e^{\int_{x_0}^{x} q(s) \frac{\hat{T}(s) - s\hat{T}'(s)}{\hat{T}(s)} ds}.$$

for every $x \in [x_0 - a, x_0]$. □

Corollary 4.0.171. *If in the last theorem* $p(x)$ *is a decreasing function in* $[x_0 - a, x_0]$ *and* $\hat{T}(x)$ *is an increasing function in* $[x_0 - a, x_0]$, *then*

$$u(x) \le p(x) e^{\int_{x}^{x_0} q(s) \frac{\hat{T}(s) - s\hat{T}'(s)}{\hat{T}(s)} ds} \quad \text{for} \quad \forall x \in [x_0 - a, x_0].$$

Proof. Since $p(x)$ is a decreasing function in $[x_0 - a, x_0]$, then

$$p(t) \le p(x) \qquad \text{for} \qquad \forall t, x \in [x_0 - a, x_0], \qquad t \ge x.$$

Because $\hat{T}(x)$ is an increasing function in $[x_0 - a, x_0]$, we have

$$\hat{T}(t) \ge \hat{T}(x) \qquad \text{for} \qquad \forall t, x \in [x_0 - a, x_0], \qquad t \ge x,$$

or

$$\frac{1}{\hat{T}(t)} \le \frac{1}{\hat{T}(x)} \qquad \text{for} \qquad \forall t, x \in [x_0 - a, x_0], \qquad t \ge x.$$

Therefore

$$\frac{p(t)}{\hat{T}(t)} \le \frac{p(x)}{\hat{T}(x)} \qquad \text{for} \qquad \forall t, x \in [x_0 - a, x_0], \qquad t \ge x.$$

From here and $(3'')$

$$u(x) \le p(x) + \hat{T}(x)\frac{p(x)}{\hat{T}(x)} \int_x^{x_0} q(t)\frac{\hat{T}(t) - t\hat{T}'(t)}{\hat{T}(t)} e^{\int_x^t q(s)\frac{\hat{T}(s) - s\hat{T}'(s)}{\hat{T}(s)} ds} dt$$

$$= p(x) + p(x) e^{\int_x^t q(s)\frac{\hat{T}(s) - s\hat{T}'(s)}{\hat{T}(s)} ds}\Big|_{t=x}^{t=x_0}$$

$$= p(x) + p(x) e^{\int_x^{x_0} q(s)\frac{\hat{T}(s) - s\hat{T}'(s)}{\hat{T}(s)} ds} - p(x)$$

$$= p(x) e^{\int_x^{x_0} q(s)\frac{\hat{T}(s) - s\hat{T}'(s)}{\hat{T}(s)} ds} \qquad \text{for} \qquad \forall x \in [x_0 - a, x_0].$$

□

Chapter 5

Existence and Uniqueness of Solutions

In this chapter $(x_0, y_0) \in \mathbb{R}^2$, D is a domain in $\mathbb{R}^2$ containing the point (x_0, y_0), J is an interval in $\mathbb{R}$ containing x_0, $\hat{T}(x) \in C^1(J)$, $\hat{T}(x) > 0$ for every $x \in J$. We begin to develop the theory of existence and uniqueness of solutions of the initial value problem

$$\left(\hat{y}^\wedge(\hat{x})\right)^{\circledast} = \hat{f}^\wedge(\hat{x}, \hat{y}^\wedge(\hat{x})), \qquad x \in J, \tag{1'}$$

$$y(x_0) = y_0, \tag{2}$$

where f will be assumed to be continuous in the domain D.

The equation $(1')$ can be rewritten in the following form

$$\frac{y'(x)\hat{T}(x) - y(x)\hat{T}'(x)}{\hat{T}(x)\left(\hat{T}(x) - x\hat{T}'(x)\right)} = \frac{f(x, y(x))}{\hat{T}(x)}, \qquad x \in J,$$

or

$$y'(x)\hat{T}(x) - y(x)\hat{T}'(x) = f(x, y(x))\left(\hat{T}(x) - x\hat{T}'(x)\right), \qquad x \in J,$$

or

$$y'(x) = y(x)\frac{\hat{T}'(x)}{\hat{T}(x)} + f(x, y)\frac{\hat{T}(x) - x\hat{T}'(x)}{\hat{T}(x)}, \qquad x \in J. \tag{1}$$

Definition 5.0.172. *We will say that a function $y(x)$ is a solution to the initial value problem* (1), (2) *if*

1. $y(x_0) = y_0$,
2. $y'(x)$ *exists for all* $x \in J$,
3. *for all* $x \in J$ *the points* $(x, y(x)) \in D$,
4. $y'(x) = y(x)\frac{\hat{T}'(x)}{\hat{T}(x)} + f(x, y(x))\frac{\hat{T}(x) - x\hat{T}'(x)}{\hat{T}(x)}$ *for all* $x \in J$.

If $f(x,y(x))$ is not continuous, then the nature of the solutions of (1) is quite arbitrary. For example, let

$$f(x,y(x)) = \frac{4(y(x)-2)}{x(1-x)} - \frac{y(x)}{1-x}, \qquad \hat{T}(x) = e^x,$$

and $(x_0, y(x_0)) = (0,0)$. Then the equation (1) admits the representation

$$\begin{aligned} y'(x) &= y(x) + \left(\tfrac{4(y(x)-2)}{x(1-x)} - \tfrac{y(x)}{1-x}\right)(1-x) \\ &= y(x) + 4\tfrac{y(x)-2}{x} - y(x) \\ &= \tfrac{4}{x}(y(x)-2), \end{aligned}$$

its general solution is

$$y(x) = 2 + Cx^4, \tag{3}$$

where C is a constant. From here, we conclude that

$$y(0) = 2 \neq 0,$$

therefore the considered initial value problem has no any solution. If we take $(x_0, y(x_0)) = (0,2)$, then every function (2) will be a solution of the considered initial value problem.

We shall need the following result to prove existence, uniqueness, and several other properties of the solutions of the initial value problem (1), (2).

Theorem 5.0.173. *Let $f(x,y(x))$ be continuous function in the domain D, then any solution of the initial value problem* (1), (2) *is also a solution of the integral equation*

$$y(x) = y_0 + \int_{x_0}^{x} \left(y(t)\frac{\hat{T}'(t)}{\hat{T}(t)} + f(t,y(t))\frac{\hat{T}(t) - t\hat{T}'(t)}{\hat{T}(t)} \right) dt \tag{4}$$

and conversely.

Proof. An integration of the equation (1) yields

$$y(x) - y(x_0) = \int_{x_0}^{x} \left(y(t)\frac{\hat{T}'(t)}{\hat{T}(t)} + f(t,y(t))\frac{\hat{T}(t) - t\hat{T}'(t)}{\hat{T}(t)} \right) dt.$$

Conversely, if $y(x)$ is any solution of (4), then

$$y(x_0) = y_0,$$

and since $f(x,y(x))$ is a continuous function in D and $\hat{T}$ is a continuous function in J, then $y(x)$ is a continuous function in J and we can differentiate (4), from where we find

$$y'(x) = y(x)\frac{\hat{T}'(x)}{\hat{T}(x)} + f(x,y(x))\frac{\hat{T}(x) - x\hat{T}'(x)}{\hat{T}(x)}.$$

□

We shall solve the integral equation (4) by using the method of successive approximations due to Picard. For this reason, let $y_0(x)$ be any continuous function, we often take $y_0(x) \equiv y_0$, which we will suppose to be initial approximation of the unknown solution of (4), then we define $y_1(x)$ as follows

$$y_1(x) = y_0 + \int_{x_0}^{x} \left(y_0(t)\frac{\hat{T}'(t)}{\hat{T}(t)} + f(t, y_0(t))\frac{\hat{T}(t) - t\hat{T}'(t)}{\hat{T}(t)} \right) dt.$$

We pick this $y_1(x)$ as our next approximation and substitute this for $y(x)$ in the right side of (4) and call it $y_2(x)$,

$$y_2(x) = y_0 + \int_{x_0}^{x} \left(y_1(t)\frac{\hat{T}'(t)}{\hat{T}(t)} + f(t, y_1(t))\frac{\hat{T}(t) - t\hat{T}'(t)}{\hat{T}(t)} \right) dt.$$

Continuing in this way, the $(m+1)$st approximation $y_{m+1}(x)$ is obtained from $y_m(x)$ by means of the relation

$$y_{m+1}(x) = y_0 + \int_{x_0}^{x} \left(y_m(t)\frac{\hat{T}'(t)}{\hat{T}(t)} + f(t, y_m(t))\frac{\hat{T}(t) - t\hat{T}'(t)}{\hat{T}(t)} \right) dt, \qquad m = 0, 1, 2, \ldots. \tag{5}$$

If the sequence $\{y_m(x)\}_{m=1}^{\infty}$ converges uniformly to a continuous function $y(x)$ in the interval J and for all $x \in J$ the points $(x, y_m(x)) \in D$, then we may pass to the limit in both sides of (5), to obtain

$$\begin{aligned} y(x) &= \lim_{m \longrightarrow \infty} y_{m+1}(x) \\ &= y_0 + \lim_{m \longrightarrow \infty} \int_{x_0}^{x} \left(y_m(t)\frac{\hat{T}'(t)}{\hat{T}(t)} + f(t, y_m(t))\frac{\hat{T}(t) - t\hat{T}'(t)}{\hat{T}(t)} \right) dt \\ &= y_0 + \int_{x_0}^{x} \left(y(t)\frac{\hat{T}'(t)}{\hat{T}(t)} + f(t, y(t))\frac{\hat{T}(t) - t\hat{T}'(t)}{\hat{T}(t)} \right) dt, \end{aligned}$$

so that $y(x)$ is the desired solution.

Example 5.0.174. *Let* $J = [-1, 1]$, $D = \{(x, y) : \quad -1 \le x \le 1, \quad y \in \mathbb{R}\}$, $\hat{T}|(x) = e^x$, $f(x, y(x)) = y(x)$, $y(0) = 1$. *We will find the first four approximation.*

We pick

$$\begin{aligned} y_0(x) &\equiv y_0 = 1, \\ y_1(x) &= 1 + \int_0^x (2-t) y_0(t) dt \\ &= 1 + \int_0^x (2-t) dt \\ &= 1 + 2x - \tfrac{x^2}{2}, \\ y_2(x) &= 1 + \int_0^1 (2-t) y_1(t) dt \\ &= 1 + \int_0^x (2-t)\left(1 + 2t - \tfrac{t^2}{2}\right) dt \\ &= 1 + \int_0^x \left(2 + 3t - 3t^2 + \tfrac{t^3}{2}\right) dt \\ &= 1 + 2x + \tfrac{3}{2}x^2 - x^3 + \tfrac{x^4}{8}, \end{aligned}$$

$$y_3(x) = 1 + \int_0^x (2-t)\left(1 + 2t + \frac{3}{2}t^2 - t^3 + \frac{t^4}{8}\right)dt$$

$$= 1 + \int_0^x \left(2 + 3t + t^2 - \frac{7}{2}t^3 + \frac{5}{4}t^4 - \frac{t^5}{8}\right)dt$$

$$= 1 + 2x + \frac{3}{2}x^2 + \frac{x^3}{3} - \frac{7}{8}x^4 + \frac{1}{4}x^5 - \frac{x^6}{48}.$$

Example 5.0.175. *Let $J = \mathbb{R}$, $D = \mathbb{R}^2$, $f(x, y(x)) = -\frac{2y(x)}{1-x}$, $\hat{T}(x) = e^x$, $y(0) = 1$. Then the equation* (1) *we can represent in the form*

$$y'(x) = y(x) - 2\frac{y(x)}{1-x}(1-x)$$

$$= y(x) - 2y(x)$$

$$= -y(x).$$

Then

$$y_0(x) \equiv y_0 = 1,$$

$$y_1(x) = 1 - \int_0^x y_0(t)dt$$

$$= 1 - x,$$

$$y_2(x) = 1 - \int_0^x y_1(t)dt$$

$$= 1 - \int_0^x (1-t)dt$$

$$= 1 - x + \frac{x^2}{2},$$

$$y_3(x) = 1 - \int_0^x \left(1 - t + \frac{t^2}{2}\right)dt$$

$$= 1 - x + \frac{x^2}{2} - \frac{x^3}{3!}.$$

We assume that

$$y_m(x) = 1 - x + \frac{x^2}{2} - \frac{x^3}{3!} + \cdots + (-1)^m \frac{x^m}{m!}.$$

Then

$$y_{m+1}(x) = 1 - \int_0^x \left(1 - t + \frac{t^2}{2} - \frac{t^3}{3!} + \cdots + (-1)^m \frac{T^m}{m!}\right)dt$$

$$= 1 - x + \frac{x^2}{2} - \frac{x^3}{3!} + \cdots + (-1)^{m+1}\frac{x^{m+1}}{(m+1)!}.$$

Therefore

$$y_m(x) = \sum_{i=0}^{m} (-1)^i \frac{x^i}{i!}, \qquad m = 0, 1, 2, \ldots.$$

From here,

$$\lim_{m \longrightarrow \infty} y_m(x) = e^{-x}.$$

Consequently

$$y(x) = e^{-x}$$

is a solution to the considered problem.

Exercise 5.0.176. *Let* $J = \mathbb{R}$, $D = \mathbb{R}^2$, $\hat{T}(x) = e^{2x}$, $f(x,y(x)) = xy(x)$, $y(0) = 1$. *Find the first three approximations of the solution of the problem* (1), (2).

Exercise 5.0.177. *Let* $J = \mathbb{R}$, $D = \mathbb{R}^2$, $\hat{T}(x) = e^x$, $f(x,y(x)) = x + y(x)$, $y(0) = 2$. *Find the first two approximations of the solution of the problem* (1), (2).

Below we will suppose that a and b are positive real numbers. Let P be positive real number such that

$$\frac{|\hat{T}'(x)|}{\hat{T}(x)} \le P, \qquad \frac{|\hat{T}(x) - x\hat{T}'(x)|}{\hat{T}(x)} \le P \qquad \text{for} \qquad \forall x \in [x_0 - a, x_0 + a].$$

Theorem 5.0.178. *Let the following conditions be satisfied*

(i) $f(x,y)$ *is continuous in the closed rectangle* $\overline{S} : |x - x_0| \le a$, $|y - y_0| \le b$ *and hence there exists a* $M > 0$ *such that* $|f(x,y)| \le M$ *for all* $(x,y) \in \overline{S}$,

(ii) $f(x,y)$ *satisfies a uniform Lipschitz condition*

$$|f(x,y_1) - f(x,y_2)| \le L|y_1 - y_2|$$

for all (x,y_1), (x,y_2) *in the closed rectangle* $\overline{S}$,

(iii) $y_0(x)$ *is continuous in* $|x - x_0| \le a$, *and* $|y_0(x) - y_0| \le b$.

Then the sequence $\{y_m(x)\}_{m=1}^{\infty}$ *generated by Picard iterative scheme* (5) *converges to the unique solution* $y(x)$ *of the initial value problem* (1), (2). *This solution is valid in the interval* $J_h : |x - x_0| \le h$, *where* $h = \min\left\{a, \frac{b}{P(b+|y_0|+M)}\right\}$. *Further, for all* $x \in J_h$ *the following error estimate holds*

$$|y(x) - y_m(x)| \le Ne^{(P+PL)h} \min\left\{1, \frac{((P+PL)h)^m}{m!}\right\}, \qquad m = 0, 1, 2, \ldots, \qquad (6)$$

where

$$\max_{x \in J_h} |y_1(x) - y_0(x)| \le N.$$

Remark 5.0.179. *This Theorem is called a local existence theorem since it guarantees a solution only in the neighborhood of the point* (x_0, y_0).

Proof. We will show that the successive approximations $y_m(x)$ defined by (5) exist as continuous function in J_h and $(x, y_m(x)) \in \overline{S}$ for all $x \in J_h$. Since $y_0(x)$ is a continuous function for all x such that $|x - x_0| \le a$, the function $F_0(x) = f(x, y_0(x))$ is continuous function in J_h, and hence $y_1(x)$ is continuous in J_h.

Also,

$$
\begin{aligned}
|y_1(x)-y_0| &= \left|\int_{x_0}^x \left(y_0(t)\frac{\hat{T}'(t)}{\hat{T}(t)}+f(t,y_0(t))\frac{\hat{T}(t)-t\hat{T}'(t)}{\hat{T}(t)}\right)dt\right| \\
&\leq \left|\int_{x_0}^x \left(|y_0(t)|\frac{|\hat{T}'(t)|}{\hat{T}(t)}+|f(t,y_0(t))|\frac{|\hat{T}(t)-t\hat{T}'(t)|}{\hat{T}(t)}\right)dt\right| \\
&\leq \left|\int_{x_0}^x \Big((b+|y_0|)P+MP\Big)dt\right| \\
&= P(b+|y_0|+M)|x-x_0| \\
&\leq P(b+|y_0|+M)h \\
&\leq b.
\end{aligned}
$$

Assuming that the assertion is true for $y_m(x)$, $m \geq 1$, then it is sufficient to prove that it is also true for $y_{m+1}(x)$. For this, since $y_m(x)$ is continuous in J_h, the function $F_m(x) = f(x,y_m(x))$ is also continuous function in J_h. Moreover,

$$
\begin{aligned}
|y_{m+1}(x)-y_0| &= \left|\int_{x_0}^x \left(y_m(t)\frac{\hat{T}'(t)}{\hat{T}(t)}+f(t,y_m(t))\frac{\hat{T}(t)-t\hat{T}'(t)}{\hat{T}(t)}\right)dt\right| \\
&\leq \left|\int_{x_0}^x \left(|y_m(t)|\frac{|\hat{T}'(t)|}{\hat{T}(t)}+|f(t,y_m(t))|\frac{|\hat{T}(t)-t\hat{T}'(t)|}{\hat{T}(t)}\right)dt\right| \\
&\leq \left|\int_{x_0}^x \Big((b+|y_0|)P+MP\Big)dt\right| \\
&\leq P(b+|y_0|+M)|x-x_0| \\
&\leq P(b+|y_0|+M)h \\
&\leq b.
\end{aligned}
$$

Now we will prove that the sequence $\{y_m(x)\}_{m=1}^{\infty}$ converges uniformly in J_h. Since $y_1(x)$ and $y_0(x)$ are continuous in J_h, there exists a constant $N > 0$ such that

$$|y_1(x)-y_0(x)| \leq N \qquad \text{for} \qquad \forall x \in J_h.$$

Also, for every $x \in J_h$, we have

$$
\begin{aligned}
|y_2(x)-y_1(x)| &= \left|\int_{x_0}^x \left((y_1(t)-y_0(t))\frac{\hat{T}'(t)}{\hat{T}(t)}+(f(t,y_1(t))-f(t,y_0(t)))\frac{\hat{T}(t)-t\hat{T}'(t)}{\hat{T}(t)}\right)dt\right| \\
&\leq \left|\int_{x_0}^x \left(|y_1(t)-y_0(t)|\frac{|\hat{T}'(t)|}{\hat{T}(t)}+|f(t,y_1(t))-f(t,y_0(t))|\frac{|\hat{T}(t)-t\hat{T}'(t)|}{\hat{T}(t)}\right)dt\right| \\
&\leq \left|\int_{x_0}^x \left(|y_1(t)-y_0(t)|\frac{|\hat{T}'(t)|}{\hat{T}(t)}+L|y_1(t)-y_0(t)|\frac{|\hat{T}(t)-t\hat{T}'(t)|}{\hat{T}(t)}\right)dt\right| \\
&\leq \left|\int_{x_0}^x (NP+LNP)dt\right| \\
&= NP(1+L)|x-x_0|.
\end{aligned}
$$

Supposing that

$$|y_m(x) - y_{m-1}(x)| \leq N \frac{\Big((P+LP)|x-x_0|\Big)^{m-1}}{(m-1)!}, \qquad x \in J_h, \tag{7}$$

for some $m \in \mathbb{N}$.

We will prove that

$$|y_{m+1}(x) - y_m(x)| \leq N \frac{\Big((P+LP)|x-x_0|\Big)^{m}}{m!}, \qquad x \in J_h.$$

Really,

$$\begin{aligned}
&|y_{m+1}(x) - y_m(x)| \\
&= \left| \int_{x_0}^{x} \left((y_m(t) - y_{m-1}(t)) \tfrac{\hat{T}'(t)}{\hat{T}(t)} + (f(t, y_m(t)) - f(t, y_{m-1}(t))) \tfrac{\hat{T}(t) - t\hat{T}'(t)}{\hat{T}(t)} \right) dt \right| \\
&\leq \left| \int_{x_0}^{x} \left(|y_m(t) - y_{m-1}(t)| \tfrac{|\hat{T}'(t)|}{\hat{T}(t)} + |f(t, y_m(t)) - f(t, y_{m-1}(t))| \tfrac{|\hat{T}(t) - t\hat{T}'(t)|}{\hat{T}(t)} \right) dt \right| \\
&\leq \left| \int_{x_0}^{x} \left(|y_m(t) - y_{m-1}(t)| \tfrac{|\hat{T}'(t)|}{\hat{T}(t)} + L|y_m(t) - y_{m-1}(t)| \tfrac{|\hat{T}(t) - t\hat{T}'(t)|}{\hat{T}(t)} \right) dt \right| \\
&\leq \left| \int_{x_0}^{x} (P+PL)|y_m(t) - y_{m-1}(t)| dt \right| \\
&\leq N(P+PL)^{m+1} \left| \int_{x_0}^{x} \tfrac{(t-x_0)^m}{m!} dt \right| \\
&= N(P+PL)^{m+1} \tfrac{|x-x_0|^{m+1}}{(m+1)!}.
\end{aligned}$$

Thus inequality (7) is true for all $m \in \mathbb{N}$.

Next, since

$$\begin{aligned}
N \textstyle\sum_{m=1}^{\infty} \dfrac{\Big((P+PL)|x-x_0|\Big)^{m-1}}{(m-1)!} &\leq N \textstyle\sum_{m=0}^{\infty} \dfrac{\Big((P+PL)h\Big)^{m}}{m!} \\
&= N e^{(P+PL)h} < \infty,
\end{aligned}$$

we have that the series

$$y_0(x) + \sum_{m=1}^{\infty} (y_m(x) - y_{m-1}(x))$$

converges absolutely and uniformly in the interval J_h, and hence its partial sums

$$y_1(x), y_2(x), \ldots, y_m(x), \ldots$$

converge to a continuous function in this interval, i.e.,

$$y(x) = \lim_{m \longrightarrow \infty} y_m(x).$$

As we have seen above we have that $y(x)$ is a solution to the problem (1), (2).

To prove that $y(x)$ is the only solution, we assume that $z(x)$ is also a solution to the initial value problem (1), (2) which exists in the interval J_h and $(x,z(x)) \in \overline{S}$ for all $x \in J_h$. The hypothesis (ii) is applicable and we have

$$|y(x)-z(x)| \leq \left| \int_{x_0}^x \left(|y(t)-z(t)| \frac{\hat{T}'(t)}{\hat{T}(t)} + |f(t,y(t)) - f(t,z(t))| \frac{|\hat{T}(t)-t\hat{T}'(t)|}{\hat{T}(t)} \right) dt \right|$$

$$\leq \left| \int_{x_0}^x \left(P|y(t)-z(t)| + LP|y(t)-z(t)| \right) dt \right|$$

$$= (P+LP) \left| \int_{x_0}^x |y(t)-z(t)| dt \right|, \qquad x \in J_h.$$

Consequently

$$|y(x)-z(x)| = 0$$

for all $x \in J_h$.

Finally, we will obtain the error bound (6).

For $n > m$ the inequality (7) gives

$$|y_n(x) - y_m(x)| = |y_n(x) - y_{n-1}(x) + y_{n-1}(x) - y_{n-2}(x) + \cdots + y_{m+1}(x) - y_m(x)|$$

$$\leq \sum_{k=m}^{n-1} |y_{k+1}(x) - y_k(x)|$$

$$\leq N \sum_{k=m}^{n-1} \frac{\left((P+LP)|x-x_0| \right)^k}{k!}$$

$$\leq N \sum_{k=m}^{n-1} \frac{\left((P+PL)h \right)^k}{k!}$$

$$= N \left((P+PL)h \right)^m \sum_{k=0}^{n--m-1} \frac{\left((P+PL)h \right)^k}{(m+k)!} \qquad \left(\frac{1}{(m+k)!} \leq \frac{1}{m!k!} \right)$$

$$\leq N \frac{\left((P+PL)h \right)^m}{m!} \sum_{k=0}^{n-m-1} \frac{\left((P+PL)h \right)^k}{k!}$$

$$\leq N \frac{\left((P+PL)h \right)^m}{m!} e^{(P+PL)h}, \tag{8}$$

and hence as $n \longrightarrow \infty$, we get

$$|y(x) - y_m(x)| \leq N \frac{\left((P+PL)h \right)^m}{m!} e^{(P+PL)h}$$

in J_h.

The inequality (8) provides

$$|y_n(x) - y_m(x)| \leq N\sum_{k=m}^{n-1} \frac{\left((P+PL)h\right)^k}{k!}$$

$$\leq N\sum_{k=0}^{\infty} \frac{\left((P+PL)h\right)^k}{k!}$$

$$= Ne^{(P+PL)h},$$

and as $n \longrightarrow \infty$, we find

$$|y(x) - y_m(x)| \leq Ne^{(P+PL)h}$$

in J_h. □

Example 5.0.180. *Let $f(x,y) = -(x + y(x))$, $\hat{T}(x) = e^x$, $y(0) = 1$. Then the equation* (1) *we can rewrite in the following form*

$$y'(x) = y(x) - x(1-x) - y(x)(1-x)$$

$$= xy(x) + x(x-1).$$

Its general solution is

$$y(x) = e^x\left(C + \int x(x-1)e^{-x}dx\right)$$

$$= e^x\left(C - (x^2 - x)e^{-x} + \int (2x-1)e^{-x}dx\right)$$

$$= e^x\left(C - (x^2 - x)e^{-x} - (2x-1)e^{-x} + 2\int e^{-x}dx\right)$$

$$= e^x\left(C - (x^2 + x - 1)e^{-x} - 2e^{-x}\right)$$

$$= Ce^x - x^2 - x - 1.$$

From here,

$$y(0) = C - 1 = 1,$$

therefore

$$C = 2.$$

Consequently

$$y(x) = 2e^x - x^2 - x - 1$$

is the solution of the considered initial value problem, which is defined for every $x \in \mathbb{R}$.

Now we will apply the last Theorem.

The function

$$f(x,y) = xy + x^2 - x$$

is a continuous function in the rectangle $\overline{S}: |x| \le a, |y-1| \le b$, *and*

$$|f(x,y)| = |xy + x^2 - x|$$

$$\le |x||y| + |x|^2 + |x|$$

$$\le a(b+1) + a^2 + a$$

$$= a^2 + 2a + ab = M.$$

For all (x,y_1), $(x,y_2) \in \overline{S}$ *we have*

$$|f(x,y_1) - f(x,y_2)| = |xy_1 + x^2 - x - xy_2 - x^2 + x|$$

$$= |xy_1 - xy_2|$$

$$= |x||y_1 - y_2|$$

$$\le a|y_1 - y_2|,$$

i.e., the function f *satisfies a uniform Lipschitz condition with* $L = a$.

Also, $y_0(x) \equiv 1$ *in* $|x| \le a$ *and*

$$|y_0(x) - 1| = 0 \le b,$$

$$\frac{|\hat{T}'(x)|}{\hat{T}(x)} = \frac{e^x}{e^x} = 1,$$

$$\frac{|\hat{T}(x) - x\hat{T}'(x)|}{\hat{T}(x)} = |1 - x|$$

$$\le |x| + 1$$

$$\le a + 1,$$

therefore $P = a + 1$.

Thus, there exists a unique solution of the considered initial value problem in the interval

$$|x| \le h = \min\left\{a, \frac{b}{(a+1)(a^2+ab+2a+b+1)}\right\}.$$

We have $y_0(x) \equiv 1$ *and*

$$|y(x) - y_0(x)| = |2e^x - x^2 - x - 1 - 1|$$

$$\le 2e^x + x^2 + |x| + 2$$

$$\le 2e^h + h^2 + h + 2 = N.$$

Further, the iterative scheme for the considered initial value problem takes the form

$$y_{m+1}(x) = 1 + \int_0^x \left(t y_m(t) + t(t-1) \right) dt$$

$$= 1 - \tfrac{x^2}{2} + \tfrac{x^3}{3} + \int_0^x t y_m(t) dt,$$

$$y_1(x) = 1 - \tfrac{x^2}{2} + \tfrac{x^3}{3} + \int_0^x t dt$$

$$= 1 + \tfrac{x^3}{3},$$

$$|y(x) - y_1(x)| \leq N e^{(P+PL)h} \min\{1, (P+PL)h\}, \qquad x \in J_h,$$

$$y_2(x) = 1 - \tfrac{x^2}{2} + \tfrac{x^3}{3} + \int_0^x t \left(1 + \tfrac{t^3}{3}\right) dt$$

$$= 1 + \tfrac{x^3}{3} + \tfrac{x^5}{15},$$

$$|y(x) - y_2(x)| \leq N e^{(P+PL)h} \min\left\{1, \tfrac{(P+PL)^2 h^2}{2}\right\}, \qquad x \in J_h.$$

Exercise 5.0.181. *Discuss the existence and uniqueness of the solutions of the initial value problem* (1), (2) *in the case when*

$$\hat{T}(x) = e^{4x} + x^2 + 1, \qquad f(x,y) = x^2 - 3y^2, \qquad y(1) = 2.$$

Definition 5.0.182. *If the solution of the initial value problem* (1), (2) *exists in the entire interval* $|x - x_0| \leq a$, *we say that the solution exists globally.*

The next result is called a global existence theorem.

Theorem 5.0.183. *Let the following conditions be satisfied*

(i) $f(x,y)$ *is continuous in the strip* $T : |x - x_0| \leq a, |y| < \infty$,

(ii) $f(x,y)$ *satisfies a uniform Lipschitz condition in* T,

(iii) $y_0(x)$ *is continuous in* $|x - x_0| \leq a$.

Then the sequence $\{y_m(x)\}_{m=1}^{\infty}$ *generated by Picard iterative scheme exists in the entire interval* $|x - x_0| \leq a$, *and converges to the unique solution* $y(x)$ *of the initial value problem* (1), (2).

Proof. For any continuous function $y_0(x)$ in $|x - x_0| \leq a$, as in the proof of the local existence Theorem, can be established the existence of each $y_m(x)$ in $|x - x_0| \leq a$ satisfying $|y_m(x)| < \infty$. Also, as in the proof of the previous Theorem we have that the sequence $\{y_m(x)(\}_{m=1}^{\infty}$ converges to $y(x)$ in $|x - x_0| \leq a$, replacing h by a throughout the proof and recalling that the function $f(x,y)$ satisfies the Lipschitz condition in the strip T. □

Corollary 5.0.184. *Let* $f(x,y)$ *be continuous in* $\mathbb{R}^2$ *and satisfies a uniform Lipschitz condition in each strip* $T_a : |x| \leq a, |y| < \infty$, *with the Lipschitz constant* L_a. *Then the initial value problem* (1), (2) *has a unique solution which exists for all x.*

Proof. For any x there exists an $a > 0$ such that $|x - x_0| \leq a$. From here and from $T \subset T_{a+|x_0|}$, it follows that the function $f(x,y)$ satisfies the conditions of the previous Theorem in the strip T. Hence, the result follows for any x. □

Exercise 5.0.185. *Discuss the existence and uniqueness of the solutions of the initial value problem* (1), (2) *in the case when*

$$\hat{T}(x) = x^2 + 1, \qquad f(x,y) = y^2, \qquad y(0) = 0.$$

We will note that there exist positive constants M_1 and M_2 such that

$$\left|\frac{\hat{T}'(x)}{\hat{T}(x)}\right| \leq M_1, \quad \left|1 - x\frac{\hat{T}'(x)}{\hat{T}(x)}\right| \leq M_2 \quad \text{for} \quad x \in [x_0 - a, x_0 + a].$$

Theorem 5.0.186. *(iso-Peano's existence theorem) Let f is defined, continuous and bounded function on the strip $T = \{(x,y) \in \mathbb{R}^2 : |x - x_0| \leq a, |y| < \infty\}$. Then the Cauchy problem* (1), (2) *has a bounded solution $y(x)$ which is defined on $|x - x_0| \leq a$ and*

$$|y(x)| \leq \left(1 + e^{aM_1}\right)(|y_0| + \sup_{(x,y)\in V} |f(x,y)|M_2 a) \quad \text{for} \quad \forall x \in [x_0 - a, x_0 + a].$$

Remark 5.0.187. *We can consider our main result as a continuation of the well - known Peano's Theorem.*

If we put

$$g(x,y) = y(x)\frac{\hat{T}'(x)}{\hat{T}(x)} + f(x, y(x))\left(1 - x\frac{\hat{T}'(x)}{\hat{T}(x)}\right),$$

then g is unbounded function on the strip T. Therefore we can not apply the classical Peano's Theorem for the Cauchy problem (1), (2), *because g has to be bounded on T.*

Proof. Since f is a bounded function on T then there exists a positive constant M such that

$$|f(x,y)| \leq M \quad \text{for} \quad (x,y) \in T.$$

We will prove our main result for $x \in [x_0, x_0 + a]$. In the same way one can prove the main result for $x \in [x_0 - a, x_0]$.

For $x \in [x_0, x_0 + a]$ we define the sequence $\{y_m(x)\}_{m=1}^{\infty}$ as follows

$$y_m(x) = y_0 \quad \text{for} \quad x \in \left[x_0, x_0 + \tfrac{a}{m}\right],$$

$$y_m(x) = y_0 + \int_{x_0}^{x-\frac{a}{m}} \left(y_m(t)\tfrac{\hat{T}'(t)}{\hat{T}(t)} + f(t, y_m(t))\left(1 - t\tfrac{\hat{T}'(t)}{\hat{T}(t)}\right)\right)dt \quad \text{for}$$

$$x \in \left[x_0 + k\tfrac{a}{m}, x_0 + (k+1)\tfrac{a}{m}\right], \quad k = 1, 2, \ldots, m-1.$$

For this sequence we have

1. Let $m \in \mathbb{N}$ is arbitrary chosen.

   If $x \in \left[x_0, x_0 + \frac{a}{m}\right]$ then

$$|y_m(x)| = |y_0|.$$

   If $x \notin \left[x_0, x_0 + \frac{a}{m}\right]$ and $x \in \left[x_0 + k\frac{a}{m}, x_0 + (k+1)\frac{a}{m}\right]$ for some $k = 1, 2, \ldots, m-1$, then

$$|y_m(x)| = \left|y_0 + \int_{x_0}^{x-\frac{a}{m}} \left(y_m(t)\frac{\hat{T}'(t)}{\hat{T}(t)} + f(t, y_m(t))\left(1 - t\frac{\hat{T}'(t)}{\hat{T}(t)}\right)\right)dt\right|$$

$$\leq |y_0| + \int_{x_0}^{x-\frac{a}{m}} \left(|y_m(t)|\left|\frac{\hat{T}'(t)}{\hat{T}(t)}\right| + |f(t, y_m(t))|\left|1 - t\frac{\hat{T}'(t)}{\hat{T}(t)}\right|\right)dt$$

$$\leq |y_0| + \int_{x_0}^{x-\frac{a}{m}} (M_1|y_m(t)| + MM_2)dt$$

$$= |y_0| + M_1\int_{x_0}^{x-\frac{a}{m}} |y_m(t)|dt + MM_2\left(x - \frac{a}{m} - x_0\right)$$

$$\leq |y_0| + M_1\int_{x_0}^{x-\frac{a}{m}} |y_m(t)|dt + MM_2\left(x_0 + (k+1)\frac{a}{m} - \frac{a}{m} - x_0\right)$$

$$\leq |y_0| + MM_2a + M_1\int_{x_0}^{x-\frac{a}{m}} |y_m(t)|dt,$$

   i.e. for $x \in \left[x_0 + k\frac{a}{m}, x_0 + (k+1)\frac{a}{m}\right]$ we have

$$|y_m(x)| \leq |y_0| + MM_2a + M_1\int_{x_0}^{x-\frac{a}{m}} |y_m(t)|dt$$

$$\leq |y_0| + MM_2a + M_1\int_{x_0}^{x} |y_m(t)|dt.$$

   From here and the Gronwall's inequality we get

$$|y_m(x)| \leq |y_0| + MM_2a + M_1\int_{x_0}^{x} (|y_0| + MM_2a)e^{M_1(x-t)}dt$$

$$= |y_0| + MM_2a + e^{M_1x}M_1(|y_0| + MM_2a)\int_{x_0}^{x} e^{-M_1t}dt$$

$$= |y_0| + MM_2a + e^{M_1x}(|y_0| + MM_2a)\left(e^{-M_1x_0} - e^{-M_1x}\right)$$

$$\leq |y_0| + MM_2a + e^{M_1(x-x_0)}(|y_0| + MM_2a)$$

$$\leq |y_0| + MM_2a + e^{aM_1}(|y_0| + MM_2a)$$

$$= (1 + e^{aM_1})(|y_0| + MM_2a) =: M_3 \quad \text{for} \quad x \in \left[x_0 + k\frac{a}{m}, x_0 + (k+1)\frac{a}{m}\right],$$

   for some $k = 1, 2, \ldots, m-1$.

   Consequently for every $x \in [x_0, x_0 + a]$ we have

$$|y_m(x)| \leq M_3 \tag{9}$$

   for every $m \in \mathbb{N}$.

   Therefore the sequence $\{y_m(x)\}_{m=1}^{\infty}$ is uniformly bounded on $[x_0, x_0 + a]$.

2. Let $x_1, x_2 \in [x_0, x_0+a]$ and $m \in \mathbb{N}$ is arbitrarily chosen. Then

1. case. $x_1, x_2 \in \left[x_0, x_0+\frac{a}{m}\right]$. Then

$$y_m(x_1) = y_m(x_2) = y_0,$$

and therefore

$$|y_m(x_2) - y_m(x_1)| = 0.$$

2. case. Let $x_1 \in \left[x_0, x_0+\frac{a}{m}\right]$, $x_2 \notin \left[x_0, x_0+\frac{a}{m}\right]$. Then there exists $k \in \{1, 2, \ldots, m-1\}$, such that $x_2 \in \left[x_0 + k\frac{a}{m}, x_0 + (k+1)\frac{a}{m}\right]$ and

$$y_m(x_1) = y_0,$$

$$y_m(x_2) = y_0 + \int_{x_0}^{x_2-\frac{a}{m}} \left(y_m(t)\frac{\hat{T}'(t)}{\hat{T}(t)} + f(t, y_m(t))\left(1 - t\frac{\hat{T}'(t)}{\hat{T}(t)}\right)\right)dt,$$

from here,

$$|y_m(x_2) - y_m(x_1)| = \left|\int_{x_0}^{x_2-\frac{a}{m}} \left(y_m(t)\frac{\hat{T}'(t)}{\hat{T}(t)} + f(t, y_m(t))\left(1 - t\frac{\hat{T}'(t)}{\hat{T}(t)}\right)\right)dt\right|$$

$$\leq \int_{x_0}^{x_2-\frac{a}{m}} \left(|y_m(t)|\left|\frac{\hat{T}'(t)}{\hat{T}(t)}\right| + |f(t, y_m(t))|\left|1 - t\frac{\hat{T}'(t)}{\hat{T}(t)}\right|\right)dt$$

$$\leq MM_2\left(x_2 - \frac{a}{m} - x_0\right) + M_1\int_{x_0}^{x_2-\frac{a}{m}} |y_m(t)|dt$$

now we use that $x_1 \in \left[x_0, x_0+\frac{a}{m}\right]$

$$\leq MM_2(x_2 - x_1) + M_1\int_{x_0}^{x_2-\frac{a}{m}} |y_m(t)|dt,$$

i.e.

$$|y_m(x_2) - y_m(x_1)| \leq MM_2(x_2 - x_1) + M_1\int_{x_0}^{x_2-\frac{a}{m}} |y_m(t)|dt.$$

From here and (9) we obtain

$$|y_m(x_2) - y_m(x_1)| \leq MM_2(x_2 - x_1) + M_1M_3\int_{x_0}^{x_2-\frac{a}{m}} dt$$

$$= MM_2(x_2 - x_1) + M_1M_3\left(x_2 - \frac{a}{m} - x_0\right)$$

$$\leq (MM_2 + M_1M_3)(x_2 - x_1).$$

3. case. Let $x_1, x_2 \notin \left[x_0, x_0+\frac{a}{m}\right]$. Without loss of generality we can suppose that $x_1 \leq x_2$. Let

$$x_1 \in \left[x_0 + k\frac{a}{m}, x_0 + (k+1)\frac{a}{m}\right], \quad x_2 \in \left[x_0 + i\frac{a}{m}, x_0 + (i+1)\frac{a}{m}\right], \quad k \leq i,$$

$k, i \in \{1, 2, \ldots, m-1\}$.
Then

$$y_m(x_2) = y_0 + \int_{x_0}^{x_2-\frac{a}{m}} \left(y_m(t)\frac{\hat{T}'(t)}{\hat{T}(t)} + f(t, y_m(t))\left(1 - t\frac{\hat{T}'(t)}{\hat{T}(t)}\right)\right)dt,$$

$$y_m(x_1) = y_0 + \int_{x_0}^{x_1-\frac{a}{m}} \left(y_m(t)\frac{\hat{T}'(t)}{\hat{T}(t)} + f(t, y_m(t))\left(1 - t\frac{\hat{T}'(t)}{\hat{T}(t)}\right)\right)dt,$$

$$y_m(x_2) - y_m(x_1) = \int_{x_1-\frac{a}{m}}^{x_2-\frac{a}{m}} \left(y_m(t)\frac{\hat{T}'(t)}{\hat{T}(t)} + f(t, y_m(t))\left(1 - t\frac{\hat{T}'(t)}{\hat{T}(t)}\right)\right)dt,$$

$$|y_m(x_2) - y_m(x_1)| = \left|\int_{x_1-\frac{a}{m}}^{x_2-\frac{a}{m}} \left(y_m(t)\frac{\hat{T}'(t)}{\hat{T}(t)} + f(t, y_m(t))\left(1 - t\frac{\hat{T}'(t)}{\hat{T}(t)}\right)\right)dt\right|$$

$$\leq \int_{x_1-\frac{a}{m}}^{x_2-\frac{a}{m}} \left(|y_m(t)|\left|\frac{\hat{T}'(t)}{\hat{T}(t)}\right| + |f(t, y_m(t))|\left|1 - t\frac{\hat{T}'(t)}{\hat{T}(t)}\right|\right)dt$$

$$\leq M_1 \int_{x_1-\frac{a}{m}}^{x_2-\frac{a}{m}} |y_m(t)|dt + MM_2 \int_{x_1-\frac{a}{m}}^{x_2-\frac{a}{m}} dt$$

$$= M_1 \int_{x_1-\frac{a}{m}}^{x_2-\frac{a}{m}} |y_m(t)|dt + MM_2(x_2 - x_1)$$

now we apply (9)

$$\leq M_1 M_3 \int_{x_1-\frac{a}{m}}^{x_2-\frac{a}{m}} dt + MM_2(x_2 - x_1)$$

$$= (M_1M_3 + MM_2)(x_2 - x_1).$$

From 1, 2, and 3 cases follows that for every $x_1, x_2 \in [x_0, x_0 + a]$ we have

$$|y_m(x_2) - y_m(x_1)| \leq (M_1M_3 + MM_2)|x_2 - x_1| \quad \text{for} \quad \forall m \in \mathbb{N}. \tag{10}$$

Let $\varepsilon > 0$ is arbitrary chosen and fixed. Let $\delta = \frac{\varepsilon}{MM_2 + M_1M_3}$. Then if $x_1, x_2 \in [x_0, x_0 + a]$, $|x_1 - x_2| < \delta$, using (10), we get

$$|y_m(x_2) - y_m(x_1)| \leq (M_1M_3 + MM_2)|x_2 - x_1|$$

$$< (M_1M_3 + MM_2)\delta = \varepsilon.$$

Consequently $\{y_m(x)\}_{m=1}^{\infty}$ is equip-continuous family on $[x_0, x_0 + a]$.

Therefore there exists a subsequence $\{y_{m_p}(x)\}_{p=1}^{\infty}$ of the sequence $\{y_m(x)\}_{m=1}^{\infty}$ which is uniformly convergent to $y(x)$ on $[x_0, x_0 + a]$.

For $y_{m_p}(x)$, $x \in [x_0, x_0 + a]$, we have

$$y_{m_p}(x) = y_0 + \int_{x_0}^{x-\frac{a}{m_p}} \left(y_{m_p}(t)\frac{\hat{T}'(t)}{\hat{T}(t)} + f(t, y_{m_p}(t))\left(1 - t\frac{\hat{T}'(t)}{\hat{T}(t)}\right)\right)dt$$

$$= y_0 + \int_{x_0}^{x} \left(y_{m_p}(t)\frac{\hat{T}'(t)}{\hat{T}(t)} + f(t, y_{m_p}(t))\left(1 - t\frac{\hat{T}'(t)}{\hat{T}(t)}\right)\right)dt \tag{11}$$

$$+ \int_{x}^{x-\frac{a}{m_p}} \left(y_{m_p}(t)\frac{\hat{T}'(t)}{\hat{T}(t)} + f(t, y_{m_p}(t))\left(1 - t\frac{\hat{T}'(t)}{\hat{T}(t)}\right)\right)dt.$$

Since f is a continuous and bounded function on T we have

$$\begin{aligned} &\lim_{p\longrightarrow\infty} \int_{x_0}^{x} \left(y_{m_p}(t)\frac{\hat{T}'(t)}{\hat{T}(t)} + f(t, y_{m_p}(t))\left(1 - t\frac{\hat{T}'(t)}{\hat{T}(t)}\right)\right)dt \\ &= \int_{x_0}^{x} \left(y(t)\frac{\hat{T}'(t)}{\hat{T}(t)} + f(t, y(t))\left(1 - t\frac{\hat{T}'(t)}{\hat{T}(t)}\right)\right)dt. \end{aligned} \tag{12}$$

Also,

$$\left|\int_{x}^{x-\frac{a}{m_p}} \left(y_{m_p}(t)\frac{\hat{T}'(t)}{\hat{T}(t)} + f(t, y_{m_p}(t))\left(1 - t\frac{\hat{T}'(t)}{\hat{T}(t)}\right)\right)dt\right|$$

$$\leq \int_{x-\frac{a}{m_p}}^{x} \left(|y_{m_p}(t)|\left|\frac{\hat{T}'(t)}{\hat{T}(t)}\right| + |f(t, y_{m_p}(t))|\left|1 - t\frac{\hat{T}'(t)}{\hat{T}(t)}\right|\right)dt$$

$$\leq M_1 \int_{x-\frac{a}{m_p}}^{x} |y_{m_p}(t)|dt + MM_2\frac{a}{m_p}$$

now we use (9)

$$\leq (M_1M_3 + MM_2)\frac{a}{m_p} \longrightarrow_{p\longrightarrow\infty} 0.$$

From here and (11), (12), when $p \longrightarrow \infty$, we get

$$y(x) = y_0 + \int_{x_0}^{x} \left(y(t)\frac{\hat{T}'(t)}{\hat{T}(t)} + f(t, y(t))\left(1 - t\frac{\hat{T}'(t)}{\hat{T}(t)}\right)\right)dt$$

for every $x \in [x_0, x_0 + a]$. Therefore y is a solution of the Cauchy problem (1), (2) which is defined on $[x_0, x_0 + a]$. From (9) follows that $|y(x)| \leq M_3$ for every $x \in [x_0, x_0 + a]$. □

Corollary 5.0.188. *Let $f(x,y)$ be continuous in $\overline{S}$, and hence there exists a $M > 0$ such that $|f(x,y)| \leq M$ for all $(x,y) \in \overline{S}$. Then the initial value problem* (1), (2) *has at least one solution in J_h.*

Proof. The proof is the same as that of the proof of iso-Peano's existence theorem with some obvious changes. □

Definition 5.0.189. *(ε-approximate solution) A function $y(x)$ defined in J is said to be an ε-approximate solution of the iso-differential equation* (1) *if*

1. *$y(x)$ is continuous for all $x \in J$,*

2. *for all $x \in J$ the points $(x, y(x)) \in D$,*

3. *$y(x)$ has piecewise continuous derivative in J which may fail to be defined only for a finite number of points, say x_1, x_2, ..., x_k,*

4. $\left|y'(x) - y(x)\frac{\hat{T}'(x)}{\hat{T}(x)} - f(x,y)\frac{\hat{T}(x) - x\hat{T}'(x)}{\hat{T}(x)}\right| \leq \varepsilon$ *for all $x \in J$, $x \neq x_i$, $i = 1, 2, \ldots, k$.*

The existence of an ε-approximate solution is provided in the following theorem.

Theorem 5.0.190. *Let $f(x,y)$ be continuous in $\overline{S}$ and hence there exists a $M > 0$ such that $|f(x,y)| \leq M$ for every $(x,y) \in \overline{S}$. Then for all $\varepsilon > 0$, there exists an ε-approximate solution $y(x)$ of the iso-differential equation* (1) *in the interval J_h such that $y(x_0) = y_0$.*

Proof. Because the function $f(x,y)$ is a continuous function in the closed rectangle $\overline{S}$, it is uniformly continuous in this rectangle. Thus, for a given $\varepsilon > 0$ there exists $\delta = \delta(\varepsilon) > 0$ so that

$$|f(x,y) - f(x_1,y_1)| \leq \varepsilon,$$

$$\left| y\frac{\hat{T}'(x)}{\hat{T}(x)} + f(x,y)\frac{\hat{T}(x)-x\hat{T}'(x)}{\hat{T}(x)} - y_1\frac{\hat{T}'(x_1)}{\hat{T}(x_1)} - f(x_1,y_1)\frac{\hat{T}(x_1)-x_1\hat{T}'(x_1)}{\hat{T}(x_1)} \right| \leq \varepsilon$$

for all (x,y), $(x_1,y_1) \in \overline{S}$ such that

$$|x - x_1| \leq \delta \qquad \text{and} \qquad |y - y_1| \leq \delta.$$

We shall construct an ε-approximate solution in the interval $[x_0, x_0 + h]$. A similar process will define it in the interval $[x_0 - h, x_0]$.

For this aim, we divide the interval $[x_0, x_0 + h]$ into m parts

$$x_0 < x_1 < x_2 \ldots < x_m = x_0 + h$$

such that

$$x_i - x_{i-1} \leq \min\left\{\delta, \frac{\delta}{P(|y_0| + b + M)}\right\}, \qquad i = 1,2,\ldots,m. \tag{13}$$

Now we define a function $y(x)$ in the interval $[x_0, x_0 + h]$ in the following manner

$$y(x) = y(x_{i-1}) + (x - x_{i-1})\left(y(x_{i-1})\frac{\hat{T}'(x_{i-1})}{\hat{T}(x_{i-1})} + f(x_{i-1}, y(x_{i-1}))\frac{\hat{T}(x_{i-1})-x_{i-1}\hat{T}'(x_{i-1})}{\hat{T}(x_{i-1})}\right),$$
$$x_{i-1} \leq x \leq x_i, \qquad i = 1,2,\ldots,m. \tag{14}$$

Obviously, this function $y(x)$ is continuous and has a piecewise continuous derivative

$$y'(x) = y(x_{i-1})\frac{\hat{T}'(x_{i-1})}{\hat{T}(x_{i-1})} + f(x_{i-1}, y(x_{i-1}))\frac{\hat{T}(x_{i-1}) - x_{i-1}\hat{T}'(x_{i-1})}{\hat{T}(x_{i-1})},$$

$x_{i-1} < x < x_i$, $i = 1,2,\ldots,m$, which fails to be defined only at the points x_i, $i = 1,2,\ldots,m-1$. Since in each subinterval $[x_{i-1}, x_i]$, $i = 1,2,\ldots,m$, the function $y(x)$ is a straight line, to prove that $(x, y(x)) \in \overline{S}$ it suffices to show that

$$|y(x_i) - y_0| \leq b$$

for all $i = 1,2,\ldots,m$.

For this reason, in (14) let $i=1$ and $x=x_1$ to obtain

$$y(x_1)=y_0+(x-x_1)\left(y_0\frac{\hat{T}'(x_0)}{\hat{T}(x_0)}+f(x_0,y_0)\frac{\hat{T}(x_0)-x_0\hat{T}'(x_0)}{\hat{T}(x_0)}\right),$$

$$|y(x_1)-y_0|=\left|(x-x_1)\left(y_0\frac{\hat{T}'(x_0)}{\hat{T}(x_0)}+f(x_0,y_0)\frac{\hat{T}(x_0)-x_0\hat{T}'(x_0)}{\hat{T}(x_0)}\right)\right|$$

$$\leq (x_1-x_0)\left(|y_0|\frac{|\hat{T}'(x_0)|}{\hat{T}(x_0)}+|f(x_0,y_0)|\frac{|\hat{T}(x_0)-x_0\hat{T}'(x_0)|}{\hat{T}(x_0)}\right)$$

$$\leq (x_1-x_0)(P|y_0|+MP)$$

$$\leq hP(M+|y_0|)$$

$$\leq hP(b+|y_0|+M)$$

$$\leq \frac{b}{P(b+|y_0|+M)}P(b+|y_0|+M)$$

$$=b.$$

Now let the assertion be true for $i=1,2,\ldots,k-1<m-1$, then from (14)

$$y(x_1)-y_0=(x_1-x_0)\left(y_0\frac{\hat{T}'(x_0)}{\hat{T}(x_0)}+f(x_0,y_0)\frac{\hat{T}(x_0)-x_0\hat{T}'(x_0)}{\hat{T}(x_0)}\right),$$

$$y(x_2)-y(x_1)=(x_2-x_1)\left(y(x_1)\frac{\hat{T}'(x_1)}{\hat{T}(x_1)}+f(x_1,y(x_1))\frac{\hat{T}(x_1)-x_1\hat{T}'(x_1)}{\hat{T}(x_1)}\right),$$

$$\ldots$$

$$y(x_k)-y(x_{k-1})=(x_k-x_{k-1})\left(y(x_{k-1})\frac{\hat{T}'(x_{k-1})}{\hat{T}(x_{k-1})}+f(x_{k-1},y(x_{k-1}))\frac{\hat{T}(x_{k-1})-x_{k-1}\hat{T}'(x_{k-1})}{\hat{T}(x_{k-1})}\right).$$

From here,

$$y(x_k)-y_0=\sum_{l=1}^{k}(x_l-x_{l-1})\left(y(x_{l-1})\frac{\hat{T}'(x_{l-1})}{\hat{T}(x_{l-1})}+f(x_{l-1},y(x_{l-1}))\frac{\hat{T}(x_{l-1})-x_{l-1}\hat{T}'(x_{l-1})}{\hat{T}(x_{l-1})}\right),$$

which gives

$$|y(x_k)-y_0|\leq \sum_{l=1}^{k}(x_l-x_{l-1})\left(|y(x_{l-1})|\frac{|\hat{T}'(x_{l-1})|}{\hat{T}(x_{l-1})}+|f(x_{l-1},y(x_{l-1}))|\frac{|\hat{T}(x_{l-1})-x_{l-1}\hat{T}'(x_{l-1})|}{\hat{T}(x_{l-1})}\right)$$

$$\leq \sum_{l=1}^{k}(x_l-x_{l-1})\Big((b+|y_0|)P+MP\Big)$$

$$=P(M+b+|y_0|)\sum_{l=1}^{k}(x_l-x_{l-1})$$

$$=P(M+b+|y_0|)(x_k-x_0)$$

$$\leq P(M+b+|y_0|)h$$

$$\leq P(M+b+|y_0|)\frac{b}{P(M+b+|y_0|)}$$

$$=b.$$

Finally, if $x_{i-1} < x < x_i$, then from (13) and (14)

$$|y(x) - y(x_{i-1})| = (x - x_i)\left|y(x_{i-1})\frac{\hat{T}'(x_{i-1})}{\hat{T}(x_{i-1})} + f(x_{i-1}, y(x_{i-1}))\frac{\hat{T}(x_{i-1}) - x_{i-1}\hat{T}'(x_{i-1})}{\hat{T}(x_{i-1})}\right|$$

$$\leq (x - x_i)\left(|y(x_{i-1})|\frac{|\hat{T}'(x_{i-1})|}{\hat{T}(x_{i-1})} + |f(x_{i-1}, y(x_{i-1}))|\frac{|\hat{T}(x_{i-1}) - x_{i-1}\hat{T}'(x_{i-1})|}{\hat{T}(x_{i-1})}\right)$$

$$\leq (x_i - x_{i-1})\Big((|y_0| + b)P + MP\Big)$$

$$\leq \frac{\delta}{P(|y_0| + b + M)}P(M + |y_0| + b)$$

$$= \delta.$$

Therefore

$$\left|y'(x) - y(x)\frac{\hat{T}'(x)}{\hat{T}(x)} - f(x, y(x))\frac{\hat{T}(x) - x\hat{T}'(x)}{\hat{T}(x)}\right|$$

$$= \left|y(x_{i-1})\frac{\hat{T}'(x_{i-1})}{\hat{T}(x_{i-1})} + f(x_{i-1}, y(x_{i-1}))\frac{\hat{T}(x_{i-1}) - x_{i-1}\hat{T}'(x_{i-1})}{\hat{T}(x_{i-1})} - y(x)\frac{\hat{T}'(x)}{\hat{T}(x)} - f(x, y(x))\frac{\hat{T}(x) - x\hat{T}'(x)}{\hat{T}(x)}\right|$$

$$\leq \varepsilon$$

for all $x \in J_h$, $x \neq x_i$, $i = 1, 2, \ldots, m-1$. This completes the proof that $y(x)$ is an ε-approximate solution of the iso-differential equation (1).

This method of constructing an approximate solution is said to be iso-Cauchy-Euler method. □

Theorem 5.0.191. *(iso-Cauchy-Peano's existence theorem) Let $f(x,y)$ be continuous in $\overline{S}$ and hence there exists a $M > 0$ such that $|f(x,y)| \leq M$ for every $(x,y) \in \overline{S}$. Then the initial value problem* (1), (2) *has at least one solution in J_h.*

Proof. We shall prove the assertion for the interval $[x_0, x_0 + h]$.

Let $\{\varepsilon_m\}_{m=1}^{\infty}$ be a monotonically decreasing sequence of positive numbers such that

$$\lim_{m \longrightarrow \infty} \varepsilon_m = 0.$$

For each ε_m we construct an ε_m-approximate solution $y_m(x)$.

As in the proof of the theorem for existence of ε-approximate solutions we have

$$|y_m(x)| \leq b + |y_0|$$

for every $m \in \mathbb{N}$ and for every $x \in J_h$. In other words, the sequence $\{y_m(x)\}_{m=1}^{\infty}$ is uniformly bounded in J_h.

Let x, $x^* \in [x_0, x_0 + h]$. Then

$$|y_m(x) - y_m(x^*)| \leq \left|\int_x^{x^*}\left(|y_m(t)|\frac{|\hat{T}'(t)|}{\hat{T}(t)} + |f(t, y_m(t))|\frac{|\hat{T}(t) - t\hat{T}'(t)|}{\hat{T}(t)}\right)dt\right|$$

$$\leq \left|\int_x^{x^*}\Big(P(|y_0| + b) + MP\Big)dt\right|$$

$$\leq P(M + b + |y_0|)|x - x^*|$$

and from this it follows that the sequence $\{y_m(x)\}_{m=1}^{\infty}$ is equip-continuous.

Consequently the sequence $\{y_m(x)\}_{m=1}^{\infty}$ contains a subsequence $\{y_{m_p}(x)\}_{p=1}^{\infty}$ which converges uniformly in $[x_0, x_0+h]$ to a continuous function $y(x)$. We define

$$e_m(x) = \begin{cases} y'_m(x) - y_m(x)\frac{\hat{T}'(x)}{\hat{T}(x)} - f(x, y_m(x))\frac{\hat{T}(x)-x\hat{T}'(x)}{\hat{T}(x)} \\ \text{at the points where } y'_m(x) \text{ exists} \\ 0 \quad \text{otherwise.} \end{cases}$$

Then

$$y_m(x) = y_0 + \int_{x_0}^{x} \left(y_m(t)\frac{\hat{T}'(t)}{\hat{T}(t)} + f(t, y_m(t)) + e_m(t) \right) dt \tag{15}$$

and

$$|e_m(x)| \le \varepsilon_m.$$

Since $f(x,y)$ is continuous in $\overline{S}$ and $y_{m_p}(x)$ converges to $y(x)$ uniformly in $[x_0, x_0+h]$, the function $f(x, y_{m_p}(x))$ converges to $f(x, y(x))$ uniformly in $[x_0, x_0+h]$. Thus, by replacing m by m_p in (15) and letting $p \longrightarrow \infty$, we become that $y(x)$ is a solution to the integral equation (4). □

Remark 5.0.192. *We suppose that all conditions of the iso-Cauchy-Peano's existence theorem are satisfied. Further, let the initial value problem* (1), (2) *has a solution* $y(x)$ *in an interval* $J = (\alpha, \beta)$. *We have*

$$|y(x_2) - y(x_1)| \le P(M + |y_0| + b)|x_2 - x_1|$$

for every x_1, $x_2 \in J$. *Therefore*

$$y(x_2) - y(x_1) \longrightarrow 0$$

as $x_1, x_2 \longrightarrow \alpha^+$. *Thus, by the Cauchy criterion of convergence we have that*

$$\lim_{x \longrightarrow \alpha^+} y(x)$$

exists.

A similar argument holds for

$$\lim_{x \longrightarrow \beta^-} y(x).$$

Theorem 5.0.193. *Let all conditions of the iso-Cauchy-Peano's existence theorem be satisfied. Let also,* $y(x)$ *be a solution of the initial value problem* (1), (2) *in the interval* $J = (\alpha, \beta)$. *Then* $y(x)$ *can be extended over the interval* $(\alpha, \beta+\gamma]$ *(*$[\alpha-\gamma, \beta)$*) for some* $\gamma > 0$.

Proof. We define the function $y_1(x)$ as follows.

$$y_1(x) = y(x) \quad \text{for} \quad x \in (\alpha, \beta),$$

$$y_1(\beta) = y(\beta - 0).$$

We observe that for all $x \in (\alpha, \beta]$ we have

$$y_1(x) = y(\beta - 0) + \int_\beta^x \left(y_1(t)\frac{\hat{T}'(t)}{\hat{T}(t)} + f(t, y_1(t))\frac{\hat{T}(t) - t\hat{T}'(t)}{\hat{T}(t)}\right)dt$$

$$= y(x_0) + \int_{x_0}^\beta \left(y_1(t)\frac{\hat{T}'(t)}{\hat{T}(t)} + f(t, y_1(t))\frac{\hat{T}(t) - t\hat{T}'(t)}{\hat{T}(t)}\right)dt$$

$$+ \int_\beta^x \left(y_1(t)\frac{\hat{T}'(t)}{\hat{T}(t)} + f(t, y_1(t))\frac{\hat{T}(t) - t\hat{T}'(t)}{\hat{T}(t)}\right)dt$$

$$= y(x_0) + \int_{x_0}^x \left(y_1(t)\frac{\hat{T}'(t)}{\hat{T}(t)} + f(t, y_1(t))\frac{\hat{T}(t) - t\hat{T}'(t)}{\hat{T}(t)}\right)dt.$$

Therefore the left-hand derivative $y_1'(\beta - 0)$ exists and

$$y_1'(\beta - 0) = y_1(\beta)\frac{\hat{T}'(\beta)}{\hat{T}(\beta)} + f(\beta, y_1(\beta))\frac{\hat{T}(\beta) - \beta\hat{T}'(\beta)}{\hat{T}(\beta)}.$$

Thus, $y_1(x)$ is a continuation of $y(x)$ in the interval $(\alpha, \beta]$.

Let $y_2(x)$ be a solution to the problem

$$y'(x) = y(x)\frac{\hat{T}'(x)}{\hat{T}(x)} + f(x, y(x))\frac{\hat{T}(x) - x\hat{T}'(x)}{\hat{T}(x)},$$

$$y(\beta) = y_1(\beta),$$

existing in the interval $[\beta, \beta + \gamma]$ for some $\gamma > 0$.

We define the function

$$y_3(x) = \begin{cases} y_1(x) & x \in (\alpha, \beta], \\ y_2(x) & x \in [\beta, \beta + \gamma], \end{cases}$$

which is a continuation of $y(x)$ in the interval $(\alpha, \beta + \gamma]$.

Also,

$$y_3(x) = y_0 + \int_{x_0}^x \left(y_3(t)\frac{\hat{T}'(t)}{\hat{T}(t)} + f(t, y_3(t))\frac{\hat{T}(t) - t\hat{T}'(t)}{\hat{T}(t)}\right)dt$$

for every $x \in (\alpha, \beta + \gamma]$, because for all $x \in [\beta, \beta + \gamma]$ we have

$$y_3(x) = y(\beta - 0) + \int_\beta^x \left(y_3(t)\frac{\hat{T}'(t)}{\hat{T}(t)} + f(t, y_3(t))\frac{\hat{T}(t) - t\hat{T}'(t)}{\hat{T}(t)}\right)dt$$

$$= y_0 + \int_{x_0}^\beta \left(y_3(t)\frac{\hat{T}'(t)}{\hat{T}(t)} + f(t, y_3(t))\frac{\hat{T}(t) - t\hat{T}'(t)}{\hat{T}x(t)}\right)dt$$

$$+ \int_\beta^x \left(y_3(t)\frac{\hat{T}'(t)}{\hat{T}(t)} + f(t, y_3(t))\frac{\hat{T}(t) - t\hat{T}'(t)}{\hat{T}(t)}\right)dt$$

$$= y_0 + \int_{x_0}^x \left(y_3(t)\frac{\hat{T}'(t)}{\hat{T}(t)} + f(t, y_3(t))\frac{\hat{T}(t) - t\hat{T}'(t)}{\hat{T}(t)}\right)dt.$$

□

Exercise 5.0.194. *Let $\hat{T}(x) = x^2+1$, $f(x,y) = x+y$. Find the maximum interval of existence of the solutions to the problem* (1), (2) *in the case when* $(x_0, y_0) = (0,0)$.

Exercise 5.0.195. *Let $\hat{T}(x) = x^2+1$, $f(x,y) = x+xy+y^2$. Find the maximum interval of existence of the solutions to the problem* (1), (2) *in the case when* $(x_0, y_0) = (0,1)$.

Exercise 5.0.196. *Let $\hat{T}(x) = x^4+1$, $f(x,y) = xy$. Find the maximum interval of existence of the solutions to the problem* (1), (2) *in the case when* $(x_0, y_0) = (1,1)$.

Theorem 5.0.197. *(iso-Lipschitz uniqueness theorem) Let $f(x,y)$ be continuous and satisfies a uniform Lipschitz condition in $\overline{S}$ with a Lipschitz constant L. Then the problem* (1), (2) *has at most one solution in $|x-x_0| \leq a$.*

Proof. We suppose that the problem (1), (2) has two solutions $y_1(x)$ and $y_2(x)$, $x \in [x_0 - a, x_0 + a]$. Then

$$y_1(x) = y_0 + \int_{x_0}^{x} \left(y_1(t) \frac{\hat{T}'(t)}{\hat{T}(t)} + f(t, y_1(t)) \frac{\hat{T}(t) - t\hat{T}'(t)}{\hat{T}(t)} \right) dt,$$

$$y_2(x) = y_0 + \int_{x_0}^{x} \left(y_2(t) \frac{\hat{T}'(t)}{\hat{T}(t)} + f(t, y_2(t)) \frac{\hat{T}(t) - t\hat{T}'(t)}{\hat{T}(t)} \right) dt,$$

whereupon

$$y_1(x) - y_2(x) = \int_{x_0}^{x} \left((y_1(t) - y_2(t)) \frac{\hat{T}'(t)}{\hat{T}(t)} + (f(t, y_1(t)) - f(t, y_2(t))) \frac{\hat{T}(t) - t\hat{T}'(t)}{\hat{T}(t)} \right) dt,$$

and

$$|y_1(x) - y_2(x)| \leq \left| \int_{x_0}^{x} \left(|y_1(t) - y_2(t)| \frac{|\hat{T}'(t)|}{\hat{T}(t)} + |f(t, y_1(t)) - f(t, y_2(t))| \frac{|\hat{T}(t) - t\hat{T}'(t)|}{\hat{T}(t)} \right) dt \right|$$

$$\leq \left| \int_{x_0}^{x} \left(P|y_1(t) - y_2(t)| + LP|y_1(t) - y_2(t)| \right) dt \right.$$

$$= P(1+L) \left| \int_{x_0}^{x} |y_1(t) - y_2(t)| dt \right|.$$

From the last inequality and Gronwall's type inequality we conclude that

$$|y_1(x) - y_2(x)| = 0 \quad \text{in} \quad [x_0 - a, x_0 + a].$$

□

Theorem 5.0.198. *(iso-Peano's uniqueness theorem) Let $f(x,y)$ be continuous in*

$$\overline{S}_+ = \{(x,y) \in \mathbb{R}^2 : x_0 \leq x \leq x_0 + a, |y - y_0| \leq b\}$$

and nonincreasing in y for all $[x_0, x_0 + a]$. Let also,

$$\hat{T}'(x) \leq 0, \qquad \hat{T}(x) - x\hat{T}'(x) \geq 0 \quad \text{for} \quad \forall x \in [x_0, x_0 + a].$$

Then the problem (1), (2) *has at most one solution in $[x_0, x_0 + a]$.*

Proof. Let the problem (1), (2) has two solutions $y_1(x)$ and $y_2(x)$ in $[x_0, x_0+a]$ which differ in $[x_0, x_0+a]$. We assume that

$$y_2(x) > y_1(x) \qquad \text{in} \qquad (x_1, x_1+\varepsilon) \subset [x_0, x_0+a],$$

while $y_1(x) = y_2(x)$ for $x \in [x_0, x_1]$, i.e., x_1 is the greatest lower bound of the set A consisting of those x for which $y_2(x) > y_1(x)$. This greatest lower bound of the set A exists because the set A is bounded below by x_0 at least. Thus for every $x \in (x_1, x_1+\varepsilon)$ we have

$$f(x, y_1(x)) \geq f(x, y_2(x)),$$

$$f(x, y_1(x)) \frac{\hat{T}(x) - x\hat{T}'(x)}{\hat{T}(x)} \geq f(x, y_2(x)) \frac{\hat{T}(x) - x\hat{T}'(x)}{\hat{T}(x)},$$

$$y_1(x) \frac{\hat{T}'(x)}{\hat{T}(x)} \geq y_2(x) \frac{\hat{T}'(x)}{\hat{T}(x)},$$

whereupon

$$y_1(x) \frac{\hat{T}'(x)}{\hat{T}(x)} + f(x, y_1(x)) \frac{\hat{T}(x) - x\hat{T}'(x)}{\hat{T}(x)}$$

$$\geq y_2(x) \frac{\hat{T}'(x)}{\hat{T}(x)} + f(x, y_2(x)) \frac{\hat{T}(x) - x\hat{T}'(x)}{\hat{T}(x)}$$

for all $x \in (x_1, x_1+\varepsilon)$, and from here

$$y_1'(x) \geq y_2'(x) \qquad \text{for} \qquad \forall x \in (x_1, x_1+\varepsilon).$$

Hence the function

$$z(x) = y_2(x) - y_1(x)$$

is nonincreasing function in $(x_1, x_1+\varepsilon)$.

Because

$$z(x_1) = y_2(x_1) - y_1(x_1) = 0$$

we obtain

$$z(x) \leq z(x_1) = 0 \qquad \text{in} \qquad (x_1, x_1+\varepsilon)$$

or

$$y_2(x) \leq y_1(x) \qquad \text{in} \qquad (x_1, x_1+\varepsilon).$$

This contradiction proves that

$$y_1(x) = y_2(x) \qquad \text{for} \qquad \forall x \in [x_0, x_0+a].$$

□

Theorem 5.0.199. *(iso-Peano's uniqueness theorem) Let $f(x,y)$ be continuous in $\overline{S}_+$ and nondecreasing in y for every $x \in [x_0, x_0+a]$. Let also,*

$$\hat{T}'(x) \leq 0, \qquad \hat{T}(x) - x\hat{T}'(x) \leq 0 \qquad \text{for} \qquad \forall x \in [x_0, x_0+a].$$

Then the problem (1), (2) has at most one solution in $[x_0, x_0+a]$.

Proof. Let the problem (1), (2) has two solutions $y_1(x)$ and $y_2(x)$ in $[x_0, x_0+a]$ which differ in $[x_0, x_0+a]$. Let

$$y_2(x) > y_1(x) \qquad \text{in} \qquad (x_1, x_1+\varepsilon) \subset [x_0, x_0+a],$$

and

$$y_2(x) = y_1(x) \qquad \text{for} \qquad \forall x \in [x_0, x_1].$$

Therefore for every $x \in (x_1, x_1+\varepsilon)$ we have

$$f(x, y_2(x)) \geq f(x, y_1(x)),$$

$$f(x, y_1(x))\frac{\hat{T}(x)-x\hat{T}'(x)}{\hat{T}(x)} \geq f(x, y_2(x))\frac{\hat{T}(x)-x\hat{T}'(x)}{\hat{T}(x)},$$

$$y_1(x)\frac{\hat{T}'(x)}{\hat{T}(x)} \geq y_2(x)\frac{\hat{T}'(x)}{\hat{T}(x)},$$

whereupon

$$y_1(x)\frac{\hat{T}'(x)}{\hat{T}(x)} + f(x, y_1(x))\frac{\hat{T}(x)-x\hat{T}'(x)}{\hat{T}(x)}$$

$$\geq y_2(x)\frac{\hat{T}'(x)}{\hat{T}(x)} + f(x, y_2(x))\frac{\hat{T}(x)-x\hat{T}'(x)}{\hat{T}(x)}$$

for every $x \in (x_1, x_1+\varepsilon)$. Consequently

$$y_1'(x) \geq y_2'(x) \qquad \text{for} \qquad \forall x \in (x_1, x_1+\varepsilon)$$

and then the function

$$z(x) = y_2(x) - y_1(x)$$

is nonincreasing function in $(x_1, x_1+\varepsilon)$, therefore

$$y_2(x) - y_1(x) \leq y_2(x_1) - y_1(x_1) = 0 \qquad \text{for} \qquad \forall x \in (x_1, x_1+\varepsilon),$$

which is a contradiction. From here we conclude that $y_1(x) = y_2(x)$ for every $x \in [x_0, x_0+a]$. □

Lemma 5.0.200. *(iso-Osgood's lemma) Let $w(z)$ be continuous function in $[0,\infty)$, $w(0) = 0$, $z+w(z) > 0$ in $(0,\infty)$, $z+w(z)$ be increasing function in $[0,\infty)$, and*

$$\lim_{\varepsilon \longrightarrow 0^+} \int_\varepsilon^a \frac{dz}{z+w(z)} = \infty. \tag{16}$$

Let $u(x)$ be a nonnegative continuous function in $[0,a]$. Then the inequality

$$u(x) \leq P\int_0^x (u(t)+w(u(t)))dt, \qquad 0 < x \leq a, \tag{17}$$

implies that $u(x) \equiv 0$ in $[0,a]$.

Proof. We define the function

$$v(x) = \max_{0 \le t \le x} u(t)$$

and assume that $v(x) > 0$ for $0 < x \le a$. Then

$$u(t) \le v(x) \qquad \text{for} \qquad \forall t \in [0, x].$$

Because $u(x)$ is a continuous function in $[0, a]$ then there exists $x_1 \in [0, x]$ such that

$$v(x) = u(x_1).$$

Therefore, using that $z + w(z)$ is an increasing function in $[0, \infty)$,

$$v(x) = u(x_1) \le P \int_0^{x_1} (u(t) + w(u(t)))dt$$

$$\le P \int_0^{x_1} (v(t) + w(v(t)))dt$$

$$\le P \int_0^{x} (v(t) + w(v(t)))dt.$$

Let

$$\overline{v}(x) = P \int_0^x (v(t) + w(v(t)))dt.$$

We have

$$\overline{v}(x) \ge 0, \qquad v(x) \le \overline{v}(x),$$

and

$$\overline{v}'(x) = P(v(x) + w(v(x)))$$

$$\le P(\overline{v}(x) + w(\overline{v}(x))),$$

and since

$$\overline{v}(x) + w(\overline{v}(x)) \ge 0,$$

then

$$\frac{\overline{v}'(x)}{P(\overline{v}(x) + w(\overline{v}(x)))}.$$

Consequently for $0 < \delta < a$ we have

$$\int_\delta^a \frac{d\overline{v}(x)}{P(\overline{v}(x) + w(\overline{v}(x)))} \le \int_\delta^a dx,$$

whereupon

$$\lim_{\delta \longrightarrow 0^+} \int_\delta^a \frac{d\overline{v}(x)}{P(\overline{v}(x)+w(\overline{v}(x)))} = \lim_{\delta \longrightarrow 0^+} \int_{\overline{v}(\delta)}^{\overline{v}(a)} \frac{dy}{P(y+w(y))}$$

$$\le a,$$

which contradicts with (16). Consequently $u(x) \equiv 0$ in $[0, a]$. □

Theorem 5.0.201. *(iso-Osgood's uniqueness theorem) Let $f(x,y)$ be continuous in $\overline{S}_+$ and for all (x,y_1), $(x,y_2) \in \overline{S}_+$ it satisfies*

$$|f(x,y_1) - f(x,y_2)| \leq w(|y_1 - y_2|),$$

where $w(z)$ satisfies all conditions of the iso-Osgood's lemma. Then the problem (1), (2) *has at most one solution in* $[x_0, x_0 + a]$.

Proof. Let $y_1(x)$ and $y_2(x)$ are two solutions of the problem (1), (2) in $[x_0, x_0 + a]$. Then, if

$$z(x) = |y_1(x) - y_2(x)|, \qquad x \in [x_0, x_0 + a],$$

we have

$$z(x) = \left| \int_{x_0}^{x} \left((y_1(t) - y_2(t)) \frac{\hat{T}'(t)}{\hat{T}(t)} + (f(t, y_1(t)) - f(t, y_2(t))) \frac{\hat{T}(t) - t\hat{T}'(t)}{\hat{T}(t)} \right) dt \right|$$

$$\leq \int_{x_0}^{x} \left(|y_1(t) - y_2(t)| \frac{|\hat{T}'(t)|}{\hat{T}(t)} + |f(t, y_1(t)) - f(t, y_2(t))| \frac{|\hat{T}(t) - t\hat{T}'(t)|}{\hat{T}(t)} \right) dt$$

$$\leq \int_{x_0} (P|y_1(t) - y_2(t)| + Pw(|y_1(t) - y_2(t)|)) dt$$

$$= P \int_{x_0}^{x} (z(t) + w(z(t))) dt.$$

Let

$$u(x) = z(x_0 + x).$$

Therefore

$$u(x) \leq P \int_{x_0}^{x_0 + x} (z(t) + w(z(t))) dt$$

$$= P \int_0^x (z(x_0 + t) + w(z(x_0 + t))) dt$$

$$= P \int_0^x (u(t) + w(u(t))) dt.$$

Consequently $u(x)$ satisfies the iso-Osgood's lemma, from where $u(x) \equiv 0$ in $[0, a]$, i.e., $y_1(x) = y_2(x)$ in $[x_0, x_0 + a]$. □

Lemma 5.0.202. *(iso-Nagumo's lemma) Let $u(x)$ be nonnegative continuous function in $[x_0, x_0 + a]$ and $u(x_0) = 0$, and let $u(x)$ be differentiable at $x = x_0$ with $u'(x_0) = 0$. Then*

$$\int_{x_0}^{x} u(t) dt \leq a \int_{x_0}^{x} \frac{u(t)}{t - x_0} dt, \qquad x \in [x_0, x_0 + a],$$

and the inequality

$$u(x) \leq \int_{x_0}^{x} \frac{u(t)}{t - x_0} dt, \qquad x \in [x_0, x_0 + a],$$

implies that $u(x) = 0$ in $[x_0, x_0 + a]$.

Proof. Let

$$g(x)=\int_{x_0}^{x}u(t)dt-a\int_{x_0}^{x}\frac{u(t)}{t-x_0}dt,\qquad x\in[x_0,x_0+a].$$

Since

$$\lim_{x\longrightarrow x_0}\frac{u(x)}{x-x_0}=u'(x_0)=0,$$

then the integral

$$\int_{x_0}^{x}\frac{u(t)}{t-x_0}dt$$

exists for $x\in[x_0,x_0+a]$.

Also,

$$g'(x)=u(x)-a\frac{u(x)}{x-x_0}=u(x)\frac{x-x_0-a}{x-x_0}\leq 0$$

for every $x\in[x_0,x_0+a]$. Therefore g is a nonincreasing function in $[x_0,x_0+a]$, whereupon

$$g(x)\leq g(x_0)\qquad\text{for}\qquad\forall x\in[x_0,x_0+a],$$

or

$$\int_{x_0}^{x}u(t)dt\leq a\int_{x_0}^{x}\frac{u(t)}{t-x_0}dt$$

for every $x\in[x_0,x_0+a]$.

Let now

$$v(x)=\int_{x_0}^{x}\frac{u(t)}{t-x_0}dt,\qquad x\in[x_0,x_0+a].$$

Then

$$u(x)\leq v(x),\qquad x\in[x_0,x_0+a],$$

and

$$\begin{aligned}v'(x)&=\tfrac{u(x)}{x-x_0}\\&\leq\tfrac{v(x)}{x-x_0},\qquad x\in[x_0,x_0+a].\end{aligned}$$

Consequently

$$\begin{aligned}\tfrac{d}{dx}\left(\tfrac{v(x)}{x-x_0}\right)&=\tfrac{v'(x)(x-x_0)-v(x)}{(x-x_0)^2}\\&\leq 0\end{aligned}$$

or the function

$$l(x)=\frac{v(x)}{x-x_0}$$

is a nonincreasing function in $[x_0,x_0+a]$ and since $l(x_0)=0$, we have that

$$v(x)\leq 0\qquad\text{in}\qquad[x_0,x_0+a],$$

from where

$$v(x)=0\qquad\text{in}\qquad[x_0,x_0+a].$$

Consequently $u(x)=0$ in $[x_0,x_0+a]$. □

Theorem 5.0.203. *(iso-Nagumo's theorem) Let $P(a+1) \leq 1$, $f(x,y)$ be continuous in $\overline{S}_+$ and for all (x,y_1), $(x,y_2) \in \overline{S}_+$ it satisfies*

$$|f(x,y_1)-f(x,y_2)| \leq k|x-x_0|^{-1}|y_1-y_2|, \qquad x \neq x_0. \qquad k \leq 1.$$

Then the problem (1), (2) *has at most one solution in* $[x_0, x_0+a]$.

Proof. Let $y_1(x)$ and $y_2(x)$ are two solutions of the problem (1), (2) in $[x_0,x_0+a]$. Then for $x \in [x_0,x_0+a]$ we have

$$|y_1(x)-y_2(x)| \leq \int_{x_0}^{x} \left(|y_1(t)-y_2(t)| \frac{|\hat{T}'(t)|}{\hat{T}(t)} + |f(t,y_1(t))-f(t,y_2(t))| \frac{|\hat{T}(t)-t\hat{T}'(t)|}{\hat{T}(t)} \right) dt$$

$$\leq \int_{x_0}^{x} \left(P|y_1(t)-y_2(t)| + k(t-y_0)^{-1}|y_1(t)-y_2(t)|P \right) dt$$

$$\leq P\int_{x_0}^{x} |y_1(t)-y_2(t)|dt + P\int_{x_0}^{x} \frac{|y_1(t)-y_2(t)|}{t-x_0} dt$$

$$\leq aP\int_{x_0}^{x} \frac{|y_1(t)-y_2(t)|}{t-x_0} dt + P\int_{x_0}^{x} \frac{|y_1(t)-y_2(t)|}{t-x_0} dt$$

$$= (a+1)P\int_{x_0}^{x} \frac{|y_1(t)-y_2(t)|}{t-x_0} dt$$

$$\leq \int_{x_0}^{x} \frac{|y_1(t)-y_2(t)|}{t-x_0} dt.$$

Let

$$u(x) = |y_1(x)-y_2(x)|, \qquad x \in [x_0,x_0+a].$$

Then $u(x_0) = 0$ and from the mean value theorem we have

$$u'(x_0) = \lim_{h \longrightarrow 0} \frac{u(x_0+h)-u(x_0)}{h}$$

$$= \lim_{h \longrightarrow 0} \frac{|y_1(x_0)+hy_1'(x_0+\theta_1 h)+y_2(x_0)-hy_2'(x_0+\theta_2 h)|}{h}$$

$$= (\mathrm{sgn} h) \lim_{h \longrightarrow 0} |y_1'(x_0+\theta_1 h) - y_2'(x_0+\theta_2 h)|$$

$$= 0, \qquad 0 < \theta_1, \theta_2 < 1.$$

Then the conditions of iso-Nagumo's lemma are satisfied and $u(x) = 0$, i.e., $y_1(x) = y_2(x)$ in $[x_0,x_0+a]$. □

Advanced Practical Exercises

Problem 5.0.204. *Let $J = \mathbb{R}$, $D = \mathbb{R}^2$, $\hat{T}(x) = e^{-x}$, $f(x,y(x)) = x^2 y(x)$, $y(0) = 1$. Find the first three approximations of the solution of the problem* (1), (2).

Problem 5.0.205. *Let $J = \mathbb{R}$, $D = \mathbb{R}^2$, $\hat{T}(x) = e^{3x}$, $f(x,y(x)) = x - 2y(x)$, $y(0) = 1$. Find the first three approximations of the solution of the problem* (1), (2).

Problem 5.0.206. *Discuss the existence and uniqueness of the solutions of the initial value problem* (1), (2) *in the case when*

1) $\hat{T}(x) = x^2 + 1, \quad f(x,y) = y^2, \quad y(0) = 0.$

2) $\hat{T}(x) = x^2 + 1, \quad f(x,y) = x - 2y, \quad y(0) = 1,$

3) $\hat{T}(x) = e^x + x^2 + 2, \quad f(x,y) = x - 3y, \quad y(0) = 2,$

4) $\hat{T}(x) = e^{-x}, \quad f(x,y) = e^{x+y}, \quad y(0) = -1,$

5) $\hat{T}(x) = 1 + x^2 + x^4, \quad f(x,y) = 2x - 6y, \quad y(0) = 0,$

6) $\hat{T}(x) = e^{-3x}, \quad f(x,y) = x^2 - y^2, \quad y(0) = 1.$

Problem 5.0.207. *Let* $(x_0, y_0) \in \mathbb{R}^2$, *a*, $b \in C(\mathbb{R})$. *Prove that the initial value problem*

$$\left(\hat{y}^\wedge(\hat{x})\right)^{\circledast} = \hat{a}^\wedge(\hat{x}) \hat{\times} \hat{y}^\wedge(\hat{x}) + \hat{b}^\wedge(\hat{x}),$$

$$y(x_0) = y_0$$

has unique continuous solution which is defined in $\mathbb{R}$.

Problem 5.0.208. *Let* $(x_0, y_0) \in \mathbb{R}^2$, *a*, $b \in C(\mathbb{R})$. *Prove that the initial value problem*

$$\left(\hat{y}^\wedge(\hat{x})\right)^{\circledast} = a^\wedge(x) \hat{\times} \hat{y}^\wedge(\hat{x}) + b^\wedge(x),$$

$$y(x_0) = y_0$$

has unique continuous solution which is defined in $\mathbb{R}$.

Problem 5.0.209. *Let* $(x_0, y_0) \in \mathbb{R}^2$, $y_0 \neq 0$, *a*, $b \in C(\mathbb{R})$. *Prove that the initial value problem*

$$\left(\hat{y}^\wedge(\hat{x})\right)^{\circledast} = \hat{a}^\wedge(\hat{x}) \hat{\times} \hat{y}^\wedge(\hat{x}) + \hat{b}^\wedge(\hat{x}) \hat{\times} \left(\hat{y}^\wedge(\hat{x})\right)^{\hat{2}},$$

$$y(x_0) = y_0$$

has unique continuous solution which is defined in $\mathbb{R}$.

Problem 5.0.210. *Let* $\hat{T}(x) = e^x$, $f(x,y) = x + y$. *Find the maximum interval of existence of the solutions to the problem* (1), (2) *in the case when* $(x_0, y_0) = (0,0)$.

Problem 5.0.211. *Let* $\hat{T}(x) = e^x$, $f(x,y) = x - y^2$. *Find the maximum interval of existence of the solutions to the problem* (1), (2) *in the case when* $(x_0, y_0) = (1,0)$.

Problem 5.0.212. *Let* $\hat{T}(x) = e^x$, $f(x,y) = xy - x^2 - y^2$. *Find the maximum interval of existence of the solutions to the problem* (1), (2) *in the case when* $(x_0, y_0) = (1,1)$.

Problem 5.0.213. *Let* $\hat{T}(x) = e^x$, $f(x,y) = y^2$. *Solve the initial value problem for* (1) *in the case when* $y(0) = 1$. *Find also the largest interval on which the solution is defined.*

Problem 5.0.214. *Let* $\hat{T}(x) = e^x$, $f(x,y) = -\frac{x+y^2}{y(1-x)}$, $y(0) = 1$. *Show that the solution of the initial value problem* (1), (2) , *can not be extended beyond the interval* $(-1,1)$.

Chapter 6

Iso-Differential Inequalities

Let D is a domain in $\mathbb{R}^2$, $a > 0$, $x_0 \in \mathbb{R}$, $J = [x_0, x_0 + a)$, $\hat{T} \in C^1(J)$, $\hat{T}(x) > 0$ in J, $f \in C(D)$.

Definition 6.0.215. *(solution of iso-differential inequality) A function $y(x)$ is said to be a solution of the iso-differential inequality*

$$\left(\hat{y}^{\wedge}(\hat{x})\right)^{\circledast} > \hat{f}^{\wedge}(\hat{x}, \hat{y}^{\wedge}(\hat{x})) \tag{1'}$$

or

$$y'(x) > y(x)\frac{\hat{T}'(x)}{\hat{T}(x)} + f(x, y(x))\frac{\hat{T}(x) - x\hat{T}'(x)}{\hat{T}(x)} \tag{1}$$

in J if

1. *$y'(x)$ exists for all $x \in J$,*

2. *for all $x \in J$ the points $(x, y(x)) \in D$,*

3. *$y'(x) > y(x)\frac{\hat{T}'(x)}{\hat{T}(x)} + f(x, y(x))\frac{\hat{T}(x) - x\hat{T}'(x)}{\hat{T}(x)}$ for all $x \in J$.*

The solutions of the iso-differential inequalities

$$y'(x) \geq y(x)\frac{\hat{T}'(x)}{\hat{T}(x)} + f(x, y(x))\frac{\hat{T}(x) - x\hat{T}'(x)}{\hat{T}(x)},$$

$$y'(x) < y(x)\frac{\hat{T}'(x)}{\hat{T}(x)} + f(x, y(x))\frac{\hat{T}(x) - x\hat{T}'(x)}{\hat{T}(x)},$$

$$y'(x) \leq y(x)\frac{\hat{T}'(x)}{\hat{T}(x)} + f(x, y(x))\frac{\hat{T}(x) - x\hat{T}'(x)}{\hat{T}(x)},$$

are defined analogously.

Example 6.0.216. *Let* $\hat{T}(x) = e^x$, $f(x,y) = 1$, $J = \left(\frac{3}{2}, 1\right)$. *Then* $y(x) = 1 - x$ *is a solution of the iso-differential inequality* (1) *in* J. *Really,*

$$\frac{\hat{T}'(x)}{\hat{T}(x)} = 1,$$

$$y(x)\frac{\hat{T}'(x)}{\hat{T}(x)} = 1 - x,$$

$$f(x,y)\frac{\hat{T}(x) - x\hat{T}'(x)}{\hat{T}(x)} = 1 - x,$$

$$y'(x) = -1.$$

Then the iso-differential inequality (1) *takes the form*

$$-1 > 1 - x + 1 - x \qquad \Longleftrightarrow$$

$$2x > 3.$$

Our first result for iso-differential inequalities is stated in the following theorem.

Theorem 6.0.217. *(basic theorem for the iso-differential inequalities) Let* $\hat{T}(x) - x\hat{T}'(x) \geq 0$ *for every* $x \in J$, $y_1(x)$ *and* $y_2(x)$ *be the solutions of the iso-differential inequalities*

$$y_1'(x) \leq y_1(x)\frac{\hat{T}'(x)}{\hat{T}(x)} + f(x, y_1(x))\frac{\hat{T}(x) - x\hat{T}'(x)}{\hat{T}(x)}, \tag{2}$$

$$y_2'(x) > y_2(x)\frac{\hat{T}'(x)}{\hat{T}(x)} + f(x, y_2(x))\frac{\hat{T}(x) - x\hat{T}'(x)}{\hat{T}(x)} \tag{3}$$

on J, *respectively. Then the inequality*

$$y_1(x_0) < y_2(x_0)$$

implies that

$$y_1(x) < y_2(x) \qquad \text{for} \qquad \forall x \in J. \tag{4}$$

Proof. We suppose that (4) is not true. Then we define the set

$$A = \{x : \qquad x \in J, \qquad y_1(x) \geq y_2(x)\}.$$

From our assumption it follows that $A \neq \emptyset$.

Let x^* be the greatest lower bound of the set A. Then $x_0 < x^*$ and

$$y_1(x^*) \geq y_2(x^*).$$

Let us assume that

$$y_1(x^*) > y_2(x^*).$$

Because $y_1(x)$ and $y_2(x)$ are continuous functions in J then there exists a $\varepsilon > 0$ such that

$$y_1(x^* - \varepsilon) \geq y_2(x^* - \varepsilon),$$

which is a contradiction with the definition of x^*. Consequently

$$y_1(x^*) = y_2(x^*).$$

Let $h < 0$. We have

$$y_1(x^* + h) < y_2(x^* + h),$$

and hence

$$y_1'(x^* - 0) = \lim_{h \longrightarrow 0} \frac{y_1(x^*+h) - y_1(x^*)}{h}$$

$$= \lim_{h \longrightarrow 0} \frac{y_1(x^*+h) - y_2(x^*)}{h}$$

$$\geq \lim_{h \longrightarrow 0} \frac{y_2(x^*+h) - y_2(x^*)}{h}$$

$$= y_2'(x^* - 0),$$

i.e.,

$$y_1'(x^* - 0) \geq y_2'(x^* - 0). \tag{5}$$

From (2) we get

$$y_1'(x^*) \leq y_1(x^*) \frac{\hat{T}'(x^*)}{\hat{T}(x^*)} + f(x^*, y_1(x^*)) \frac{\hat{T}(x^*) - x^* \hat{T}'(x^*)}{\hat{T}(x^*)},$$

from where, using (5),

$$y_1(x^*) \frac{\hat{T}'(x^*)}{\hat{T}(x^*)} + f(x^*, y_1(x^*)) \frac{\hat{T}(x^*) - x^* \hat{T}'(x^*)}{\hat{T}(x^*)} \geq y_2'(x^* - 0). \tag{6'}$$

On the other hand, from (3) we have

$$y_2'(x^* - 0) > y_2(x^*) \frac{\hat{T}'(x^*)}{\hat{T}(x^*)} + f(x^*, y_2(x^*)) \frac{\hat{T}(x^*) - x^* \hat{T}'(x^*)}{\hat{T}(x^*)},$$

whereupon, using (6′),

$$y_2(x^*) \frac{\hat{T}'(x^*)}{\hat{T}(x^*)} + f(x^*, y_2(x^*)) \frac{\hat{T}(x^*) - x^* \hat{T}'(x^*)}{\hat{T}(x^*)}$$

$$< y_1(x^*) \frac{\hat{T}'(x^*)}{\hat{T}(x^*)} + f(x^*, y_1(x^*)) \frac{\hat{T}(x^*) - x^* \hat{T}'(x^*)}{\hat{T}(x^*)}$$

$$= y_2(x^*) \frac{\hat{T}'(x^*)}{\hat{T}(x^*)} + f(x^*, y_2(x^*)) \frac{\hat{T}(x^*) - x^* \hat{T}'(x^*)}{\hat{T}(x^*)},$$

and since $\hat{T}(x^*) - x^* \hat{T}'(x^*) \geq 0$ we get the contradiction

$$f(x^*, y_2(x^*)) < f(x^*, y_2(x^*)).$$

Consequently

$$A = \emptyset,$$

from where we conclude that

$$y_1(x) < y_2(x) \quad \text{in} \quad J.$$

□

Exercise 6.0.218. *Let $\hat{T}(x) - x\hat{T}'(x) \leq 0$ for every $x \in J$, $y_1(x)$ and $y_2(x)$ be the solutions of the iso-differential inequalities (2) and (3) on J, respectively. Prove that the inequality*

$$y_1(x_0) < y_2(x_0)$$

implies that

$$y_1(x) < y_2(x) \qquad \text{for} \qquad \forall x \in J.$$

Corollary 6.0.219. *Let $\hat{T}(x) - x\hat{T}'(x) \geq 0$ in the interval J. Let also,*

(i) *$y(x)$ be a solution of the initial value problem*

$$\begin{aligned} & y'(x) = -y(x)\tfrac{\hat{T}'(x)}{\hat{T}(x)} + f(x,y(x))\tfrac{\hat{T}(x)-x\hat{T}'(x)}{\hat{T}(x)} \qquad \text{in} \qquad (x_0, x_0+a), \\ & y(x_0) = y_0, \end{aligned} \tag{6}$$

(ii) *$y_1(x)$ and $y_2(x)$ be the solutions of the iso-differential inequalities*

$$y_1'(x) < y_1(x)\frac{\hat{T}'(x)}{\hat{T}(x)} + f(x,y_1(x))\frac{\hat{T}(x)-x\hat{T}'(x)}{\hat{T}(x)}, \tag{7}$$

$$y_2'(x) > y_2(x)\frac{\hat{T}'(x)}{\hat{T}(x)} + f(x,y_2(x))\frac{\hat{T}(x)-x\hat{T}'(x)}{\hat{T}(x)} \tag{8}$$

in J, respectively,

(iii) *$y_1(x_0) \leq y_0 \leq y_2(x_0)$.*

Then

$$y_1(x) < y(x) < y_2(x)$$

for all $x \in (x_0, x_0+a)$.

Proof. We shall prove that

$$y(x) < y_2(x) \qquad \text{for} \qquad \forall x \in (x_0, x_0+a).$$

1. case $y_0 < y_2(x_0)$. Then from the last theorem we have that

$$y(x) < y_2(x) \qquad \text{in} \qquad (x_0, x_0+a).$$

2. case $y_0 = y_2(x_0)$.

Let

$$z(x) = y_2(x) - y(x).$$

Then

$$z(x_0) = y_2(x_0) - y(x_0) = 0,$$

$$z'(x) = y_2'(x) - y'(x),$$

$$z'(x_0) = y_2'(x_0) - y'(x_0)$$

$$> y_2(x_0)\frac{\hat{T}'(x_0)}{\hat{T}(x_0)} + f(x_0, y_2(x_0))\frac{\hat{T}(x_0) - x_0\hat{T}'(x_0)}{\hat{T}(x_0)}$$

$$-y(x_0)\frac{\hat{T}'(x_0)}{\hat{T}(x_0)} - f(x_0, y(x_0))\frac{\hat{T}(x_0) - x_0\hat{T}'(x_0)}{\hat{T}(x_0)}$$

$$= 0,$$

therefore the function z is an increasing function to the right of x_0 in a sufficiently small interval $[x_0, x_0 + \delta]$. Consequently $y(x) < y_2(x)$ for all $x \in (x_0, x_0 + \delta]$, from where

$$y(x_0 + \delta) < y_2(x_0 + \delta).$$

Now the last theorem gives that

$$y(x) < y_2(x) \qquad \text{in} \qquad [x_0 + \delta, x_0 + a).$$

Since δ can be chosen sufficiently small, then

$$y(x) < y_2(x) \qquad \text{in} \qquad (x_0, x_0 + a).$$

□

Exercise 6.0.220. *Prove that*

$$y_1(x) < y(x) \qquad \text{in} \qquad (x_0, x_0 + a).$$

Exercise 6.0.221. *Let $\hat{T}(x) - x\hat{T}'(x) \geq 0$ in the interval J. Let also,*

(i) *$y(x)$ be a solution of the initial value problem (6),*

(ii) *$y_1(x)$ and $y_2(x)$ be the solutions of the iso-differential inequalities (7) and (8) in J, respectively,*

(iii) *$y_1(x_0) \leq y_0 \leq y_2(x_0)$.*

Prove that

$$y_1(x) < y(x) < y_2(x)$$

for all $x \in (x_0, x_0 + a)$.

Exercise 6.0.222. *Let $\hat{T}(x) = e^x$, $f(x,y) = \frac{-y+y^2+x^2}{1-x}$, $x \in (0,1)$, $y(0) = 1$. Prove that*

$$1 + \frac{x^3}{3} < y(x) < \tan\left(x + \frac{\pi}{4}\right), \qquad x \in (0,1),$$

where $y(x)$ is the solution of the initial value problem (6) in $(0,1)$.

Theorem 6.0.223. *Let $\hat{T}(x) - x\hat{T}'(x) \geq 0$, $\hat{T}'(x) \leq 0$, $\frac{\hat{T}(x)-x\hat{T}'(x)}{\hat{T}(x)} \leq P$ in J for some positive constant P, and for all (x,y), $(x,z) \in D$ such that $x \geq x_0$, $y \geq z$, we have*

$$f(x,y) - f(x,z) \leq L(y-z),$$

for some positive constant L. Let also,

(i) *$y(x)$ be a solution to the initial value problem (6),*

(ii) *$y_1(x)$ and $y_2(x)$ be solutions to the iso-differential inequalities (2) and (3) on J, respectively.*

(iii) *$y_1(x_0) \leq y_0 \leq y_2(x_0)$.*

Then

$$y_1(x) \leq y(x) \leq y_2(x) \qquad \text{for} \qquad \forall x \in J.$$

Proof. Let $\varepsilon > 0$, $\lambda > LP$. Let also,

$$z_1(x) = y_1(x) - \varepsilon e^{\lambda(x-x_0)}, \qquad x \in J.$$

Then

$$z_1(x_0) = y_1(x_0) - \varepsilon < y_1(x_0)$$

and

$$\begin{aligned} z_1'(x) &= y_1'(x) - \varepsilon\lambda e^{\lambda(x-x_0)} \\ &\leq y_1(x)\frac{\hat{T}'(x)}{\hat{T}(x)} + f(x,y_1(x))\frac{\hat{T}(x)-x\hat{T}'(x)}{\hat{T}(x)} - \varepsilon\lambda e^{\lambda(x-x_0)}. \end{aligned} \tag{9}$$

On the other hand, from the definition of the function $z_1(x)$ we have

$$z_1(x) \leq y_1(x) \qquad \text{in} \qquad J.$$

Then

$$f(x,y_1(x)) - f(x,z_1(x)) \leq L(y_1(x) - z_1(x))$$

or

$$f(x,y_1(x)) \leq f(x,z_1(x)) + L(y_1(x) - z_1(x)) \qquad \text{in} \qquad J.$$

From the last inequality and (9) we become

$$\begin{aligned} z_1'(x) &\leq z_1(x)\frac{\hat{T}'(x)}{\hat{T}(x)} + (f(x,z_1(x)) + L(y_1(x) - z_1(x)))\frac{\hat{T}(x)-x\hat{T}'(x)}{\hat{T}(x)} - \varepsilon\lambda e^{\lambda(x-x_0)} \\ &= z_1(x)\frac{\hat{T}'(x)}{\hat{T}(x)} + f(x,z_1(x))\frac{\hat{T}(x)-x\hat{T}'(x)}{\hat{T}(x)} + L\varepsilon e^{\lambda(x-x_0)}\frac{\hat{T}(x)-x\hat{T}'(x)}{\hat{T}(x)} - \varepsilon\lambda e^{\lambda(x-x_0)} \\ &\leq z_1(x)\frac{\hat{T}'(x)}{\hat{T}(x)} + f(x,z_1(x))\frac{\hat{T}(x)-x\hat{T}'(x)}{\hat{T}(x)} + LPe^{\varepsilon(x-x_0)} - \varepsilon\lambda e^{\lambda(x-x_0)} \\ &< z_1(x)\frac{\hat{T}'(x)}{\hat{T}(x)} + f(x,z_1(x))\frac{\hat{T}(x)-x\hat{T}'(x)}{\hat{T}(x)}, \end{aligned}$$

i.e.,

$$z_1'(x) < z_1(x)\frac{\hat{T}'(x)}{\hat{T}(x)} + f(x,z_1(x))\frac{\hat{T}(x)-x\hat{T}'(x)}{\hat{T}(x)} \quad \text{in} \quad J, \tag{10}$$

$$z_1(x_0) < y(x_0).$$

Let now

$$z_2(x) = y_2(x) + \varepsilon e^{\lambda(x-x_0)}, \qquad x \in J.$$

Then

$$z_2(x) > y_2(x) \quad \text{in} \quad J.$$

Therefore

$$f(x,z_2(x)) - f(x,y_2(x)) \le L(z_2(x) - y_2(x)) \quad \text{in} \quad J,$$

from where

$$f(x,y_2(x)) \ge f(x,z_2(x)) + L(y_2(x) - z_2(x)) \quad \text{in} \quad J.$$

Also, using the last inequality,

$$\begin{aligned}
z_2'(x) &= y_2'(x) + \varepsilon\lambda e^{\lambda(x-x_0)} \\
&> y_2(x)\frac{\hat{T}'(x)}{\hat{T}(x)} + f(x,y_2(x))\frac{\hat{T}(x)-x\hat{T}'(x)}{\hat{T}(x)} + \varepsilon\lambda e^{\lambda(x-x_0)} \\
&\ge y_2(x)\frac{\hat{T}'(x)}{\hat{T}(x)} + (f(x,z_2(x)) + L(y_2(x) - z_2(x)))\frac{\hat{T}(x)-x\hat{T}'(x)}{\hat{T}(x)} + \varepsilon\lambda e^{\lambda(x-x_0)} \\
&\ge z_2(x)\frac{\hat{T}'(x)}{\hat{T}(x)} + f(x,z_2(x))\frac{\hat{T}(x)-x\hat{T}'(x)}{\hat{T}(x)} - L\varepsilon e^{\lambda(x-x_0)}\frac{\hat{T}(x)-x\hat{T}'(x)}{\hat{T}(x)} + \varepsilon\lambda e^{\lambda(x-x_0)} \\
&\ge z_2(x)\frac{\hat{T}'(x)}{\hat{T}(x)} + f(x,z_2(x))\frac{\hat{T}(x)-x\hat{T}'(x)}{\hat{T}(x)} - L\varepsilon P e^{\lambda(x-x_0)} + \varepsilon\lambda e^{\lambda(x-x_0)} \\
&> z_2(x)\frac{\hat{T}'(x)}{\hat{T}(x)} + f(x,z_2(x))\frac{\hat{T}(x)-x\hat{T}'(x)}{\hat{T}(x)},
\end{aligned}$$

i.e.,

$$z_2'(x) > z_2(x)\frac{\hat{T}'(x)}{\hat{T}(x)} + f(x,z_2(x))\frac{\hat{T}(x)-x\hat{T}'(x)}{\hat{T}(x)} \quad \text{in} \quad J, \tag{11}$$

$$z_2(x_0) > y_2(x_0).$$

From (10) and (11) it follows that the functions $z_1(x)$ and $z_2(x)$ satisfy all conditions of the basic theorem for the iso-differential inequalities. Therefore

$$z_1(x) < y(x) < z_2(x) \quad \text{in} \quad (x_0, x_0 + a),$$

i.e.

$$y_1(x) - \varepsilon e^{\lambda(x-x_0)} < y(x) < y_2(x) + \varepsilon e^{\lambda(x-x_0)} \quad \text{in} \quad (x_0, x_0 + a),$$

from here, when $\varepsilon \longrightarrow 0$,

$$y_1(x) \le y(x) \le y_2(x) \quad \text{in} \quad J.$$

□

Corollary 6.0.224. *Let for every points (x,y), $(x,z) \in D$ such that $x \geq x_0$, we have*

$$|f(x,y) - f(x,z)| \leq L|y-z| \tag{12}$$

for some positive constant L, $-P \leq \frac{\hat{T}'(x)}{\hat{T}(x)} \leq 0$, $0 \leq \frac{\hat{T}(x)-x\hat{T}'(x)}{\hat{T}(x)} \leq P$ in J for some positive constant P.

Let also,

(i) *y be a solution to the initial value problem (6),*

(ii) *$y_1(x)$ and $y_2(x)$ be solutions to the iso-differential inequalities (2) and (3) on J, respectively,*

(iii) *$y_1(x_0) = y_0 = y_2(x_0)$.*

Then for every $x_1 \in J$, $x_1 > x_0$, either $y_1(x_1) < y(x_1)$ ($y(x_1) < y_2(x_1)$) or $y_1(x) = y(x)$ ($y_2(x) = y(x)$) for $\forall x \in [x_0, x_1]$.

Proof. From (12) we have that if $y \geq z$ then

$$-L(y-z) \leq f(x,y) - f(x,z) \leq L(y-z).$$

Therefore all conditions of the last theorem are fulfilled. Consequently

$$y_1(x) \leq y(x) \leq y_2(x) \qquad \text{for} \qquad \forall x \in J.$$

Also, we have

$$y'(x) - y_1'(x) = y(x)\frac{\hat{T}'(x)}{\hat{T}(x)} + f(x,y(x))\frac{\hat{T}(x)-x\hat{T}'(x)}{\hat{T}(x)} - y_1'(x)$$

$$\geq y(x)\frac{\hat{T}'(x)}{\hat{T}(x)} + f(x,y(x))\frac{\hat{T}(x)-x\hat{T}'(x)}{\hat{T}(x)}$$

$$-y_1'(x)\frac{\hat{T}'(x)}{\hat{T}(x)} - f(x,y_1(x))\frac{\hat{T}(x)-x\hat{T}'(x)}{\hat{T}(x)}$$

$$= (y(x)-y_1(x))\frac{\hat{T}'(x)}{\hat{T}(x)} + (f(x,y(x)) - f(x,y_1(x)))\frac{\hat{T}(x)-x\hat{T}'(x)}{\hat{T}(x)}$$

$$\geq -(y(x)-y_1(x))P - LP(y(x)-y_1(x))$$

$$= -P(1+L)(y(x)-y_1(x)),$$

from where

$$(y(x)-y_1(x))' + P(1+L)(y(x)-y_1(x)) \geq 0,$$

and

$$\left(e^{P(1+L)x}(y(x)\right)' \geq 0,$$

From the last inequality, when $x \leq x_1$, we get

$$\int_{x_1}^{x} \left(e^{P(1+L)x}(y(x)\right)' dx \leq 0,$$

or

$$e^{P(1+L)x}(y(x)-y_1(x)) \leq e^{P(1+L)x_1}(y(x_1)-y_1(x_1)). \tag{13}$$

Then, if $y(x_1) = y_1(x_1)$, using (13), we have that for every $x \in [x_0, x_1]$

$$y(x) \leq y_1(x),$$

whereupon

$$y(x) = y_1(x) \qquad \text{for} \qquad \forall x \in [x_0, x_1].$$

□

Definition 6.0.225. *A solution $r(x)$ ($\rho(x)$) of the initial value problem* (6) *which exists in $J = [x_0, x_0 + a)$ is said to be maximal(minimal) if for an arbitrary solution $y(x)$ of* (6) *existing in J, the inequality $y(x) \leq r(x)$ ($\rho(x) \leq y(x)$) holds for all $x \in J$.*

Theorem 6.0.226. *Let $f(x,y)$ be continuous in $\overline{S}_+ = \{(x,y): \quad x_0 \leq x \leq x_0 + a, \quad |y - y_0| \leq b\}$ and hence there exists a $M > 0$ such that $|f(x,y)| \leq M$ for all $(x,y) \in \overline{S}_+$. Let also, $\hat{T}(x) - x\hat{T}'(x) \geq 0$ in $[x_0, x_0 + a)$, $\frac{|\hat{T}'(x)|}{\hat{T}(x)} \leq P$, $\frac{\hat{T}(x) - x\hat{T}'(x)}{\hat{T}(x)} \leq P$ in $[x_0, x_0 + a)$. Then there exists a maximal solution $r(x)$ and a minimal solution $\rho(x)$ of the initial value problem* (6) *in the interval $[x_0, x_0 + \alpha)$, where*

$$\alpha = \left\{a, \frac{2b}{2P(b + |y_0| + M) + b}\right\}.$$

Proof. We will prove the existence of a maximal solution.

Let

$$0 < \varepsilon \leq \frac{b}{2}.$$

Let us consider the initial value problem

$$\begin{aligned} &y'(x) = y(x)\tfrac{\hat{T}'(x)}{\hat{T}(x)} + f(x, y(x))\tfrac{\hat{T}(x) - x\hat{T}'(x)}{\hat{T}(x)} + \varepsilon \qquad \text{in} \qquad [x_0, x_0 + a), \\ &y(x_0) = y_0. \end{aligned} \tag{14}$$

We define

$$\overline{S}_\varepsilon = \{(x,y) \in \mathbb{R}^2: \quad x_0 \leq x \leq x_0 + a, \quad |y - (y_0 + \varepsilon)| \leq \frac{b}{2}\}.$$

We have that

$$\overline{S}_\varepsilon \subset \overline{S}_+,$$

because

$$\begin{aligned} \tfrac{b}{2} &\geq |y - (y_0 + \varepsilon)| \\ &= |y - y_0 - \varepsilon| \\ &\geq |y - y_0| - \varepsilon, \end{aligned}$$

or

$$|y-y_0| \le \frac{b}{2}+\varepsilon$$

$$\le \frac{b}{2}+\frac{b}{2}$$

$$= b.$$

Also, for every $(x,y) \in \overline{S}_+$ we have

$$\left|y(x)\frac{\hat{T}'(x)}{\hat{T}(x)} + f(x,y)\frac{\hat{T}(x)-x\hat{T}'(x)}{\hat{T}(x)} + \varepsilon\right| \le |y(x)|\frac{|\hat{T}'(x)|}{\hat{T}(x)} + |f(x,y)|\frac{|\hat{T}(x)-x\hat{T}'(x)|}{\hat{T}(x)} + \varepsilon$$

$$\le P(b+|y_0|)+MP+\varepsilon$$

$$\le P(b+|y_0|+M)+\frac{b}{2}.$$

From here and from the iso-Cauchy-Peano's existence theorem it follows that the problem (14) has a solution $y(x,\varepsilon)$ which is defined in $[x_0, x_0+\alpha)$.

Let now

$$0 < \varepsilon_2 < \varepsilon_1 < \varepsilon.$$

We have

$$y(x_0,\varepsilon_2) = y_0+\varepsilon_2 < y_0+\varepsilon_1 = y(x_0,\varepsilon_1),$$

$$y'(x,\varepsilon_2) = y(x)\frac{\hat{T}'(x)}{\hat{T}(x)} + f(x,y(x))\frac{\hat{T}(x)-x\hat{T}'(x)}{\hat{T}(x)} + \varepsilon_2,$$

$$y'(x,\varepsilon_1) = y(x)\frac{\hat{T}'(x)}{\hat{T}(x)} + f(x,y(x))\frac{\hat{T}(x)-x\hat{T}'(x)}{\hat{T}(x)} + \varepsilon_1$$

$$> y(x)\frac{\hat{T}'(x)}{\hat{T}(x)} + f(x,y(x))\frac{\hat{T}(x)-x\hat{T}'(x)}{\hat{T}(x)} + \varepsilon_2.$$

From here and from the basic theorem for the iso-differential inequalities it follows that

$$y(x,\varepsilon_2) < y(x,\varepsilon_1) \qquad \text{for} \qquad \forall x \in [x_0, x_0+\alpha).$$

Using the proof of the iso-Cauchy-Peano's existence theorem we have that the sequence $\{y(x,\varepsilon)\}_{\varepsilon>0}$ is equip-continuous and uniformly bounded.

Let $\{\varepsilon_n\}_{n=1}^{\infty}$ be a sequence of positive real numbers such that

$$\lim_{n\longrightarrow\infty} \varepsilon_n = 0$$

and the corresponding sequence $\{y(x,\varepsilon_n)\}_{n=1}^{\infty}$ of solutions of (14) is defined in $[x_0, x_0+\alpha)$.

We have

$$y(x,\varepsilon_n) = y_0+\varepsilon_n + \int_{x_0}^{x}\left(y(t)\frac{\hat{T}'(t)}{\hat{T}(t)} + f(t,y(t))\frac{\hat{T}(t)-t\hat{T}'(t)}{\hat{T}(t)}\right)dt,$$

$$y_0 = y(x_0,0) < y_0+\varepsilon_n,$$

$$y'(x,\varepsilon_n) = y(x)\frac{\hat{T}'(x)}{\hat{T}(x)} + f(x,y(x))\frac{\hat{T}(x)-x\hat{T}'(x)}{\hat{T}(x)} + \varepsilon_n$$

$$> y(x)\frac{\hat{T}'(x)}{\hat{T}(x)} + f(x,y(x))\frac{\hat{T}(x)-x\hat{T}'(x)}{\hat{T}(x)}.$$

From here and from the basic theorem for the iso-differential inequalities it follows that

$$y(x) < y(x,\varepsilon_n) \qquad \text{in} \qquad [x_0, x_0+\alpha).$$

Consequently

$$y(x) \le \lim_{n \longrightarrow \infty} y(x,\varepsilon_n) := r(x) \qquad \text{for} \qquad \forall x \in [x_0, x_0+\alpha).$$

□

Exercise 6.0.227. *Let* $\hat{T}(x) = e^x$, $f(x,y) = \frac{|y|^{\frac{1}{2}} - y}{1-x}$, $(x_0, y_0) = (0,0)$. *Prove that the functions*

$$r(x) = \begin{cases} \frac{x^2}{4} & \text{for} \quad x \in [0,1), \\ 0 & \text{for} \quad x \in (-1,0], \end{cases}$$

$$\rho(x) = \begin{cases} 0 & \text{for} \quad x \in [0,1), \\ -\frac{x^2}{4} & \text{for} \quad x \in (-1,0] \end{cases}$$

are a maximal and a minimal solution, respectively, to the initial value problem (6) *in* $(-1,1)$.

Theorem 6.0.228. *Let* $r(x)$ *be a maximal solution to the initial value problem* (6) *in J,* $J = [x_0, x_0+a)$. *Let also,* $y(x)$ *be a solution to the iso-differential inequality* (2) *in J. If*

$$y(x_0) \le y_0$$

then

$$y(x) \le r(x) \qquad \text{in} \qquad J.$$

Proof. Let $x_1 \in [x_0, x_0+a)$. Let also $\varepsilon > 0$ be chosen enough small. We consider the problem

$$y'(x) = y(x)\frac{\hat{T}'(x)}{\hat{T}(x)} + f(x,y(x)) + \varepsilon \qquad \text{in} \qquad J,$$
$$y(x_0) = y_0. \tag{15}$$

Let $r(x,\varepsilon)$ be a maximal solution of the problem (15) in the interval J. We have that

$$\lim_{\varepsilon \longrightarrow 0} r(x,\varepsilon) = r(x)$$

uniformly in $[x_0, x_1]$.

Since

$$y(x_0) \le y_0 < y_0 + \varepsilon,$$

$$y'(x) \le y(x)\frac{\hat{T}'(x)}{\hat{T}(x)} + f(x,y(x))\frac{\hat{T}(x) - x\hat{T}'(x)}{\hat{T}(x)}$$

$$< y(x)\frac{\hat{T}'(x)}{\hat{T}(x)} + f(x,y(x))\frac{\hat{T}(x) - x\hat{T}'(x)}{\hat{T}(x)} + \varepsilon$$

and

$$r'(x,\varepsilon) = r(x,\varepsilon)\frac{\hat{T}'(x)}{\hat{T}(x)} + f(x, r(x,\varepsilon))\frac{\hat{T}(x) - x\hat{T}'(x)}{\hat{T}(x)},$$

then from the basic theorem for the iso-differential inequalities it follows that

$$y(x) < r(x,\varepsilon) \quad \text{in} \quad [x_0, x_1],$$

whereupon

$$y(x) \leq \lim_{\varepsilon \longrightarrow 0} r(x,\varepsilon) = r(x).$$

□

Advanced Practical Exercises

Problem 6.0.229. *Let* $\hat{T}(x) = e^x$, $f(x,y) = \frac{y^2}{x-1}$, $y(x)$ *be a solution of the initial value problem* (6) *with* $y(0) = y_0$, $0 < y_0 < 1$. *Prove that*

$$y_0 < y(x) \leq 1 \quad \text{for} \quad \forall x \in (0, \infty).$$

Problem 6.0.230. *Let* $\hat{T}(x) = e^x$, $f(x,y) = \frac{x+y-y^2}{x-1}$, y *be a solution to the initial value problem* (6) *with* $y(0) = 1$, $J = (0,1)$. *Prove that*

$$1 + x < y(x) < \frac{1}{1-x} \quad \text{in} \quad J.$$

Problem 6.0.231. *Let* $\hat{T}(x) = 1 + x^2$, $f(x,y) = \frac{2|y|^{\frac{1}{2}}(1+x^2) - 2xy}{1-x^2}$, $(x_0, y_0) = (0,0)$. *Prove that the functions*

$$r(x) = \begin{cases} x^2 & \text{for} \quad x \in [0,1), \\ 0 & \text{for} \quad (-1, 0], \end{cases}$$

$$\rho(x) = \begin{cases} 0 & \text{for} \quad x \in [0,1), \\ -x^2 & \text{for} \quad x \in (-1, 0], \end{cases}$$

are a maximal and a minimal solution, respectively, in $(-1,1)$, *to the initial value problem* (6).

Chapter 7

Continuous Dependence on Initial Conditions

Let D be a domain in $\mathbb{R}^2$ containing the points (x_0, y_0) and (x_1, y_1), J be an interval in $\mathbb{R}$, $\hat{T} \in \mathcal{C}^1(J)$, $\hat{T}(x) > 0$ for every $x \in J$, $f \in \mathcal{C}(D)$.

Theorem 7.0.232. *Let the following conditions be satisfied.*

(i) *$f(x,y)$ is bounded by M in the domain D, where M is some positive constant,*

(ii) *$f(x,y)$ satisfies a uniform Lipschitz condition*

$$|f(x,y) - f(x,z)| \leq L|y-z| \qquad \text{for} \qquad \forall (x,y), (x,z) \in D,$$

for some positive constant L.

(iii) *$g(x,y)$ is continuous and bounded by M_1 in the domain D, where M_1 is some positive constant,*

(iv) *$y(x)$ is the solution of the initial value problem*

$$y'(x) = y(x)\frac{\hat{T}'(x)}{\hat{T}(x)} + f(x,y(x))\frac{\hat{T}(x) - x\hat{T}';(x)}{\hat{T}(x)} \qquad \text{in} \qquad J, \tag{1}$$

$$y(x_0) = y_0, \tag{2}$$

(v) *$z(x)$ is the solution of the initial value problem*

$$z'(x) = z(x)\frac{\hat{T}'(x)}{\hat{T}(x)} + (f(x,z(x)) + g(x,z(x)))\frac{\hat{T}(x) - x\hat{T}'(x)}{\hat{T}(x)} \qquad \text{in} \qquad J,$$

$$z(x_1) = y_1,$$

(vi) $\frac{|\hat{T}'(x)|}{\hat{T}(x)} \leq P$, $\frac{|\hat{T}(x) - x\hat{T}'(x)|}{\hat{T}(x)} \leq P$ *for every $x \in J$.*

Then

$$|y(x)-z(x)| \leq \Big(|y_0-y_1|+(M+M_1)P|x_0-x_1|+P\Big|\int_{x_1}^{x_0}|z(t)|dt\Big|$$

$$+\frac{M_1}{1+L}\Big)e^{P(1+L)|x-x_0|}-\frac{M_1}{1+L}$$

for all $x \in J$.

Proof. We have

$$y(x) = y_0+\int_{x_0}^{x}\Big(y(t)\frac{\hat{T}'(t)}{\hat{T}(t)}+f(t,y(t))\frac{\hat{T}(t)-t\hat{T}'(t)}{\hat{T}(t)}\Big)dt,$$

$$z(x) = y_1+\int_{x_1}^{x}\Big(z(t)\frac{\hat{T}'(t)}{\hat{T}(t)}+(f(t,z(t))+g(t,z(t)))\frac{\hat{T}(t)-t\hat{T}'(t)}{\hat{T}(t)}\Big)dt$$

$$= y_1+\int_{x_1}^{x}g(t,z(t))\frac{\hat{T}(t)-t\hat{T}'(t)}{\hat{T}(t)}dt+\int_{x_0}^{x}z(t)\frac{\hat{T}'(t)}{\hat{T}(t)}dt+\int_{x_1}^{x_0}z(t)\frac{\hat{T}'(t)}{\hat{T}(t)}dt$$

$$+\int_{x_0}^{x}f(t,z(t))\frac{\hat{T}(t)-t\hat{T}'(t)}{\hat{T}(t)}dt+\int_{x_1}^{x_0}f(t,z(t))\frac{\hat{T}(t)-t\hat{T}'(t)}{\hat{T}(t)}dt$$

for every $x \in J$.

Therefore

$$|z(x)-y(x)| = \Big|(y_1-y_0)$$

$$+\int_{x_0}^{x}\Big((y(t)-z(t))\frac{\hat{T}'(t)}{\hat{T}(t)}+(f(t,y(t))-f(t,z(t)))\frac{\hat{T}(t)-t\hat{T}'(t)}{\hat{T}(t)}\Big)dt$$

$$+\int_{x_1}^{x_0}z(t)\frac{\hat{T}'(t)}{\hat{T}(t)}dt+\int_{x_1}^{x_0}f(t,z(t))\frac{\hat{T}(t)-t\hat{T}'(t)}{\hat{T}(t)}dt+\int_{x_1}^{x}g(t,z(t))\frac{\hat{T}(t)-t\hat{T}'(t)}{\hat{T}(t)}dt\Big|$$

$$\leq |y_0-y_1|+\Big|\int_{x_0}^{x}\Big(|y(t)-z(t)|\frac{|\hat{T}'(t)|}{\hat{T}(t)}+|f(t,y(t))-f(t,z(t))|\frac{|\hat{T}(t)-t\hat{T}'(t)|}{\hat{T}(t)}\Big)dt\Big|$$

$$+\Big|\int_{x_1}^{x_0}|z(t)|\frac{|\hat{T}'(t)|}{\hat{T}(t)}dt\Big|+\Big|\int_{x_1}^{x_0}|f(t,z(t))|\frac{|\hat{T}(t)-t\hat{T}'(t)|}{\hat{T}(t)}dt\Big|$$

$$+\Big|\int_{x_1}^{x}|g(t,z(t))|\frac{|\hat{T}(t)-t\hat{T}'(t)|}{\hat{T}(t)}dt\Big|$$

$$\leq \Big(|y_0-y_1|+(M+M_1)P|x_0-x_1|+P\Big|\int_{x_1}^{x_0}|z(t)|dt\Big|\Big)$$

$$+M_1P|x-x_0|+P(1+L)\Big|\int_{x_0}^{x}|y(t)-z(t)|dt\Big|,$$

i.e.,

$$|z(x)-y(x)| \leq \Big(|y_0-y_1|+(M+M_1)P|x_0-x_1|+P\Big|\int_{x_1}^{x_0}|z(t)|dt\Big|\Big)$$

$$+M_1P|x-x_0|+P(1+L)\Big|\int_{x_0}^{x}|y(t)-z(t)|dt\Big|$$

for all $x \in J$.

From the last inequality and the classical Gronwall's-type inequality we conclude the assertion. □

Theorem 7.0.233. *Let the following conditions be satisfied.*

(i) $f(x,y)$ *is continuous and bounded by* M *in a domain* D *containing the point* (x_0,y_0)*,*

(ii) $\frac{\partial}{\partial y}f(x,y)$ *exists, continuous and bounded by L in D,*

(iii) $\frac{|\hat{T}'(x)|}{\hat{T}(x)} \leq P$, $\frac{|\hat{T}(x)-x\hat{T}'(x)|}{\hat{T}(x)} \leq P$ *in an interval* J*, containing the point* x_0*,*

(iv) *the solution* $y(x,x_0,y_0)$ *of the initial value problem* (1)*,* (2) *exists in the interval* J*.*

Then

(i) *the solution* $y(x,x_0,y_0)$ *is differentiable with respect to* y_0 *and*

$$z(x) = \frac{\partial y}{\partial y_0}(x,x_0,y_0)$$

is the solution of the initial value problem

$$\begin{aligned} z'(x) &= z(x)\frac{\hat{T}'(x)}{\hat{T}(x)} + \frac{\partial f}{\partial y}(x,y(x,x_0,y_0))z(x) \\ z(x_0) &= z_0. \end{aligned} \tag{3}$$

(ii) *the solution* $y(x,x_0,y_0)$ *is differentiable with respect to* x_0 *and* $z(x) = \frac{\partial y(x,x_0,y_0)}{\partial x_0}$ *is the solution of the equation* (3)*, satisfying the initial data*

$$z(x_0) = -y(x_0)\frac{\hat{T}'(x_0)}{\hat{T}(x_0)} - f(x_0,y_0)\frac{\hat{T}(x_0)-x_0\hat{T}'(x_0)}{\hat{T}(x_0)}.$$

The equation (3) *is called the variational equation corresponding to the solution* $y(x,x_0,y_0)$*.*

Proof. **(i)** Let $(x_0,y_1) \in D$ be such that the solution $y(x,x_0,y_1)$ of the initial value problem for the equation (1) with initial data $y(x_0) = y_1$ exists in the interval J_1. Then for all $x \in J_2 = J \cap J_1$ the previous theorem implies that

$$|y(x,x_0,y_0) - y(x,x_0,y_1)| \leq |y_0 - y_1| e^{P(1+L)|x-x_0|},$$

i.e.,

$$|y(x,x_0,y_0) - y(x,x_0,y_1)| \longrightarrow 0$$

when

$$|y_0 - y_1| \longrightarrow 0.$$

Now for all $x \in J_2$ we have

$$y(x,x_0,y_0)-y(x,x_0,y_1)-z(x)(y_0-y_1)$$

$$=\int_{x_0}^{x}\Big(y(t,x_0,y_0)\tfrac{\hat{T}'(t)}{\hat{T}(t)}+f(t,y(t,x_0,y_0))\tfrac{\hat{T}(t)-t\hat{T}'(t)}{\hat{T}(t)}$$

$$-y(t,x_0,y_1)\tfrac{\hat{T}'(t)}{\hat{T}(t)}-f(t,y(t,x_0,y_1))\tfrac{\hat{T}(t)-t\hat{T}'(t)}{\hat{T}(t)}$$

$$-z(t)(y_0-y_1)\tfrac{\hat{T}'(t)}{\hat{T}(t)}-\tfrac{\partial f}{\partial y}(y,y(t,x_0,y_0))\tfrac{\hat{T}(t)-t\hat{T}'(t)}{\hat{T}(t)}z(t)(y_0-y_1)\Big)dt$$

$$=\int_{x_0}^{x}\Big(\big(y(t,x_0,y_0)-y(t,x_0,y_1)\big)\tfrac{\hat{T}'(t)}{\hat{T}(t)}+\Big(f(t,y(t,x_0,y_0))-f(t,y(t,x_0,y_1))\Big)\tfrac{\hat{T}(t)-t\hat{T}'(t)}{\hat{T}(t)}$$

$$-z(t)(y_0-y_1)\tfrac{\hat{T}'(t)}{\hat{T}(t)}-\tfrac{\partial f}{\partial y}(t,y(t,x_0,y_0))\tfrac{\hat{T}(t)-t\hat{T}'(t)}{\hat{T}(t)}z(t)(y_0-y_1)\Big)dt$$

$$=\int_{x_0}^{x}\Big(\tfrac{\partial f}{\partial y}(t,y(t,x_0,y_0))\Big(y(t,x_0,y_0)-y(t,x_0,y_1)-z(t)(y_0-y_1)\Big)\tfrac{\hat{T}(t)-t\hat{T}'(t)}{\hat{T}(t)}$$

$$+\Big(y(t,x_0,y_0)-y(t,x_0,y_1)-z(t)(y_0-y_1)\Big)\tfrac{\hat{T}'(t)}{\hat{T}(t)}\Big)dt$$

$$=\int_{x_0}^{x}\Big(y(t,x_0,y_0)-y(t,x_0,y_1)-z(t)(y_0-y_1)\Big)\Big(\tfrac{\partial f}{\partial y}(t,y(t,x_0,y_0))\tfrac{\hat{T}(t)-t\hat{T}'(t)}{\hat{T}(t)}+\tfrac{\hat{T}'(t)}{\hat{T}(t)}\Big)dt$$

$$+\int_{x_0}^{x}\delta\{y(t,x_0,y_0),y(t,x_0,y_1)\}dt,$$

where

$$\delta\{y(t,x_0,y_0),y(t,x_0,y_1)\}\longrightarrow 0$$

when

$$|y(t,x_0,y_0)-y(t,x_0,y_1)|\longrightarrow 0,$$

i.e., as

$$|y_0-y_1|\longrightarrow 0.$$

Hence,

$$|y(x,x_0,y_0)-y(x,x_0,y_1)-z(x)(y_0-y_1)|$$

$$\le P(L+1)\Big|\int_{x_0}^{x}|y(t,x_0,y_0)-y(t,x_0,y_1)-z(t)(y_0-y_1)|dt\Big|$$

$$+o(|y_0-y_1|),$$

from where

$$|y(x,x_0,y_0)-y(x,x_0,y_1)-z(x)(y_0-y_1)|\le o(|y_0-y_1|)e^{(1+L)P|x-x_0|}.$$

Thus

$$|y(x,x_0,y_0)-y(x,x_0,y_1)-z(x)(y_0-y_1)|\longrightarrow 0$$

as

$$|y_0-y_1|\longrightarrow 0.$$

(ii) Let $(x_1, y_0) \in D$ be such that $x_1 \in J$ and the solution $y(x, x_1, y_0)$ of the initial value problem for the equation (1) with initial data $y(x_1) = y_0$ exists in an interval J_2. Then for all $x \in J_3 = J \cap J_2$, using the last theorem, we have

$$|y(x,x_0,y_0) - y(x,x_1,y_0)| \leq \left(MP|x_0 - x_1| + P\left|\int_{x_1}^{x_0} |y(t,x_1,y_0)|dt\right|\right) e^{P(1+L)|x-x_0|},$$

from here

$$|y(x,x_0,y_0) - y(x,x_1,y_0)| \longrightarrow 0$$

when

$$|x_0 - x_1| \longrightarrow 0.$$

Now for all $x \in J_3$ we have

$$y(x,x_0,y_0) - y(x,x_1,y_0) = \int_{x_0}^{x} \left(y(t,x_0,y_0)\frac{\hat{T}'(t)}{\hat{T}(t)} + f(t,y(t,x_0,y_0))\frac{\hat{T}(t)-t\hat{T}'(t)}{\hat{T}(t)} \right) dt$$

$$- \int_{x_1}^{x} \left(y(t,x_1,y_0)\frac{\hat{T}'(t)}{\hat{T}(t)} + f(t,y(t,x_1,y_0))\frac{\hat{T}(t)-t\hat{T}'(t)}{\hat{T}(t)} \right) dt$$

$$= \int_{x_0}^{x} \left((y(t,x_0,y_0) - y(t,x_1,y_0))\frac{\hat{T}'(t)}{\hat{T}(t)} \right.$$

$$\left. + (f(t,y(t,x_0,y_0)) - f(t,y(t,x_1,y_0)))\frac{\hat{T}(t)-t\hat{T}'(t)}{\hat{T}(t)} \right) dt$$

$$+ \int_{x_0}^{x_1} \left(y(t,x_1,y_0)\frac{\hat{T}'(t)}{\hat{T}(t)} + f(t,y(t,x_1,y_0))\frac{\hat{T}(t)-t\hat{T}'(t)}{\hat{T}(t)} \right) dt$$

$$= \int_{x_0}^{x} \left((y(t,x_0,y_0) - y(t,x_1,y_0))\frac{\hat{T}'(t)}{\hat{T}(t)} \right.$$

$$\left. + \frac{\partial f}{\partial y}(t,y(t_0,x_0,y_0))(y(t,x_0,y_0) - y(t,x_1,y_0))\frac{\hat{T}(t)-t\hat{T}'(t)}{\hat{T}(t)} \right) dt$$

$$+ \int_{x_0}^{x_1} \left(y(t,x_1,y_0)\frac{\hat{T}'(t)}{\hat{T}(t)} + f(t,y(t,x_1,y_0))\frac{\hat{T}(t)-t\hat{T}'(t)}{\hat{T}(t)} \right) dt$$

$$+ \int_{x_0}^{x} \delta\{y(t,x_0,y_0), y(t,x_1,y_0)\}$$

$$= \int_{x_0}^{x} (y(t,x_0,y_0) - y(t,x_1,y_0)) \left(\frac{\hat{T}'(t)}{\hat{T}(t)} + \frac{\partial f}{\partial y}(t,y(t,x_0,y_0))\frac{\hat{T}(t)-t\hat{T}'(t)}{\hat{T}(t)} \right) dt$$

$$+ \int_{x_0}^{x_1} \left(y(t,x_1,y_0)\frac{\hat{T}'(t)}{\hat{T}(t)} + f(t,y(t,x_1,y_0))\frac{\hat{T}(t)-t\hat{T}'(t)}{\hat{T}(t)} \right) dt$$

$$+ \int_{x_0}^{x} \delta\{y(t,x_0,y_0), y(t,x_1,y_0)\}dt,$$

where

$$\delta\{y(t,x_0,y_0), y(t,x_1,y_0)\} \longrightarrow 0$$

when

$$|x_0 - x_1| \longrightarrow 0.$$

Consequently

$$\int_{x_0}^{x}(y(t,x_0,y_0)-y(t,x_1,y_0))\left(\frac{\hat{T}'(t)}{\hat{T}(t)}+\frac{\partial f}{\partial y}(t,y(t,x_0,y_0))\frac{\hat{T}(t)-t\hat{T}'(t)}{\hat{T}(t)}\right)dt$$

$$+\int_{x_0}^{x_1}\left(y(t,x_1,y_0)\frac{\hat{T}'(t)}{\hat{T}(t)}+f(t,y(t,x_1,y_0))\frac{\hat{T}(t)-t\hat{T}'(t)}{\hat{T}(t)}\right)dt$$

$$+\int_{x_0}^{x}\delta\{y(t,x_0,y_0),y(t,x_1,y_0)\}dt$$

$$+\int_{x_0}^{x_1}\left(y_0\frac{\hat{T}'(x_0)}{\hat{T}(x_0)}+f(x_0,y_0)\frac{\hat{T}(x_0)-x_0\hat{T}'(x_0)}{\hat{T}(x_0)}\right)dt$$

$$-(x_0-x_1)\int_{x_0}^{x}\left(z(t)\frac{\hat{T}'(t)}{\hat{T}(t)}+\frac{\partial f}{\partial y}(t,y(t,x_0,y_0))z(t)\right)dt$$

$$=\int_{x_0}^{x}\Big(y(t,x_0,y_0)-y(t,x_1,y_0)-(x_0-x_1)z(t)\Big)\left(\frac{\hat{T}'(t)}{\hat{T}(t)}+\frac{\partial f}{\partial y}(t,x_0,y_0)\frac{\hat{T}(t)-t\hat{T}'(t)}{\hat{T}(t)}\right)dt$$

$$+\int_{x)}^{x_1}\left((y(t,x_1,y_0)+y_0)\frac{\hat{T}'(t)}{\hat{T}(t)}+(f(t,y(t,x_0,y_0))+f(x_0,y_0))\frac{\hat{T}(t)-t\hat{T}'(t)}{\hat{T}(t)}\right)dt$$

$$+\int_{x_0}^{x}\delta\{y(t,x_0,y_0),y(t,x_1,y_0)\}dt$$

for all $x\in J_3$.

From the last inequality it follows that there exist $C_1=C_1(P,L,M)>0$, $C_2=C_2(P,L,M)$ such that

$$|y(t,x_0,y_0)-y(t,x_1,y_0)-z(x)(x_0-x_1)|$$

$$\leq C_1\left|\int_{x_0}^{x}|y(t,x_0,y_0)-y(t,x_1,y_0)-(x_0-x_1)z(t)|dt+C_2o(|x_0-x_1|)\right)$$

for all $x\in J_3$, from where it follows that there exists $C_3=C_3(P,L,M)>0$ such that

$$|y(t,x_0,y_0)-y(t,x_1,y_0)-(x_0-x_1)z(x)|\leq C_3o(|x_0-x_1|).$$

□

Example 7.0.234. *Let* $\hat{T}(x)=e^x$, $f(x,y)=y$, $(x_0,y_0)=(0,1)$. *Then the initial value problem* (1), (2)*admits the following representation*

$$y'(x)=y(x)(2-x),$$

$$y(0)=1,$$

its solution is

$$y(x,x_0,y_0)=e^{2x-\frac{x^2}{2}}.$$

The derivative $z(x)=\frac{\partial y}{\partial x_0}(y,x_0,y_0)$ *satisfies the initial value problem*

$$z'(x)=2z(x)$$

$$z(0)=1,$$

therefore

$$z(x) = e^{2x}.$$

Exercise 7.0.235. *Let $\hat{T}(x) = e^x$, $f(x,y) = 2y$, $(x_0, y_0) = (0,1)$, $y(x, x_0, y_0)$ be the solution of the initial value problem* (1), (2) *in $J = [0,1)$. Find*

$$z(x) = \frac{\partial y}{\partial x_0}(x, x_0, y_0).$$

Answer. e^{3x}.

Finally, we shall consider the initial value problem

$$\begin{aligned} y'(x,\lambda) &= y(x,\lambda)\frac{\hat{T}'(x)}{\hat{T}(x)} + f(x, y(x,\lambda),\lambda)\frac{\hat{T}(x) - x\hat{T}'(x)}{\hat{T}(x)}, \\ y(x_0,\lambda) &= y_0(\lambda), \end{aligned} \tag{4}$$

where $\lambda \in \mathbb{R}$ is a parameter.

The proof of the following theorem is very similar to earlier results.

Theorem 7.0.236. *Let the following conditions be satisfied.*

(i) *$f(x,y,\lambda)$ is continuous and bounded by M in a domain $D \subset \mathbb{R}^3$ containing the point (x_0, y_0, λ_0).*

(ii) $\frac{\partial f(x,y,\lambda)}{\partial y}$, $\frac{\partial f(x,y,\lambda)}{\partial \lambda}$ *exist, continuous and bounded, respectively, by L and L_1 in D.*

Then

(i) *there exist positive numbers h and ε such that given any λ in the interval $|\lambda - \lambda_0| \leq \varepsilon$, there exists a unique solution $y(x,\lambda)$ of the initial value problem* (4) *in the interval $|x - x_0| \leq h$.*

(ii) *the solution $y(x,\lambda)$ is differentiable with respect to λ and*

$$z(x,\lambda) = \frac{\partial y(x,\lambda)}{\partial \lambda}$$

is the solution of the initial value problem

$$\begin{aligned} z'(x,\lambda) &= z(x,\lambda)\frac{\hat{T}'(x)}{\hat{T}(x)} + \left(\frac{\partial f}{\partial y}(x, y(x,\lambda),\lambda) z(x,\lambda) + \frac{\partial f}{\partial \lambda}(x, y(x,\lambda),\lambda)\right)\frac{\hat{T}(x) - x\hat{T}'(x)}{\hat{T}(x)}, \\ z(x_0,\lambda) &= y_0'(\lambda). \end{aligned}$$

If λ is such that $|\lambda - \lambda_0|$is sufficiently small then we have a first-order approximation of the solution $y(x,\lambda)$ given by

$$\begin{aligned} y(x,\lambda) &\simeq y(x,\lambda_0) + (\lambda - \lambda_0)\left(\frac{\partial y}{\partial \lambda}(x,\lambda)\right)\Big|_{\lambda=\lambda_0} \\ &= y(x,\lambda_0) + (\lambda - \lambda_0) z(x,\lambda_0). \end{aligned}$$

Example 7.0.237. *Let* $\hat{T}(x) = e^x$, $f(x,y,\lambda) = \frac{\lambda y^2 - y + 1}{1-x}$, $\lambda \geq 0$, $(x_0, y_0) = (0,0)$. *Then*

$$y\frac{\hat{T}'(x)}{\hat{T}(x)} = y,$$

$$f(x,y,\lambda)\frac{\hat{T}(x) - x\hat{T}'(x)}{\hat{T}(x)} = \frac{\lambda y^2 - y + 1}{1-x}(1-x)$$

$$= \lambda y^2 - y + 1,$$

from here

$$y' = \lambda y^2 + 1,$$

$$y(0) = 0.$$

Its general solution is

$$y(x,\lambda) = \frac{1}{\sqrt{\lambda}}\tan(\sqrt{\lambda}x).$$

We will find $z(x,0)$. *It satisfies the initial value problem*

$$z'(x,0) = x^2,$$

$$z(0,0) = 0.$$

Consequently

$$z(x,0) = \frac{x^3}{3}.$$

Exercise 7.0.238. *Let* $\hat{T}(x) = e^x$, $f(x,y,\lambda) = \frac{\lambda(x+y^2)}{1-x}$, $(x_0, y_0) = (0,1)$. *Find*

$$\frac{\partial y}{\partial \lambda}(x,y,\lambda)\Big|_{\lambda=0}.$$

Answer. $e^{2x} - x - 1$.

Exercise 7.0.239. *Let* $\hat{T}(x) = e^x$, $f(x,y,\lambda) = \frac{2x + \lambda y^2 - y}{1-x}$, $(x_0, y_0) = (0, \lambda - 1)$. *Find*

$$\frac{\partial y}{\partial \lambda}(x,y,\lambda)\Big|_{\lambda=0}.$$

Answer. $\frac{x^5}{5} - \frac{2}{3}x^3 + x + 1$.

Exercise 7.0.240. *Let* $\hat{T}(x) = e^x$, $f(x,y,\lambda) = \frac{y^2 + xy^3}{1-x}$, $(x_0, y_0) = (2, \lambda)$. *Find*

$$\frac{\partial y}{\partial \lambda}(x,y,\lambda)\Big|_{\lambda=0}.$$

Answer. e^{x-2}.

Exercise 7.0.241. *Let* $\hat{T}(x) = e^x$, $f(x,y,\lambda) = \frac{y+\lambda x^2 e^{-y}}{x(1-x)}$, $(x_0, y_0) = (1,1)$. *Find*

$$\frac{\partial y}{\partial \lambda}(x,y,\lambda)\bigg|_{\lambda=0}.$$

Answer. $x(e^{-1} - e^{-x})$.

Exercise 7.0.242. *Let* $\hat{T}(x) = e^x$, $f(x,y,\lambda) = \frac{-y+y^2+\lambda xy^3}{1-x}$, $(x_0, y_0) = (0, 1+\lambda)$. *Find*

$$\frac{\partial y}{\partial \lambda}(x,y,\lambda)\bigg|_{\lambda=0}.$$

Answer. $\frac{1-x-\log(1-x)}{(1-x)^2}$.

Chapter 8

Existence and Uniqueness of Solutions of Systems

Here we suppose that $\hat{T}^1 \in C(J)$, $\hat{T}(x) > 0$ for every $x \in J$, where J is an interval.

We consider a system of first-order iso-differential equations of the form

$$
\begin{aligned}
\left(\hat{u}_1^{\wedge}(\hat{x})\right)^{\circledast} &= \hat{g}_1^{\wedge}(\hat{x}, \hat{u}_1^{\wedge}(\hat{x}), \hat{u}_2^{\wedge}(\hat{x}), \ldots, \hat{u}_n^{\wedge}(\hat{x}))\\
\left(\hat{u}_2^{\wedge}(\hat{x})\right)^{\circledast} &= \hat{g}_2^{\wedge}(\hat{x}, \hat{u}_1^{\wedge}(\hat{x}), \hat{u}_2^{\wedge}(\hat{x}), \ldots, \hat{u}_n^{\wedge}(\hat{x}))\\
&\ldots\\
\left(\hat{u}_n^{\wedge}(\hat{x})\right)^{\circledast} &= \hat{g}_n^{\wedge}(\hat{x}, \hat{u}_1^{\wedge}(\hat{x}), \hat{u}_2^{\wedge}(\hat{x}), \ldots, \hat{u}_n^{\wedge}(\hat{x})),
\end{aligned}
\qquad (1')
$$

which we can rewrite in the form

$$
\begin{aligned}
u_1'(x) &= u_1(x)\frac{\hat{T}'(x)}{\hat{T}(x)} + g_1(x_1, u_1(x), u_2(x), \ldots, u_n(x))\frac{\hat{T}(x) - x\hat{T}'(x)}{\hat{T}(x)}\\
u_2'(x) &= u_2(x)\frac{\hat{T}'(x)}{\hat{T}(x)} + g_2(x_1, u_1(x), u_2(x), \ldots, u_n(x))\frac{\hat{T}(x) - x\hat{T}'(x)}{\hat{T}(x)}\\
&\ldots\\
u_n'(x) &= u_n(x)\frac{\hat{T}'(x)}{\hat{T}(x)} + g_n(x_1, u_1(x), u_2(x), \ldots, u_n(x))\frac{\hat{T}(x) - x\hat{T}'(x)}{\hat{T}(x)}.
\end{aligned}
\qquad (1)
$$

Throughout, we shall assume that the functions g_1, g_2, …, g_n, are continuous in some domain E of $(n+1)$-dimensional space $\mathbb{R}^{n+1}$.

Definition 8.0.243. *By a solution of the iso-differential system* (1) *in the interval* J *we mean a set of n-functions* $u_1(x)$, $u_2(x)$, …, $u_n(x)$ *such that*

1. $u_1'(x)$, $u_2'(x)$, …, $u_n'(x)$ *exist for all* $x \in J$,

2. *for all $x \in J$ the points*

$$(x, u_1(x), u_2(x), \ldots, u_n(x)) \in E,$$

3.

$$u_i'(x) = u_i(x)\frac{\hat{T}'(x)}{\hat{T}(x)} + g_i(x, u_1(x), u_2(x), \ldots, u_n(x))\frac{\hat{T}(x) - x\hat{T}'(x)}{\hat{T}(x)}$$

for all $x \in J$, $i = 1, 2, \ldots, n$.

In addition to the iso-differential system (1) there may also be given initial conditions of the form

$$u_1(x_0) = u_1^0, \quad u_2(x_0) = u_2^0, \ldots, u_n(x_0) = u_n^0, \tag{2}$$

where x_0 is a specified value of x in the interval J, and

$$u_1^0, \quad u_2^0, \ldots, u_n^0$$

are prescribed numbers such that

$$(x_0, u_1^0, u_2^0, \ldots, u_n^0) \in E.$$

The iso-differential system (1) together with the initial conditions (2) forms an initial value problem.

To study the existence and uniqueness of the solutions of (1), (2) there are two possible approaches, either directly imposing sufficient conditions on the functions g_1, g_2, $\ldots$, g_n, and proving the results, or alternatively using vector notations to write (1), (2) in a compact form and then proving the results. We shall prefer to use the second approach since then the proofs are very similar to the scalar case.

By setting

$$u(x) = (u_1(x), u_2(x), \ldots, u_n(x)),$$

$$\frac{\hat{T}'(x)}{\hat{T}(x)}u(x) = \left(\frac{\hat{T}'(x)}{\hat{T}(x)}u_1(x), \frac{\hat{T}'(x)}{\hat{T}(x)}u_2(x), \ldots, \frac{\hat{T}'(x)}{\hat{T}(x)}u_n(x)\right),$$

$$g(x,u) = (g_1(x,u), g_2(x,u), \ldots, g_n(x,u)),$$

$$\frac{\hat{T}(x)-x\hat{T}'(x)}{\hat{T}(x)}g(x,u) = \left(\frac{\hat{T}(x)-x\hat{T}'(x)}{\hat{T}(x)}g_1(x,u), \frac{\hat{T}(x)-x\hat{T}'(x)}{\hat{T}(x)}g_2(x,u),\right.$$

$$\left.\ldots, \frac{\hat{T}(x)-x\hat{T}'(x)}{\hat{T}(x)}g_n(x,u)\right),$$

and agreeing that differentiation and integration to be performed component-wise, i.e.,

$$u'(x) = (u_1'(x), u_2'(x), \ldots, u_n'(x)),$$

$$\int_a^b u(x)dx = \left(\int_a^b u_1(x)dx, \int_a^b u_2(x)dx, \ldots, \int_a^b u_n(x)dx\right),$$

the problem (1), (2) can be written as

$$u'(x) = \frac{\hat{T}'(x)}{\hat{T}(x)}u(x) + g(x,u)\frac{\hat{T}(x) - x\hat{T}'(x)}{\hat{T}(x)}, \tag{3}$$

$$u(x_0) = u^0. \tag{4}$$

We define

$$||u|| = \max_J \sum_{i=1}^n |u_i(x)|.$$

Definition 8.0.244. *The function $g(x,u)$ is said to be continuous in E if each of its components is continuous in E.*

Definition 8.0.245. *The function $g(x,u)$ is defined to be uniformly Lipschitz continuous in E if there exists a nonnegative constant L(Lipschitz constant) such that*

$$||g(x,u) - g(x,v)|| \leq L||u-v|| \tag{5}$$

for all (x,u) and (x,v) in the domain E.

Example 8.0.246. *Let*

$$g(x,u) = (a_{11}u_1 + a_{12}u_2, a_{21}u_1 + a_{22}u_2)$$

and $E = \mathbb{R}^3$. Then, for (x,u), $(x,v) \in E$ we have

$$g(x,u) - g(x,v) = (a_{11}(u_1 - v_1) + a_{12}(u_2 - v_2), a_{21}(u_1 - v_1) + a_{22}(u_2 - v_2)),$$

$$||g(x,u) - g(x,v)|| = ||(a_{11}(u_1 - v_1) + a_{12}(u_2 - v_2), a_{21}(u_1 - v_1) + a_{22}(u_2 - v_2))||$$

$$\leq \max\nolimits_{x\in J}\Big(|a_{11}(u_1 - v_1) + a_{12}(u_2 - v_2)| + |a_{21}(u_1 - v_1) + a_{22}(u_2 - v_2)|\Big)$$

$$\leq \max\nolimits_{x\in J}\Big(|a_{11}||u_1 - v_1| + |a_{12}||u_2 - v_2| + |a_{21}||u_1 - v_1| + |a_{22}||u_2 - v_2|\Big)$$

$$= \max\nolimits_{x\in J}\Big((|a_{11}| + |a_{21}|)|u_1 - v_1| + (|a_{12}| + |a_{22}|)|u_2 - v_2|\Big)$$

$$\leq \max\nolimits_{x\in J}\{|a_{11} + |a_{21}|, |a_{12}| + |a_{22}|\} \max\nolimits_{x\in J}(|u_1 - v_1| + |u_2 - v_2|)$$

$$= \max\nolimits_{x\in J}\{|a_{11} + |a_{21}|, |a_{12}| + |a_{22}|\}||u - v||,$$

therefore

$$L = \max_{x\in J}\{|a_{11} + |a_{21}|, |a_{12}| + |a_{22}|\}.$$

The following result provides sufficient conditions for the function $g(x,u)$ to satisfy the Lipschitz condition.

Theorem 8.0.247. *Let the domain E be convex and for all $(x,u) \in E$ the partial derivatives $\frac{\partial g}{\partial u_k}$, $k = 1,2,\ldots,n$, exist and*

$$\left|\left|\frac{\partial g}{\partial u}\right|\right| \leq L,$$

where

$$\left|\left|\frac{\partial g}{\partial u}\right|\right| = \max_j \max_E \sum_{i=1}^n \left|\frac{\partial g_i}{\partial u_j}(x,u)\right|.$$

Then the function $g(x,u)$ satisfies the Lipschitz condition (5) in E with Lipschitz constant L.

Proof. Let (x,u) and (x,v) be fixed points in E. Then since E is convex for all $0 \le t \le 1$ the points

$$(x, v+t(u-v)) \in E.$$

Then the vector-valued function

$$G(t) = g(x, v+t(u-v)), \qquad 0 \le t \le 1,$$

is well defined, also

$$G'(t) = (u_1 - v_1)\tfrac{\partial g}{\partial u_1}(x, v+t(u-v))$$

$$+(u_2 - v_2)\tfrac{\partial g}{\partial u_2}(x, v+t(u-v))$$

$$+\dots$$

$$+(u_n - v_n)\tfrac{\partial g}{\partial u_n}(x, v+t(u-v))$$

$$= \Big((u_1 - v_1)\tfrac{\partial g_1}{\partial u_1}(x, v+t(u-v)), (u_1 - v_1)\tfrac{\partial g_2}{\partial u_1}(x, v+t(u-v)), \dots, (u_1 - v_1)\tfrac{\partial g_n}{\partial u_1}(x, v+t(u-v))\Big)$$

$$+\Big((u_2 - v_2)\tfrac{\partial g_1}{\partial u_2}(x, v+t(u-v)), (u_2 - v_2)\tfrac{\partial g_2}{\partial u_2}(x, v+t(u-v)), \dots, (u_2 - v_2)\tfrac{\partial g_n}{\partial u_2}(x, v+t(u-v))\Big)$$

$$+\cdots$$

$$+\Big((u_n - v_n)\tfrac{\partial g_1}{\partial u_n}(x, v+t(u-v)), (u_n - v_n)\tfrac{\partial g_2}{\partial u_n}(x, v+t(u-v)), \dots, (u_n - v_n)\tfrac{\partial g_n}{\partial u_n}(x, v+t(u-v))\Big),$$

$$||G'(t)|| = \max_E \textstyle\sum_{i=1}^n \Big|\sum_{j=1}^n (u_j - v_j)\tfrac{\partial g_i}{\partial u_j}(x, v+t(u-v))\Big|$$

$$\le \max_E \textstyle\sum_{i=1}^n \sum_{j=1}^n |u_j - v_j| \Big|\tfrac{\partial g_i}{\partial u_j}(x, v+t(u-v))\Big|$$

$$\le \Big(\max_j \max_E \textstyle\sum_{i=1}^n \Big|\tfrac{\partial g_i}{\partial u_j}(x, v+t(u-v))\Big|\Big) \max_J \sum_{j=1}^n |u_j - v_j|$$

$$= \Big|\Big|\tfrac{\partial g}{\partial u}\Big|\Big| ||u-v||$$

$$\le L||u-v||.$$

Now from the relation

$$g(x,u) - g(x,v) = G(1) - G(0) = \int_0^1 G'(t)dt,$$

we find that

$$||g(x,u)-g(x,v)|| = \left|\left|\int_0^1 G'(t)dt\right|\right|$$

$$\leq \int_0^1 ||G'(t)||dt$$

$$\leq \int_0^1 L||u-v||dt$$

$$= L||u-v||.$$

□

Example 8.0.248. *Let*

$$g(x,u) = (g_1(x,u), g_2(x,u))$$

$$= (a_{11}(x)u_1(x) + a_{12}(x)u_2(x), a_{21}(x)u_1(x) + a_{22}(x)u_2(x)).$$

Then

$$\frac{\partial g}{\partial u_1}(x,u) = (a_{11}(x), a_{21}(x)),$$

$$\frac{\partial g}{\partial u_2}(x,u) = (a_{12}(x), a_{22}(x)).$$

Then

$$\left|\left|\frac{\partial g}{\partial u}\right|\right| = \max\left\{\max_J(|a_{11}(x)| + |a_{21}(x)|), \max_J(|a_{21}(x)| + |a_{22}(x)|)\right\}.$$

Therefore the function $g(x,u)$ satisfies the Lipschitz condition with the Lipschitz constant

$$L = \max\left\{\max_J(|a_{11}(x)| + |a_{21}(x)|), \max_J(|a_{21}(x)| + |a_{22}(x)|)\right\}.$$

If $g(x,u)$ is continuous in the domain E, then any solution of the initial value problem (3), (4) is also a solution of the integral equation

$$u(x) = u^0 + \int_{x_0}^x \left(\frac{\hat{T}'(t)}{\hat{T}(t)}u(t) + g(t,u(t))\frac{\hat{T}(t) - t\hat{T}'(t)}{\hat{T}(t)}\right)dt \tag{6}$$

and conversely.

To find a solution of the integral equation (6) the iso-Picard method of successive approximations is equally useful. Let $u^0(x)$ be any continuous function which we assume to be an initial approximation of the solution, then we define approximations successively by

$$u^{m+1}(x) = u^0 + \int_{x_0}^x \left(\frac{\hat{T}'(t)}{\hat{T}(t)}u^m(t) + g(t,u^m(t))\frac{\hat{T}(t) - t\hat{T}'(t)}{\hat{T}(t)}\right)dt, \quad m = 0,1,2,\ldots, \tag{7}$$

and, as before, if the sequence $\{u^m(x)\}_{m=1}^\infty$ converges uniformly to a continuous function $u(x)$ in some interval J containing the point x_0 and for all $x \in J$ the points $(x,u(x)) \in E$, then this function $u(x)$ will be a solution of the integral equation (6).

Below we will suppose that

$$\frac{|\hat{T}'(x)|}{\hat{T}(x)} \leq P, \qquad \frac{|\hat{T}(x) - x\hat{T}'(x)|}{\hat{T}(x)} \leq P$$

for all x in the interval J.

Now we shall state several results for the initial value problem (3), (4) which are analogous to those proved in the earlier chapters for the scalar case.

Theorem 8.0.249. *(local existence theorem) Let the following conditions hold.*

(i) $g(x,u)$ *is continuous in*

$$\Omega = \{(x,u) \in \mathbb{R}^{n+1} : |x - x_0| \leq a, \qquad ||u - u^0|| \leq b\}$$

and hence there exists a $M > 0$ *such that*

$$||g(x,u)|| \leq M$$

for all $(x,u) \in \Omega$*,*

(ii) $g(x,u)$ *satisfies the uniform Lipschitz condition* (5) *in* Ω*,*

(iii) $u^0(x)$ *is a continuous function in the interval* $|x - x_0| \leq a$ *and*

$$||u^0(x) - u^0|| \leq b.$$

Then the sequence $\{u^m(x)\}_{m=1}^{\infty}$ *generated by the iso-Picard iterative scheme* (7) *converges to the unique solution* $u(x)$ *of the problem* (3), (4). *This solution is valid in the interval*

$$J_h = \{x \in \mathbb{R} : \quad |x - x_0| \leq h\},$$

where

$$h = \min\left\{a, \frac{b}{P(b + |y_0| + M)}\right\}.$$

Further, for all $x \in J_h$*, the following error estimate holds*

$$||u(x) - u^m(x)|| \leq Ne^{(P+PL)h} \min\left\{1, \frac{((P+PL)h)^m}{m!}\right\}, \qquad m = 0, 1, 2, \ldots,$$

where

$$||u^1(x) - u^0(x)|| \leq N.$$

Theorem 8.0.250. *(global existence theorem) Let the following conditions hold.*

(i) $g(x,u)$ *is continuous in*

$$\Delta = \{(x,u) \in \mathbb{R}^{n+1} : \qquad |x - x_0| \leq a, \qquad ||u|| < \infty\},$$

(ii) $g(x,u)$ *satisfies the uniform Lipschitz condition* (5) *in* Δ*,*

(iii) *$u^0(x)$ is continuous in $|x-x^0| \leq a$.*

Then the sequence $\{u^m(x)\}_{m=1}^{\infty}$ generated by the iso-Picard iterative scheme (7) exists in the entire interval $|x-x_0| \leq a$, and converges to the unique solution $u(x)$ of the problem (3), (4).

Corollary 8.0.251. *Let $g(x,u)$ be continuous function in $\mathbb{R}^{n+1}$ and satisfies the uniform Lipschitz condition (5) in each*

$$\Delta_a = \{(x,u) \in \mathbb{R}^{n+1} : \qquad |x| \leq a, \qquad ||u|| < \infty\}$$

with the Lipschitz constant L_a. Then the problem (3), (4) has a unique solution, which exists for all $x \in \mathbb{R}$.

Theorem 8.0.252. *(iso-Peano's existence theorem) Let $g(x,u)$ be continuous and bounded in Δ. Then the problem (3), (4) has at least one solution in the interval $|x-x_0| \leq a$.*

Definition 8.0.253. *(ε-approximate solution) Let $g(x,u)$ be continuous in the domain E. A function $u(x)$ defined in the interval J is said to be an ε-approximate solution of the iso-differential system (3) if*

1. *$u(x)$ is a continuous function in the interval J,*

2. *for all $x \in J$ the points $(x,u(x)) \in E$,*

3. *$u(x)$ has a piecewise continuous derivative in the interval J which may fail to be defined only for a finite number of points, say, x_1, x_2, ..., x_k,*

4.

$$\left|\left|u'(x) - u(x)\frac{\hat{T}'(x)}{\hat{T}(x)} - g(x,u(x))\frac{\hat{T}(x) - x\hat{T}'(x)}{\hat{T}(x)}\right|\right| \leq \varepsilon$$

for all $x \in J$, $x \neq x_i$, $i = 1,2,\ldots,k$.

Theorem 8.0.254. *Let $g(x,u)$ be continuous in Ω, and hence there exists a $M > 0$ such that*

$$||g(x,u)|| \leq M$$

for all $(x,u) \in \Omega$. Then for any $\varepsilon > 0$ there exists an ε-approximate solution $u(x)$ of the iso-differential system (3) in the interval J_h such that

$$u(x_0) = u^0.$$

Theorem 8.0.255. *(iso-Cauchy-Peano's existence theorem) Let $g(x,u)$ be continuous in Ω and hence there exists a $M > 0$ such that $|g(x,u)| \leq M$ for every point $(x,y) \in \Omega$. Then the initial value problem (3), (4) has at least one solution in J_h.*

Theorem 8.0.256. *(continuation of solutions) Assume that $g(x,u)$ is continuous in E and $u(x)$ is a solution of the problem (3), (4) in the interval J. Then $u(x)$ can be extended as a solution of the initial value problem (3), (4) to the boundary of E.*

Corollary 8.0.257. *Assume that $g(x,u)$ is continuous in*

$$E_1 = \{(x,u) \in E : \quad x_0 \leq x < x_0 + a, \quad a < \infty, \quad ||u|| < \infty\}.$$

If $u(x)$ is any solution of the initial value problem (3), (4), then the largest interval of the existence of the solution $u(x)$ is either $[x_0, x_0 + a]$ or $[x_0, x_0 + \alpha)$, $\alpha < a$, and

$$||u(x)|| \longrightarrow \infty$$

as $x \longrightarrow x_0 + \alpha$.

Theorem 8.0.258. *(continuous dependence on initial conditions) Let the following conditions hold.*

(i) *$g(x,u)$ is continuous and bounded by M in the domain E containing the points (x_0, u^0) and (x_1, u^1),*

(ii) *$g(x,u)$ satisfies the uniform Lipschitz condition (5) in E,*

(iii) *$h(x,u)$ is continuous and bounded by M_1 in E,*

(iv) *$u(x)$ and $v(x)$ are the solutions of the initial value problem (3), (4) and*

$$v'(x) = v(x)\frac{\hat{T}'(x)}{\hat{T}(x)} + g(x, v(x))\frac{\hat{T}(x) - x\hat{T}'(x)}{\hat{T}(x)} + h(x, v(x))\frac{\hat{T}(x) - x\hat{T}'(x)}{\hat{T}(x)},$$

$$v(x_1) = u^1,$$

respectively, which exist in the interval J containing the points x_0 and x_1.

Then for all $x \in J$, the following inequality holds

$$||u(x) - v(x)|| \leq \left(||u^0 - u^1|| + (M + M_1)P|x_0 - x_1| + P\left|\left| \int_{x_1}^{x_0} |v(t)|dt \right|\right| \right.$$

$$\left. + \frac{M_1}{1+L} \right) e^{P(1+L)|x - x_0|} - \frac{M_1}{1+L}.$$

Theorem 8.0.259. *(differentiation with respect to initial conditions) Let the following conditions be satisfied.*

(i) *$g(x,u)$ is continuous and bounded by M in the domain E containing the point (x_0, u^0),*

(ii) *the matrix $\frac{\partial g}{\partial u}(x,u)$ exists and is continuous, and bounded by L in the domain E,*

(iii) *the solution $u(x, x_0, u^0)$ of the initial value problem (3), (4) exists in an interval J containing x_0.*

Then the following hold.

(i) *The solution $u(x, x_0, u^0)$ is differentiable with respect to u^0, and for all j, $1 \leq j \leq n$, $v^j(x) = \frac{\partial u}{\partial u_j^0}(x, x_0, u^0)$ is the solution of the following initial value problem*

$$\begin{aligned} v'(x) &= v(x)\frac{\hat{T}'(x)}{\hat{T}(x)} + \frac{\partial g}{\partial u}(x, u(x, x_0, u^0))v(x), \\ v(x_0) &= e^j = (0, 0, \ldots, 0, 1, 0, \ldots, 0). \end{aligned} \tag{8}$$

(ii) *The solution $u(x,x_0,u^0)$ is differentiable with respect to x_0 and $v(x) = \frac{\partial u(x,x_0,u^0)}{\partial x_0}$ is the solution of the iso-differential system* (8)*, satisfying the initial condition*

$$v(x_0) = -u^0 \frac{\hat{T}'(x_0)}{\hat{T}(x_0)} - g(x_0,u^0)\frac{\hat{T}(x_0) - x_0\hat{T}'(x_0)}{\hat{T}(x_0)}.$$

Theorem 8.0.260. *Let the following conditions be satisfied.*

(i) *$g(x,u,\lambda)$ is continuous and bounded by M in a domain $E \subset \mathbb{R}^{n+m+1}$ containing the point (x_0,u^0,λ^0),*

(ii) *the matrix $\frac{\partial g(x,u,\lambda)}{\partial u}$ exists and is continuous, an bounded by L in E*

(iii) *the $n \times m$ matrix $\frac{\partial g(x,u,\lambda)}{\partial u}$ exists and is continuous, an bounded by L_1 in E,*

Then the following hold.

(i) *There exist positive numbers h and ε such that for any λsatisfying*

$$||\lambda - \lambda^0|| \leq \varepsilon,$$

there exists a unique solution $u(x,\lambda)$ to the following initial value problem

$$u'(x,\lambda) = u(x,\lambda)\frac{\hat{T}'(x)}{\hat{T}(x)} + f(x,u(x,\lambda))\frac{\hat{T}(x)-x\hat{T}'(x)}{\hat{T}(x)},$$
$$u(x_0,\lambda) = u^0, \tag{9}$$

in the interval $|x - x_0| \leq h$.

(ii) *The solution $u(x,\lambda)$ is differentiable with respect to λ and for each j, $1 \leq j \leq m$, $v^j(x,\lambda) = \frac{\partial u(x,\lambda)}{\partial \lambda_j}$ is the solution of the following initial value problem*

$$v'(x,\lambda) = v(x,\lambda)\frac{\hat{T}'(x)}{\hat{T}(x)} + \frac{\partial g}{\partial u}(x,u(x,\lambda),\lambda)v(x,\lambda)\frac{\hat{T}(x)-x\hat{T}'(x)}{\hat{T}(x)}$$

$$+\frac{\partial g}{\partial \lambda_j}(x,u(x,\lambda),\lambda)\frac{\hat{T}(x)-x\hat{T}'(x)}{\hat{T}(x)},$$

$$v(x_0,\lambda) = 0.$$

Chapter 9

General Properties of Linear Systems

Here we will suppose that $\hat{T} \in \mathcal{C}^1(J)$, $\hat{T}(x) > 0$ for every $x \in J$, where J is an interval.

We will consider the system

$$u'(x) = \left(A(x)\frac{\hat{T}(x) - x\hat{T}'(x)}{\hat{T}(x)} + \frac{\hat{T}'(x)}{\hat{T}(x)}\right)u(x) + b(x)\frac{\hat{T}(x) - x\hat{T}'(x)}{\hat{T}(x)}, \tag{1}$$

where $A(x)$ is a $n \times n$ matrix with elements $a_{ij}(x)$, $b(x)$ is a $n \times 1$ vector with components $b_i(x)$, and $u(x)$ is $n \times 1$ unknown vector with components $u - i(x)$.

The existence and uniqueness of the solutions of the iso-differential system (1) together with the initial condition

$$u(x_0) = u^0 \tag{2}$$

in the interval J containing x_0 follows from the previous chapter provided the functions $a_{ij}(x)$, $b_i(x)$, $1 \le i, j \le n$, are continuous in the interval J which we shall assume throughout.

The principle of the superposition is stated as follows.

Let $u(x)$ is a solution to the iso-differential system

$$u'(x) = \left(A(x)\frac{\hat{T}(x) - x\hat{T}'(x)}{\hat{T}(x)} + \frac{\hat{T}'(x)}{\hat{T}(x)}\right)u(x) + b^1(x)\frac{\hat{T}(x) - x\hat{T}'(x)}{\hat{T}(x)}$$

and $v(x)$ is a solution to the system

$$v'(x) = \left(A(x)\frac{\hat{T}(x) - x\hat{T}'(x)}{\hat{T}(x)} + \frac{\hat{T}'(x)}{\hat{T}(x)}\right)v(x) + b^2(x)\frac{\hat{T}(x) - x\hat{T}'(x)}{\hat{T}(x)},$$

then

$$\phi(x) = c_1 u(x) + c_2 v(x),$$

where c_1 and c_2 are real constants, is a solution of the following iso-differential system

$$\phi'(x) = \left(A(x)\frac{\hat{T}(x) - x\hat{T}'(x)}{\hat{T}(x)} + \frac{\hat{T}'(x)}{\hat{T}(x)}\right)\phi(x) + \left(c_1 b^1(x) + c^2 b^2(x)\right)\frac{\hat{T}(x) - x\hat{T}'(x)}{\hat{T}(x)}.$$

For this we have

$$\begin{aligned}
\phi'(x) &= c_1u'(x)+c_2v'(x)\\
&= c_1\left(A(x)\tfrac{\hat{T}(x)-x\hat{T}'(x)}{\hat{T}(x)}+\tfrac{\hat{T}'(x)}{\hat{T}(x)}\right)u(x)+c_1b^1(x)\tfrac{\hat{T}(x)-x\hat{T}'(x)}{\hat{T}(x)}\\
&+c_2\left(A(x)\tfrac{\hat{T}(x)-x\hat{T}'(x)}{\hat{T}(x)}+\tfrac{\hat{T}'(x)}{\hat{T}(x)}\right)v(x)+c_2b^2(x)\tfrac{\hat{T}(x)-x\hat{T}'(x)}{\hat{T}(x)}\\
&= \left(A(x)\tfrac{\hat{T}(x)-x\hat{T}'(x)}{\hat{T}(x)}+\tfrac{\hat{T}'(x)}{\hat{T}(x)}\right)(c_1u(x)+c_2v(x))\\
&+\left(c_1b^1(x)+c_2b^2(x)\right)\tfrac{\hat{T}(x)-x\hat{T}'(x)}{\hat{T}(x)}\\
&= \left(A(x)\tfrac{\hat{T}(x)-x\hat{T}'(x)}{\hat{T}(x)}+\tfrac{\hat{T}'(x)}{\hat{T}(x)}\right)\phi(x)+\left(c_1b^1(x)+c_2b^2(x)\right)\tfrac{\hat{T}(x)-x\hat{T}'(x)}{\hat{T}(x)}.
\end{aligned}$$

In particular, if

$$b^1(x)=b^2(x)\equiv 0,$$

i.e., $u(x)$ and $v(x)$ are solutions of the homogeneous iso-differential system

$$u'(x)=\left(A(x)\frac{\hat{T}(x)-x\hat{T}'(x)}{\hat{T}(x)}+\frac{\hat{T}'(x)}{\hat{T}(x)}\right)u(x), \tag{3}$$

then

$$c_1u(x)+c_2v(x)$$

is also a solution.

Thus, solutions of the homogeneous iso-differential system (3) form a vector space.

Further, if

$$b^1(x)=b^2(x), \qquad c_1=1, c_2=-1,$$

and $u(x)$ is a solution of (1), then $v(x)$ is also a solution of (1) if and only if

$$u(x)-v(x)$$

is a solution of (3).

Thus, the general solution of (1) is obtained by adding to a particular solution of (1) the general solution of the corresponding homogeneous system (3).

To find the dimension of the vector space of the solutions of (3) we need to define the concept for linear independence and dependence of vector-valued functions.

Definition 9.0.261. *The vector-valued functions*

$$u^1(x), u^2(x), \ldots, u^m(x)$$

defined in the interval J are said to be linearly independent in J, if the relation

$$c_1u^1(x)+c_2u^2(x)+\cdots+c_mu^m(x)=0$$

for all $x \in J$ implies that

$$c_1 = c_2 = \cdots = c_m = 0.$$

Conversely, these functions are said to be linearly independent if there exist constants

$$c_1, c_2, \ldots, c_m$$

not all zero such that

$$c_1 u^1(x) + c_2 u^2(x) + \cdots + c_m u^m(x) = 0$$

for all $x \in J$.

Let the functions

$$u^1(x), u^2(x), \ldots, u^m(x)$$

be linearly dependent in the interval J and $c_k \neq 0$ for some $k \in \{1, 2, \ldots, m\}$. Then we have

$$u^k(x) = -\frac{c_1}{c_k} u^1(x) - \cdots - \frac{c_{k-1}}{c_1} u^{k-1}(x) - \frac{c_{k+1}}{c_k} u^{k+1}(x) - \cdots - \frac{c_m}{c_k} u^m(x),$$

i.e., $u^k(x)$ (and hence at least one of these functions) can be expressed as a linear combination of the remaining $m-1$ functions, so that

$$u^k(x) = c_1 u^1(x) + \cdots + c_{k-1} u^{k-1}(x) + c_{k+1} u^{k+1}(x) + \cdots + c_m u^m(x),$$

then obviously these functions are linearly dependent. Hence, if two functions are linearly dependent in the interval J, then each one of these functions is identically equal to a constant times the other function, while if two functions are linearly independent, then it is impossible to express either function as a constant times the other.

Example 9.0.262. *The functions*

$$1, \quad x, \quad x^2, \quad \cdots, \quad x^{m-1}$$

are linearly independent in the interval J.

Really, we suppose that

$$c_1 + c_2 x + c_3 x^2 + \cdots + c_m x^{m-1} = 0 \tag{4}$$

in an interval J and any c_k were not zero. Then the equation (4) could hold for at most $m-1$ values of x, whereas it must hold for all $x \in J$.

Therefore the considered functions are linearly independent.

Example 9.0.263. *The functions*

$$u^1(x) = \begin{pmatrix} e^x \\ e^x \end{pmatrix}, \qquad u^2(x) = \begin{pmatrix} e^{2x} \\ 3e^{2x} \end{pmatrix}$$

are linearly independent in every interval J. Indeed,

$$c_1 \begin{pmatrix} e^x \\ e^x \end{pmatrix} + c_2 \begin{pmatrix} e^{2x} \\ 3e^{2x} \end{pmatrix} = 0$$

implies that

$$c_1 e^x + c_2 e^{2x} = 0$$

$$c_1 e^x + 3c_2 e^{2x} = 0,$$

which is possible if

$$c_1 = c_2 = 0.$$

Example 9.0.264. *The functions*

$$u^1(x) = \begin{pmatrix} \sin x \\ \cos x \end{pmatrix}, \qquad u^2(x) = \begin{pmatrix} 0 \\ 0 \end{pmatrix}$$

are linearly dependent in every interval J. Let $c_1 = 0$, $c_2 = 1$. Then

$$0 \begin{pmatrix} \sin x \\ \cos x \end{pmatrix} + 1 \begin{pmatrix} 0 \\ 0 \end{pmatrix} = 0$$

in every interval J.

Definition 9.0.265. *For the given n vector-valued functions*

$$u^1(x), \quad u^2(x), \quad \cdots, \quad u^n(x)$$

the determinant $W(u^1, u^2, \ldots, u^n)(x)$ or $W(x)$, defined by

$$\begin{vmatrix} u_1^1(x) & u_1^2(x) & \cdots & u_1^n(x) \\ u_2^1(x) & u_2^2(x) & \cdots & u_2^n(x) \\ \cdots & \cdots & \cdots & \cdots \\ u_n^1(x) & u_n^2(x) & \cdots & u_n^n(x) \end{vmatrix}$$

is called the Wronskian of these functions.

Theorem 9.0.266. *If the Wronskian $W(x)$ of n vector-valued functions*

$$u^1(x), \quad u^2(x), \quad ,\cdots, \quad u^n(x)$$

is different from zero for at least one point in an interval J, then these functions are linearly independent in J.

Proof. Let

$$u^1(x), \quad u^2(x), \quad ,\cdots, \quad u^n(x)$$

be linearly dependent in J, then there exist n constants

$$c_1, \quad c_2, \quad \ldots, \quad c_n,$$

such that

$$\sum_{i=1}^{n} c_i u^i(x) = 0 \qquad \text{in} \qquad J.$$

From here

$$\sum_{i=1}^{n} u_k^i(x)c_i = 0 \qquad \text{in} \qquad J$$

for all k, $1 \le k \le n$, has a nontrivial solution, which is possible if and only if $W(x) = 0$ for every $x \in J$. But $W(x) \neq 0$ for at least one point x in the interval J, and, therefore

$$u^1(x), \quad u^2(x), \quad ,\cdots, \quad u^n(x)$$

can not be linearly independent. □

Theorem 9.0.267. *Let*

$$u^1(x), \quad u^2(x),\ldots, \quad u^n(x)$$

be linearly independent solutions of the iso-differential systems (3) *in the interval* J. *Then* $W(x) \neq 0$ *for all* $x \in J$.

Proof. Let x_0 be a point in J where $W(x_0) = 0$. Then there exists constants

$$c_1, \quad c_2, \quad \ldots, \quad c_n$$

not all zero such that

$$\sum_{i=1}^{n} c_i u^i(x_0) = 0.$$

Since

$$u(x) = \sum_{i=1}^{n} c_i u^i(x)$$

is a solution of (3) , and $u(x_0) = 0$, from the uniqueness of the solutions it follows that

$$u(x) = \sum_{i=1}^{n} c_i u^i(x) = 0$$

in the interval J. However, the functions

$$u^1(x), \quad u^2(x), \quad ,\cdots, \quad u^n(x)$$

are linearly independent J so we must have

$$c_1 = c_2 = \cdots = c_n = 0,$$

which is a contradiction. □

Theorem 9.0.268. *(iso-Abel's formula) Let*

$$u^1(x), \quad u^2(x), \quad ,\cdots, \quad u^n(x)$$

be the solutions of the iso-differential system (3) *in the interval* J *and* $x_0 \in J$. *Then for all* $x \in J$ *we have*

$$W(x) = \frac{W(x_0)}{\hat{T}^n(x_0)} \hat{T}^n(x) e^{\int_{x_0}^{x} \mathrm{Tr}A(t) \frac{\hat{T}(t) - t\hat{T}'(t)}{\hat{T}(t)} dt}.$$

Proof. The derivative of the Wronskian $W(x)$ can be written as

$$W'(x) = \sum_{i=1}^{n} \begin{vmatrix} u_1^1(x) & \cdots & u_1^n(x) \\ \cdots & \cdots & \cdots \\ u_{i-1}^1(x) & \cdots & u_{i-1}^n(x) \\ u_i'^1(x) & \cdots & u_i'^n(x) \\ u_{i+1}^1(x) & \cdots & u_{i+1}^n(x) \\ \cdots & \cdots & \cdots \\ u_n^1(x) & \cdots & u_n^n(x) \end{vmatrix}. \tag{5}$$

In the ith determinant of the right hand side of (5) we use the iso-differential system (3) to replace $u_i^j(x)$ by

$$\sum_{k=1}^{n} \left(a_{ik}(x) \frac{\hat{T}(x) - x\hat{T}'(x)}{\hat{T}(x)} + \frac{\hat{T}'(x)}{\hat{T}(x)} \right) u_k^j(x)$$

and multiply the first row by

$$a_{i1}(x) \frac{\hat{T}(x) - x\hat{T}'(x)}{\hat{T}(x)} + \frac{\hat{T}'(x)}{\hat{T}(x)},$$

the second row by

$$a_{i2}(x) \frac{\hat{T}(x) - x\hat{T}'(x)}{\hat{T}(x)} + \frac{\hat{T}'(x)}{\hat{T}(x)},$$

and so on, then subtract their sum from the ith row, to get

$$W'(x) = \left(\mathrm{Tr}A(x) \frac{\hat{T}(x) - x\hat{T}'(x)}{\hat{T}(x)} + n \frac{\hat{T}'(x)}{\hat{T}(x)} \right) W(x). \tag{6}$$

Integration of the first-order iso-differential equation (6) from x_0 to x gives the required relation. □

Example 9.0.269. *Let* $\hat{T}(x) = e^x$,

$$A(x) = \begin{pmatrix} \frac{1}{x-1} & 0 \\ \frac{x^2+2x+1}{(x-1)(x^2+2x-1)} & \frac{x^2-3}{(x-1)(x^2+2x-1)} \end{pmatrix}, \qquad x \neq 1, -1 \pm \sqrt{2}.$$

Then

$$\hat{T}'(x) = e^x,$$

$$\frac{\hat{T}(x)-x\hat{T}'(x)}{\hat{T}(x)} = \frac{e^x-xe^x}{e^x}$$

$$= 1-x,$$

$$\frac{\hat{T}'(x)}{\hat{T}(x)} = \frac{e^x}{e^x}$$

$$= 1,$$

$$a_{11}(x)\frac{\hat{T}(x)-x\hat{T}'(x)}{\hat{T}(x)} + \frac{\hat{T}'(x)}{\hat{T}(x)} = \frac{1}{x-1}(1-x)+1$$

$$= 0,$$

$$a_{12}(x)\frac{\hat{T}(x)-x\hat{T}'(x)}{\hat{T}(x)} + \frac{\hat{T}'(x)}{\hat{T}(x)} = 0(1-x)+1$$

$$= 1,$$

$$a_{21}(x)\frac{\hat{T}(x)-x\hat{T}'(x)}{\hat{T}(x)} + \frac{\hat{T}'(x)}{\hat{T}(x)} = \frac{x^2+2x+1}{(x-1)(x^2+2x-1)}(1-x)+1$$

$$= -\frac{x^2+2x+1}{x^2+2x-1}+1$$

$$= -\frac{2}{x^2+2x-1},$$

$$a_{22}(x)\frac{\hat{T}(x)-x\hat{T}'(x)}{\hat{T}(x)} + \frac{\hat{T}'(x)}{\hat{T}(x)} = \frac{x^2-3}{(x-1)(x^2+2x_1)}(1-x)+1$$

$$= \frac{-x^2+3}{x^2+2x-1}+1$$

$$= \frac{2(x+1)}{x^2+2x-1}.$$

In this way we obtain the system

$$u_1'(x) = u_2(x)$$

$$u_2'(x) = -\frac{2}{x^2+2x-1}u_1(x) + \frac{2(x+1)}{x^2+2x-1}u_2(x).$$

From here

$$(x^2+2x-1)u_2'(x) = -2u_1(x)+2(x+1)u_2(x)$$

or

$$u_1(x) = -\frac{x^2+2x-1}{2}u_2'(x)+(x+1)u_2(x),$$

whereupon

$$u_1'(x) = -\tfrac{2x+2}{2}u_2'(x) - \tfrac{x^2+2x-1}{2}u_2''(x) + u_2(x) + (x+1)u_2'(x)$$

$$= -\tfrac{x^2+2x-1}{2}u_2''(x) + u_2(x).$$

Therefore

$$-\frac{x^2+2x-1}{2}u_2''(x) + u_2(x) = u_2(x)$$

or

$$u_2''(x) = 0.$$

Consequently

$$u_2(x) = c_1 + c_2 x,$$

where c_1 and c_2 are real constants.

Also,

$$u_1(x) = -\tfrac{x^2+2x-1}{2}c_2 + (x+1)(c_1 + c_2 x)$$

$$= \left(\tfrac{-x^2-2x+1}{2} + x^2 + x\right)c_2 + c_1(x+1)$$

$$= c_1(x+1) + c_2\tfrac{x^2+1}{2},$$

whereas

$$u(x) = \begin{pmatrix} c_1(x+1) + c_2\frac{x^2+1}{2} \\ c_1 + c_2 x \end{pmatrix}.$$

For

$$c_1 = 1, \qquad c_2 = 0$$

we get

$$u^1(x) = \begin{pmatrix} x+1 \\ 1 \end{pmatrix},$$

for

$$c_1 = 0, \qquad c_2 = 2$$

we have

$$u^2(x) = \begin{pmatrix} x^2+1 \\ 2x \end{pmatrix}.$$

Therefore

$$W(u^1, u^2)(x) = \begin{vmatrix} x+1 & x^2+1 \\ 1 & 2x \end{vmatrix}$$

$$= 2x(x+1) - (x^2+1)$$

$$= x^2 + 2x - 1.$$

From iso-Abel's formula we have

$$W(x) = W(x_0)e^{\int_{x_0}^{x}\left(\mathrm{Tr}A(t)\frac{\hat{T}(t)-t\hat{T}'(t)}{\hat{T}(t)}+2\frac{\hat{T}'(t)}{\hat{T}(t)}\right)dt}$$

$$= (x_0^2+2x_0-1)e^{\int_{x_0}^{x}\frac{2(t+1)}{t^2+2t-1}dt}$$

$$= (x_0^2+2x_0-1)e^{\int_{x_0}^{x}\frac{d(t+1)^2}{(t+1)^2-2}dt}$$

$$= (x_0^2+2x_0-1)\frac{x^2+2x-1}{x_0^2+2x_0-1}$$

$$= x^2+2x-1.$$

Exercise 9.0.270. *Let* $\hat{T}(x) = e^x$,

$$A(x) = \begin{pmatrix} \frac{1}{x-1} & \frac{1-e^{-x}}{x-1} \\ \frac{1}{x-1} & 0 \end{pmatrix}.$$

Find two linearly independent solutions $u^1(x)$, $u^2(x)$ *of the system* (3) *and* $W(u^1,u^2)(x)$. *Check your answer using the definition of the Wronskian and the iso-Abel's formula.*

Exercise 9.0.271. *Let the solutions*

$$u^1(x), \qquad u^2(x), \qquad \ldots, \qquad u^n(x)$$

of the system (3) *satisfy the initial conditions*

$$u^i(x_0) = e^i, \qquad i = 1,2,\ldots,n,$$

which are defined in an interval J, where

$$e^i = (0,0,\ldots,0,1,0,\ldots,0), \qquad i = 1,2,\ldots,n.$$

Prove that they are linearly independent in the interval J.

Exercise 9.0.272. *The Wronskian of n functions*

$$y_1(x), \quad y_2(x), \quad , \quad \ldots, \quad y_n(x)$$

which are $(n-1)$*-times differentiable in an interval J is defined by the determinant*

$$W(y_1,y_2,\ldots,y_n)(x) = \begin{vmatrix} y_1(x) & y_2(x) & \cdots & y_n(x) \\ y_1'(x) & y_2'(x) & \cdots & y_n'(x) \\ \cdots & \cdots & \cdots & \cdots \\ y_1^{(n-1)}(x) & y_2^{(n-1)}(x) & \cdots & y_n^{(n-1)}(x) \end{vmatrix}.$$

Prove the following.

(i) *If $W(y_1, y_2, \ldots, y_n)(x)$ is different from zero for at least one point in J, then the functions*

$$y_1(x), \quad y_2(x), \quad , \ldots, \quad y_n(x)$$

are linearly independent.

(ii) *If the functions*

$$y_1(x), \quad y_2(x), \quad , \ldots, \quad y_n(x)$$

are linearly dependent in J, then the Wronskian

$$W(y_1, y_2, \ldots, y_n)(x) = 0 \quad \text{in} \quad J.$$

(iii) *The converse of (i) as well as of (ii) is not necessary true.*

(iv) *If*

$$W(y_1, y_2, \ldots, y_n)(x) = 0 \quad \text{in} \quad J,$$

but for some set of $(n-1)$, y's(say, without loss of generality, all but $y_n(x)$)

$$W(y_1, y_2, \ldots, y_{n-1}(x)) \neq 0$$

for all $x \in J$, then the functions

$$y_1(x), \quad y_2(x), \quad , \ldots, \quad y_n(x)$$

are linearly dependent in J.

Exercise 9.0.273. *Prove that any solution $u(x)$ of the iso-differential system (3) satisfying the initial data*

$$u(x_0) = u^0$$

can be written as

$$\sum_{i=1}^{n} u_i^0 u^i(x),$$

where $u^i(x)$, $1 \leq i \leq n$, is the solution of the initial value problem for the iso-differential system (3) with initial conditions

$$u^i(x_0) = e^i, \qquad i = 1, 2, \ldots, n,$$

$$e^i = (0, 0, \ldots, 0, 1, 0, \ldots, 0)$$

for $i = 1, 2, \ldots, n$.

Exercise 9.0.274. *Let $\hat{T}(x) = e^x$,*

$$A(x) = \begin{pmatrix} \frac{1}{x-1} & 0 & \frac{1}{x-1} \\ \frac{1}{x-1} & \frac{1}{x-1} & 0 \\ 0 & \frac{1}{x-1} & \frac{1}{x-1} \end{pmatrix},$$

$x \neq 1$. *Let* $u(x)$, $v(x)$ *and* $w(x)$ *be the solutions of the iso-differential system* (3) *satisfying*

$$u(0) = 1, \quad u'(0) = 0, \quad u''(0) = 0,$$

$$v(0) = 0, \quad v'(0) = 1, \quad v''(0) = 0,$$

$$w(0) = 0, \quad w'(0) = 0, \quad w''(0) = 1.$$

Without solving the iso-differential system (3), *show that*

(i) $u'(x) = -w(x)$,

(ii) $v'(x) = u(x)$,

(iii) $w'(x) = v(x)$,

(iv) $W(u, v, w) = u^3 - v^3 + w^3 + 3uvw = 1$.

Chapter 10

Fundamental Matrix Solution

Here we will suppose that $\hat{T} \in C^1(J)$, $\hat{T}(x) > 0$ for every $x \in J$, where J is an interval in $\mathbb{R}$.

We will investigate the system

$$u'(x) = \left(A(x)\frac{\hat{T}(x) - x\hat{T}'(x)}{\hat{T}(x)} + \frac{\hat{T}'(x)}{\hat{T}(x)}\right)u(x), \tag{1}$$

where $A(x)$ is $n \times n$ matrix with elements $a_{ij}(x)$, which are continuous functions in the interval J, $1 \leq i, j \leq n$.

We have that any solution $u(x)$ of the iso-differential system (1) for which

$$u(x_0) = u^0$$

can be written as

$$u(x) = \sum_{i=1}^{n} u_i^0 u^i(x),$$

where $u^i(x)$ is the solution of the initial value problem (1),

$$u^i(x_0) = u^i, \qquad i = 1, 2, \ldots, n, \tag{1'}$$

$$e^i = (0, 0, \ldots, 0, 1, 0, \ldots, 0),$$

$i = 1, 2, \ldots, n$.

In matrix notation this solution can be written as

$$u(x) = \Phi(x, x_0)u^0,$$

where $\Phi(x, x_0)$ is an $n \times n$ matrix whose ith column is $u^i(x)$.

Definition 10.0.275. *The matrix $\Phi(x, x_0)$ is said to be the principal fundamental matrix or evolution matrix or transition matrix.*

We have that $\Phi(x, x_0)$ satisfies the initial value problem

$$\Phi'(x, x_0) = \left(A(x)\frac{\hat{T}(x) - x\hat{T}'(x)}{\hat{T}(x)} + \frac{\hat{T}'(x)}{\hat{T}(x)}\right)\Phi(x, x_0), \tag{2}$$

$$\Phi(x_0,x_0)=I. \tag{3}$$

The fact that the initial value problem (2), (3) has a unique solution $\Phi(x,x_0)$ in the interval J can be proved exactly as in the previous chapters. Moreover, the iterative scheme

$$\Phi^{m+1}(x)=I+\int_{x_0}^{x}\left(A(t)\frac{\hat{T}(t)-t\hat{T}'(t)}{\hat{T}(t)}+\frac{\hat{T}'(t)}{\hat{T}(t)}\right)\Phi^{m}(t)dt,$$

$$\Phi^{0}(x)=I,$$

converges to $\Phi(x,x_0)$, and

$$\Phi^{1}(x)=I+\int_{x_0}^{x}\left(A(t)\frac{\hat{T}(t)-t\hat{T}'(t)}{\hat{T}(t)}+\frac{\hat{T}'(t)}{\hat{T}(t)}\right)\Phi^{0}(t)dt$$

$$=I+\int_{x_0}^{x}\left(A(t)\frac{\hat{T}(t)-t\hat{T}'(t)}{\hat{T}(t)}+\frac{\hat{T}'(t)}{\hat{T}(t)}\right)dt,$$

$$\Phi^{2}(x)=I+\int_{x_0}^{x}\left(A(t)\frac{\hat{T}(t)-t\hat{T}'(t)}{\hat{T}(t)}+\frac{\hat{T}'(t)}{\hat{T}(t)}\right)\Phi^{1}(t)dt$$

$$=I+\int_{x_0}^{x}\left(A(t)\frac{\hat{T}(t)-t\hat{T}'(t)}{\hat{T}(t)}+\frac{\hat{T}'(t)}{\hat{T}(t)}\right)\left(I+\int_{x_0}^{t}\left(A(s)\frac{\hat{T}(s)-s\hat{T}'(s)}{\hat{T}(s)}+\frac{\hat{T}'(s)}{\hat{T}(s)}\right)ds\right)dt$$

$$=I+\int_{x_0}^{x}\left(A(t)\frac{\hat{T}(t)-t\hat{T}'(t)}{\hat{T}(t)}+\frac{\hat{T}'(t)}{\hat{T}(t)}\right)dt$$

$$+\int_{x_0}^{x}\left(A(t)\frac{\hat{T}(t)-t\hat{T}'(t)}{\hat{T}(t)}+\frac{\hat{T}'(t)}{\hat{T}(t)}\right)\int_{x_0}^{t}\left(A(s)\frac{\hat{T}(s)-s\hat{T}'(s)}{\hat{T}(s)}+\frac{\hat{T}'(s)}{\hat{T}(s)}\right)dsdt,$$

and so on,

$$\begin{aligned}\Phi(x,x_0)&=I+\int_{x_0}^{x}\left(A(t)\frac{\hat{T}(t)-t\hat{T}'(t)}{\hat{T}(t)}+\frac{\hat{T}'(t)}{\hat{T}(t)}\right)dt\\&+\int_{x_0}^{x}\int_{x_0}^{t}\left(A(t)\frac{\hat{T}(t)-t\hat{T}'(t)}{\hat{T}(t)}+\frac{\hat{T}'(t)}{\hat{T}(t)}\right)\left(A(s)\frac{\hat{T}(s)-s\hat{T}'(s)}{\hat{T}(s)}+\frac{\hat{T}'(s)}{\hat{T}(s)}\right)dsdt+\cdots.\end{aligned}\tag{4}$$

Definition 10.0.276. *The series* (4) *is said to be iso-Peano-Baker series of the initial value problem* (2), (3).

Theorem 10.0.277. *If $\Psi(x)$ is a fundamental matrix of the iso-differential system* (1), *then for any constant nonsingular $n\times n$ matrix C, the matrix $\Psi(x)C$ is also a fundamental matrix of the iso-differential system* (1), *and any fundamental matrix of* (1) *is of the form $\Psi(x)C$ for some constant nonsingular $n\times n$ matrix C.*

Proof. We have

$$\Psi'(x)=\left(A(x)\frac{\hat{T}(x)-x\hat{T}'(x)}{\hat{T}(x)}+\frac{\hat{T}'(x)}{\hat{T}(x)}\right)\Psi(x),$$

and hence

$$\Psi'(x)C=\left(A(x)\frac{\hat{T}(x)-x\hat{T}'(x)}{\hat{T}(x)}+\frac{\hat{T}'(x)}{\hat{T}(x)}\right)\Psi(x)C,$$

which is the same as

$$(\Psi(x)C)' = \left(A(x)\frac{\hat{T}(x) - x\hat{T}'(x)}{\hat{T}(x)} + \frac{\hat{T}'(x)}{\hat{T}(x)}\right)(\Psi(x)C),$$

i.e., $\Psi(x)$ and $\Psi(x)C$ both are solutions of the same matrix iso-differential system (1).

Further, since

$$\det\Psi(x) \neq 0$$

and

$$\det C \neq 0$$

it follows that

$$\det(\Psi(x)C) \neq 0$$

and hence $\Psi(x)C$ is also a fundamental matrix solution of the iso-differential system (1).

Conversely, let $\Psi_1(x)$ and $\Psi_2(x)$ be two fundamental matrix solutions of (1). If

$$\Psi_2^{-1}(x)\Psi_1(x) = C(x)$$

or

$$\Psi_1(x) = \Psi_2(x)C(x),$$

then we have

$$\Psi_1'(x) = \Psi_2'(x)C(x) + \Psi_2(x)C'(x)$$

or

$$\left(A(x)\tfrac{\hat{T}(x)-x\hat{T}'(x)}{\hat{T}(x)} + \tfrac{\hat{T}'(x)}{\hat{T}(x)}\right)\Psi_1(x) = \left(A(x)\tfrac{\hat{T}(x)-x\hat{T}'(x)}{\hat{T}(x)} + \tfrac{\hat{T}'(x)}{\hat{T}(x)}\right)\Psi_2(x)C(c)$$

$$+\Psi_2(x)C'(x),$$

whereupon

$$\Psi_2(x)C'(x) = 0$$

or

$$C'(x) = 0.$$

Therefore $C(x)$ is a constant matrix.

Moreover, since $\Psi_1(x)$ and $\Psi_2(x)$ are nonsingular matrices, this constant matrix is also nonsingular. □

As a consequence of this theorem we find

$$\Phi(x, x_0) = \Psi(x)\Psi^{-1}(x_0)$$

and the solution of the initial value problem for (1) with

$$u(x_0) = u^0$$

can be written in the following form

$$u(x) = \Psi(x)\Psi^{-1}(x_0)u^0.$$

Since the product of matrices is not commutative, for a given constant nonsingular matrix C, the matrix $C\Psi(x)$ need not be a fundamental matrix solution of the iso-differential system (1).

Further, two different homogeneous systems cannot have the same fundamental matrix, i.e. $\Psi(x)$ determines the matrix

$$A(x)\frac{\hat{T}(x)-x\hat{T}'(x)}{\hat{T}(x)}+\frac{\hat{T}'(x)}{\hat{T}(x)}I$$

uniquely by the relation

$$A(x)\frac{\hat{T}(x)-x\hat{T}'(x)}{\hat{T}(x)}+\frac{\hat{T}'(x)}{\hat{T}(x)}I=\Psi'(x)\Psi^{-1}(x).$$

Now we differentiate the relation

$$\Psi(x)\Psi^{-1}(x)=I,$$

and we obtain

$$\Psi'(x)\Psi^{-1}(x)+\Psi(x)\left(\Psi^{-1}(x)\right)'=0,$$

and hence

$$\begin{aligned}\left(\Psi^{-1}(x)\right)' &= -\Psi^{-1}(x)\Psi'(x)\Psi^{-1}(x)\\ &= -\Psi^{-1}(x)\left(A(x)\tfrac{\hat{T}(x)-x\hat{T}'(x)}{\hat{T}(x)}+\tfrac{\hat{T}'(x)}{\hat{T}(x)}I\right),\end{aligned}$$

which is the same as

$$(4')\qquad \left(\left(\Psi^{-1}(x)\right)^T\right)'=-\left(A^T(x)\tfrac{\hat{T}(x)-x\hat{T}'(x)}{\hat{T}(x)}+\tfrac{\hat{T}'(x)}{\hat{T}(x)}\right)\left(\Psi^{-1}(x)\right)^T.$$

Therefore, $\left(\Psi^{-1}(x)\right)^T$ is a fundamental matrix of the following iso-differential system

$$(4)\qquad u'(x)=-\left(A^T(x)\tfrac{\hat{T}(x)-x\hat{T}'(x)}{\hat{T}(x)}+\tfrac{\hat{T}'(x)}{\hat{T}(x)}\right)u(x).$$

Definition 10.0.278. *The system* (4) *is said to be the iso-adjoint system of the iso-differential system* (1).

Theorem 10.0.279. *If* $\Psi(x)$ *is a fundamental matrix of the iso-differential system* (1), *then* $\chi(x)$ *is a fundamental matrix of its iso-adjoint system* (4) *if and only if*

$$\chi^T(x)\Psi(x)=C, \tag{5}$$

where C *is a constant nonsingular* $n\times n$ *matrix.*

Proof. If $\Psi(x)$ is a fundamental matrix of the iso-differential system (1), then from $(4')$ it follows that $\left(\Psi^{-1}(x)\right)^T$ is a fundamental matrix of the iso-differential system (4). Therefore there exists a constant nonsingular matrix D such that

$$\chi(x)=\left(\Psi^{-1}(x)\right)^T D,$$

from where

$$\Psi^T(x)\chi(x) = D,$$

which is the same as

$$\chi^T(x)\Psi(x) = D^T.$$

Therefore (5) holds with $C = D^T$.

Conversely, if $\Psi(x)$ is a fundamental matrix of (1) satisfying (5), then

$$\Psi^T(x)\chi(x) = C^T$$

and hence

$$\chi(x) = \left(\Psi^T(x)\right)^{-1} C^T.$$

Consequently, $\chi(x)$ is a fundamental matrix of the iso-adjoint system (4). □

Now we consider the system

$$u'(x) = \left(A(x)\frac{\hat{T}(x) - x\hat{T}'(x)}{\hat{T}(x)} + \frac{\hat{T}'(x)}{\hat{T}(x)}\right)u(x) + \frac{\hat{T}(x) - x\hat{T}'(x)}{\hat{T}(x)}b(x). \tag{6}$$

We seek a vector-valued function $v(x)$ such that

$$\Phi(x, x_0)v(x)$$

is a solution of (6).

We have

$$\Phi'(x,x_0)v(x) + \Phi(x,x_0)v'(x) = \left(A(x)\frac{\hat{T}(x)-x\hat{T}'(x)}{\hat{T}(x)} + \frac{\hat{T}'(x)}{\hat{T}(x)}\right)\Phi(x,x_0)v(x)$$

$$+b(x)\frac{\hat{T}(x)-x\hat{T}'(x)}{\hat{T}(x)},$$

from where

$$\Phi(x,x_0)v'(x) = b(x)\frac{\hat{T}(x) - x\hat{T}'(x)}{\hat{T}(x)}$$

or

$$v'(x) = \Phi^{-1}(x,x_0)b(x)\frac{\hat{T}(x) - x\hat{T}'(x)}{\hat{T}(x)}.$$

Consequently

$$v(x) = v(x_0) + \int_{x_0}^{x} \Phi^{-1}(t,x_0)b(t)\frac{\hat{T}(t) - t\hat{T}'(t)}{\hat{T}(t)}dt.$$

In this way the solution of (6) takes the form

$$u(x) = \Phi(x,x_0)v(x_0) + \Phi(x,x_0)\int_{x_0}^{x} \Phi^{-1}(t,x_0)b(t)\frac{\hat{T}(t) - t\hat{T}'(t)}{\hat{T}(t)}dt.$$

Exercise 10.0.280. *Let $\hat{T}(x) = e^x$. Solve the system* (1) *in the cases*

$$1)\quad A(x) = \begin{pmatrix} \frac{1}{1-x} & 0 \\ \frac{2}{1-x} & \frac{3}{1-x} \end{pmatrix}, \quad 2)\quad A(x) = \begin{pmatrix} 0 & \frac{2}{1-x} \\ \frac{5}{1-x} & 0 \end{pmatrix}.$$

Answer.

$$1)\quad \begin{aligned} u_1(x) &= C_1 e^x + C_2 e^{5x} \\ u_2(x) &= -C_1 e^x + 3C_2 e^{5x}, \end{aligned} \qquad 2)\quad \begin{aligned} u_1(x) &= C_1 e^{-x} + C_2 e^{3x} \\ u_2(x) &= 2C_1 e^{-x} - 2C_2 e^{3x}, \end{aligned}$$

where C_1 and C_2 are real constants.

Exercise 10.0.281. *Let $\hat{T}(x) = x^2 + 1$, $x \neq \pm 1$. Solve the system* (1) *in the cases*

$$1)\quad A(x) = \begin{pmatrix} \frac{x+1}{x-1} & 2\frac{4x^2-x+4}{1-x^2} \\ \frac{1-x}{1+x} & \frac{1-x}{1+x} \end{pmatrix}$$

$$2)\quad A(x) = \begin{pmatrix} \frac{1-x}{1+x} & \frac{1-x}{1+x} \\ -2\frac{1+x+x^2}{1+x^2} & \frac{3x^2-2x+3}{1-x^2} \end{pmatrix}.$$

Answer.

$$1)\quad \begin{aligned} u_1(x) &= 2C_1 e^{3x} - 4C_2 e^{-3x} \\ u_2(x) &= C_1 e^{3x} + C_2 e^{-3x}, \end{aligned}$$

$$2)\quad \begin{aligned} u_1(x) &= e^{2x}\left(C_1 \cos x + C_2 \sin x\right) \\ u_2(x) &= e^{2x}\left((C_1 + C_2)\cos x + (C_2 - C_1)\sin x\right), \end{aligned}$$

where C_1 and C_2 are real constants.

Exercise 10.0.282. *Let $\hat{T}(x) = e^{3x}$, $x \neq \frac{1}{3}$. Solve the system* (1) *in the cases*

$$1)\quad A(x) = \begin{pmatrix} \frac{2}{3x-1} & \frac{6}{3x-1} \\ 0 & \frac{2}{3x-1} \end{pmatrix},$$

$$2)\quad A(x) = \begin{pmatrix} \frac{4}{3x-1} & \frac{8}{3x-1} \\ \frac{2}{3x-1} & \frac{2}{3x-1} \end{pmatrix}.$$

Answer.

$$1)\quad \begin{aligned} u_1(x) &= e^x\left(C_1 \cos(3x) + C_2 \sin(3x)\right) \\ u_2(x) &= e^x\left(C_1 \sin(3x) - C_2 \cos(3x)\right), \end{aligned}$$

$$2)\quad \begin{aligned} u_1(x) &= \left(2C_2 - C_1\right)\cos(2x) - \left(2C_1 + C_2\right)\sin(2x) \\ u_2(x) &= C_1 \cos(2x) + C_2 \sin(2x), \end{aligned}$$

where C_1 and C_2 are real constants.

Exercise 10.0.283. *Let* $\hat{T}(x) = e^{x^2}$, $x \neq \pm\frac{1}{\sqrt{2}}$. *Solve the system* (1) *in the cases*

$$1)\quad A(x) = \begin{pmatrix} \frac{2-2x}{1-2x^2} & \frac{1-2x}{1-2x^2} \\ -\frac{1+2x}{1-2x^2} & \frac{4-2x}{1-2x^2} \end{pmatrix},$$

$$2)\quad A(x) = \begin{pmatrix} \frac{3-2x}{1-2x^2} & -\frac{1+2x}{1-2x^2} \\ \frac{4-2x}{1-2x^2} & -\frac{1+2x}{1-2x^2} \end{pmatrix}.$$

Answer.

$$1)\quad \begin{aligned} u_1(x) &= \Big(C_1 + C_2x\Big)e^{3x} \\ u_2(x) &= \Big(C_1 + C_2 + C_2x\Big)e^{3x}, \end{aligned}$$

$$2)\quad \begin{aligned} u_1(x) &= \Big(C_1 + C_2x\Big)e^{3x} \\ u_2(x) &= \Big(2C_1 - C_2 + 2C_2x\Big)e^{x}, \end{aligned}$$

where C_1 and C_2 are real constants.

Exercise 10.0.284. *Let* $\hat{T}(x) = (x^2+1)e^x$, $x^3+x^2+x-1 \neq 0$. *Solve the system* (1) *in the cases*

$$1)\quad A(x) = \begin{pmatrix} \frac{4x^2+2x+4}{x^3+x^2+x-1} & -\frac{x^2-2x+1}{x^3+x^2+x-1} \\ \frac{3x^2+2x+3}{x^3+x^2+x-1} & \frac{2x}{x^3+x^2+x-1} \end{pmatrix},$$

$$2)\quad A(x) = \begin{pmatrix} -\frac{4x^2-2x+4}{x^3+x^2+x-1} & -2\frac{x^2-x+1}{x^3+x^2+x-1} \\ \frac{4x^2+2x+4}{x^3+x^2+x-1} & 2\frac{x^2+x+1}{x^3+x^2+x-1} \end{pmatrix}.$$

Answer.

$$1)\quad \begin{aligned} u_1(x) &= \Big(C_1 + 2C_2x\Big)e^{-x} \\ u_2(x) &= \Big(C_1 + C_2 + 2C_2x\Big)e^{-x}, \end{aligned}$$

$$2)\quad \begin{aligned} u_1(x) &= \Big(C_1 + 3C_2x\Big)e^{2x} \\ u_2(x) &= \Big(C_2 - C_1 - 3C_2x\Big)e^{2x}, \end{aligned}$$

where C_1 and C_2 are real constants.

Exercise 10.0.285. *Let* $\hat{T}(x) = e^x$, $x \neq 1$. *Solve the system* (6) *in the cases*

$$1)\quad A(x) = \begin{pmatrix} \frac{3}{1-x} & \frac{4}{x-1} \\ \frac{1}{1-x} & \frac{2}{1-x} \end{pmatrix}, \quad b(x) = \begin{pmatrix} \frac{\sin x}{1-x} \\ -2\frac{\cos x}{1-x} \end{pmatrix},$$

$$2)\quad A(x) = \begin{pmatrix} \frac{1}{1-x} & 0 \\ 0 & \frac{1}{1-x} \end{pmatrix}, \quad b(x) = \begin{pmatrix} 2\frac{e^x}{1-x} \\ -3\frac{e^{4x}}{1-x} \end{pmatrix}.$$

Answer.

1)
$$u_1(x) = C_1 e^x + 3C_2 e^{2x} + \cos x - 2\sin x$$
$$u_2(x) = C_1 e^x + 2C_2 e^{2x} + 2\cos x - 2\sin x,$$

2)
$$u_1(x) = C_1 e^x + C_2 e^{3x} + x e^x - e^{4x}$$
$$u_2(x) = -C_1 e^x + C_2 e^{3x} - (x+1)e^x - 2e^{4x},$$

where C_1 and C_2 are real constants.

Chapter 11

Periodic Linear Systems

Definition 11.0.286. *A function $y(x)$ is said to be periodic of period $\omega > 0$ if for all x in the domain of the function $y(x)$ we have*

$$y(x+\omega) = y(x). \tag{1}$$

For example, the functions

$$2\sin x + 1, \qquad 3\cos^2 x + 2$$

are periodic functions of period 2π.

Here we will suppose that $\omega > 0$ is the smallest positive number for which (1) holds.

Definition 11.0.287. *If all components $u_i(x)$, $1 \le i \le n$, of a vector $u(x)$ are periodic of period ω, then the vector $u(x)$ will be called periodic vector of period ω.*

Definition 11.0.288. *If all elements $a_{ij}(x)$, $1 \le i, j \le n$, of a $n \times n$ matrix $A(x)$ are periodic of period ω, then the matrix $A(x)$ will be called periodic matrix of period ω.*

Below we will assume that $\hat{T} \in C^1(\mathbb{R})$, $\hat{T}(x) > 0$ for every $x \in \mathbb{R}$ and $\hat{T}(x)$ is a periodic function of period ω.

We will investigate the linear iso-differential system

$$\left(\hat{u}^\wedge(\hat{x})\right)^{\circledast} = \hat{A}^\wedge(\hat{x}) \hat{\times} \hat{u}^\wedge(\hat{x}) + \hat{b}^\wedge(\hat{x}), \tag{2'}$$

where $A(x)$ is a $n \times n$ matrix with continuous elements, $u(x)$ and $b(x)$ are $n \times 1$ vector.

The system $(2')$ we can rewrite in the form

$$u'(x) = \left(A(x)\frac{\hat{T}(x) - x\hat{T}'(x)}{\hat{T}(x)} + \frac{\hat{T}'(x)}{\hat{T}(x)}\right)u(x) + b(x)\frac{\hat{T}(x) - x\hat{T}'(x)}{\hat{T}(x)}. \tag{2}$$

Here we will provide certain characterizations for the existence of periodic solutions of period ω of the system (2). We suppose that the matrix $A(x)$ satisfies the conditions

$$\begin{aligned} &A(x)\frac{\hat{T}(x)-x\hat{T}'(x)}{\hat{T}(x)} \in C(\mathbb{R}), \\ &A(x)\frac{\hat{T}(x)-x\hat{T}'(x)}{\hat{T}(x)} = A(x+\omega)\frac{\hat{T}(x+\omega)-(x+\omega)\hat{T}'(x+\omega)}{\hat{T}(x+\omega)} \end{aligned} \tag{3}$$

for all $x \in \mathbb{R}$, and $b(x)$ satisfies the conditions

$$
\begin{aligned}
& b(x)\frac{\hat{T}(x)-x\hat{T}'(x)}{\hat{T}(x)} \in C(\mathbb{R}), \\
& b(x)\frac{\hat{T}(x)-x\hat{T}'(x)}{\hat{T}(x)} = b(x+\omega)\frac{\hat{T}(x+\omega)-(x+\omega)\hat{T}'(x+\omega)}{\hat{T}(x+\omega)}
\end{aligned}
\tag{4}
$$

for all $x \in \mathbb{R}$.

To begin with we will provide necessary and sufficient conditions for the iso-differential system (2) to have a periodic solution of period ω.

Theorem 11.0.289. *Let the matrix $A(x)$ and the function $b(x)$ satisfy (3) and (4), respectively. Then the iso-differential system (2) has a periodic solution of period ω if and only if*

$$u(0) = u(\omega).$$

Proof. Let $u(x)$ be a periodic solution of period ω of the iso-differential system (2). Then from the definition for periodic function we get

$$u(0) = u(0+\omega) = u(\omega). \tag{5}$$

Now we assume that $u(x)$ is a solution to (2) for which (5) holds.

Let

$$v(x) = u(x+\omega).$$

Then it follows that

$$
\begin{aligned}
v'(x) &= u'(x+\omega) \\
&= \left(A(x+\omega)\frac{\hat{T}(x+\omega)-(x+\omega)\hat{T}(x+\omega)}{\hat{T}(x+\omega)} + \frac{\hat{T}'(x+\omega)}{\hat{T}(x+\omega)}\right)u(x+\omega) + b(x+\omega)\frac{\hat{T}(x+\omega)-(x+\omega)\hat{T}'(x+\omega)}{\hat{T}(x+\omega)} \\
&= \left(A(x)\frac{\hat{T}(x)-x\hat{T}(x)}{\hat{T}(x)} + \frac{\hat{T}'(x)}{\hat{T}(x)}\right)u(x+\omega) + b(x)\frac{\hat{T}(x)-x\hat{T}'(x)}{\hat{T}(x)} \\
&= \left(A(x)\frac{\hat{T}(x)-x\hat{T}(x)}{\hat{T}(x)} + \frac{\hat{T}'(x)}{\hat{T}(x)}\right)v(x) + b(x)\frac{\hat{T}(x)-x\hat{T}'(x)}{\hat{T}(x)},
\end{aligned}
$$

i.e., $v(x)$ is a solution of the iso-differential system (2). However, since

$$v(0) = u(\omega) = u(0),$$

the uniqueness of the initial value problems implies that

$$v(x) = u(x)$$

and hence $u(x)$ is periodic of period ω. □

Corollary 11.0.290. *Let the matrix $A(x)$ satisfies the conditions (3). Further, let $\Psi(x)$ be a fundamental matrix of the iso-differential system*

$$u'(x) = \left(A(x)\frac{\hat{T}(x)-x\hat{T}'(x)}{\hat{T}(x)} + \frac{\hat{T}'(x)}{\hat{T}(x)}\right)u(x). \tag{6}$$

Then the iso-differential system (6) *has a nontrivial periodic solution* $u(x)$ *of period* ω *if and only if*

$$\det\Big(\Psi(0)-\Psi(\omega)\Big)=0.$$

Proof. We have that the general solution of the iso-differential system (6) is

$$u(x)=\Psi(x)c,$$

where c is an arbitrary constant vector.

This $u(x)$ is periodic if and only if

$$u(x+\omega)=u(x) \qquad \Longleftrightarrow$$

$$\Psi(x+\omega)c=\Psi(x)c \qquad \text{for} \qquad \forall x\in\mathbb{R},$$

i.e., the system

$$\Big(\Psi(0)-\Psi(\omega)\Big)c=0$$

has a nontrivial solution vector c. Consequently

$$\det\Big(\Psi(0)-\Psi(\omega)\Big)=0.$$

□

Corollary 11.0.291. *Let the matrix* $A(x)$ *and the vector* $b(x)$ *satisfy the conditions* (3) *and* (4), *respectively. Then the iso-differential system* (2) *has a unique periodic solution of period* ω *if and only if the iso-differential system* (6) *does not have a periodic solution of period* ω *other than the trivial one.*

Proof. Let $\Psi(x)$ be a fundamental matrix of the iso-differential system (6). Then the general solution of the iso-differential system (2) can be written in the following way

$$u(x)=\Psi(x)c+\int_0^x\Psi(x)\Psi^{-1}(t)b(t)\frac{\hat{T}(t)-t\hat{T}'(t)}{\hat{T}(t)}dt,$$

where c is an arbitrary constant.

This $u(x)$ is a periodic solution of period ω if and only if

$$u(0)=u(\omega)$$

or

$$\Psi(0)c=\Psi(\omega)c+\int_0^\omega\Psi(x)\Psi^{-1}(t)b(t)\frac{\hat{T}(t)-t\hat{T}'(t)}{\hat{T}(t)}dt,$$

i.e., the system

$$\Big(\Psi(0)-\Psi(\omega)\Big)c=\int_0^\omega\Psi(x)\Psi^{-1}(t)b(t)\frac{\hat{T}(t)-t\hat{T}'(t)}{\hat{T}(t)}dt$$

has a unique solution vector c, which is possible if and only if

$$\det\Big(\Psi(0)-\Psi(\omega)\Big)\neq 0.$$

Now the conclusion follows from the last corollary. □

Theorem 11.0.292. *(iso-Floquet's theorem) Let the matrix $A(x)$ satisfies the conditions (3). Further, let $\Psi(x)$ be a fundamental matrix of the iso-differential system (6). Then the following hold.*

(i) *The matrix*

$$\chi(x) = \Psi(x+\omega)$$

is also a fundamental matrix of the iso-differential system (6).

(ii) *There exists a periodic nonsingular matrix $P(x)$ of period ω and a constant matrix R such that*

$$\Psi(x) = P(x)e^{Rx}.$$

Proof. **(i)** Since $\Psi(x)$ is a fundamental matrix of the iso-differential system (6) then

$$\begin{aligned}
\chi'(x) &= \Psi'(x+\omega) \\
&= \left(A(x+\omega)\frac{\hat{T}(x+\omega)-(x+\omega)\hat{T}'(x+\omega)}{\hat{T}(x+\omega)} + \frac{\hat{T}'(x+\omega)}{\hat{T}(x+\omega)}\right)\Psi(x+\omega) \\
&= \left(A(x)\frac{\hat{T}(x)-x\hat{T}'(x)}{\hat{T}(x)} + \frac{\hat{T}'(x)}{\hat{T}(x)}\right)\Psi(x+\omega) \\
&= \left(A(x)\frac{\hat{T}(x)-x\hat{T}'(x)}{\hat{T}(x)} + \frac{\hat{T}'(x)}{\hat{T}(x)}\right)\chi(x),
\end{aligned}$$

i.e., $\chi(x)$ is a solution matrix of the iso-differential system (6).

Further, since

$$\det\Psi(x+\omega) \neq 0$$

for all x, we have that

$$\det\chi(x) \neq 0$$

for all x.

Hence, we conclude that $\chi(x)$ is a fundamental matrix of the iso-differential system (6).

(ii) Since $\Psi(x)$ and $\Psi(x+\omega)$ are both fundamental matrices of the iso-differential system (6) then there exists a nonsingular constant matrix C such that

$$\Psi(x+\omega) = \Psi(x)C.$$

Also, there exists a constant matrix R such that

$$C = e^{R\omega}.$$

Consequently

$$\Psi(x+\omega) = \Psi(x)e^{R\omega}.$$

Let

$$P(x) = \Psi(x)e^{-Rx}.$$

Then

$$P(x+\omega) = \Psi(x+\omega)e^{-R(x+\omega)}$$

$$= \Psi(x+\omega)e^{-R\omega}e^{-Rx}$$

$$= \Psi(x)e^{-Rx}$$

$$= P(x).$$

Hence, $P(x)$ is periodic of period ω .

Since $\Psi(x)$ and e^{-Rx} are nonsingular, then

$$\det P(x) \neq 0 \qquad \text{in} \qquad \mathbb{R}.$$

□

Theorem 11.0.293. *Let $P(x)$ and R be the matrices obtained in the iso-Floquet's theorem. Then the transformation*

$$u(x) = P(x)v(x)$$

reduces the iso-differential system (6) *to the system*

$$v'(x) = Rv(x).$$

Proof. We have

$$\Psi'(x) = \Big(A(x)\frac{\hat{T}(x) - x\hat{T}'(x)}{\hat{T}(x)} + \frac{\hat{T}'(x)}{\hat{T}(x)}\Big)\Psi(x),$$

from where

$$\Big(P(x)e^{Rx}\Big)' = \Big(A(x)\frac{\hat{T}(x) - x\hat{T}'(x)}{\hat{T}(x)} + \frac{\hat{T}'(x)}{\hat{T}(x)}\Big)P(x)e^{Rx},$$

or

$$\Big(P'(x) + P(x)R\Big)e^{Rx} = \Big(A(x)\frac{\hat{T}(x) - x\hat{T}'(x)}{\hat{T}(x)} + \frac{\hat{T}'(x)}{\hat{T}(x)}\Big)P(x)e^{Rx},$$

or

$$P'(x) + P(x)R = \Big(A(x)\frac{\hat{T}(x) - x\hat{T}'(x)}{\hat{T}(x)} + \frac{\hat{T}'(x)}{\hat{T}(x)}\Big)P(x),$$

or

$$P'(x) + P(x)R - \Big(A(x)\frac{\hat{T}(x) - x\hat{T}'(x)}{\hat{T}(x)} + \frac{\hat{T}'(x)}{\hat{T}(x)}\Big)P(x) = 0. \tag{7}$$

Using the transformation

$$u(x) = P(x)v(x)$$

in the iso-differential system (6) we get

$$P'(x)v(x) + P(x)v'(x) = \Big(A(x)\frac{\hat{T}(x) - x\hat{T}'(x)}{\hat{T}(x)} + \frac{\hat{T}'(x)}{\hat{T}(x)}\Big)P(x)v(x)$$

or

$$P(x)v'(x) + \Big(P'(x) - \Big(A(x)\frac{\hat{T}(x) - x\hat{T}'(x)}{\hat{T}(x)} + \frac{\hat{T}'(x)}{\hat{T}(x)}\Big)P(x)\Big)v(x) = 0. \tag{8}$$

We multiply the equation (7) by $v(x)$ and we get

$$P'(x)v(x) + P(x)Rv(x) - \Big(A(x)\frac{\hat{T}(x) - x\hat{T}'(x)}{\hat{T}(x)} + \frac{\hat{T}'(x)}{\hat{T}(x)}\Big)P(x)v(x) = 0.$$

Now we compare the last equality with (8) and we find

$$v'(x) = Rv(x).$$

□

Let now the matrix $A(x)$ satisfies the conditions (3), let $\Psi(x)$ be a fundamental matrix of the iso-differential system (6). Then there exists a nonsingular matrix M such that

$$\Psi(x) = \Psi_1(X)M. \tag{9}$$

Since

$$\Psi(x+\omega) = \Psi(x)e^{R\omega},$$

then

$$\Psi(x+\omega) = \Psi_1(x)Me^{R\omega}. \tag{10}$$

Also, using (9),

$$\Psi(x+\omega) = \Psi_1(x+\omega)M.$$

From the last equality and (10) it follows that

$$\Psi_1(x+\omega)M = \Psi_1(x)Me^{R\omega}$$

or

$$\Psi_1(x+\omega) = \Psi_1(x)Me^{R\omega}M^{-1}. \tag{11}$$

Hence, we conclude that every fundamental matrix $\Psi_1(x)$ of the iso-differential system (3) determines a matrix

$$Me^{R\omega}M^{-1},$$

which is a similar to the matrix $e^{R\omega}$.

Conversely, if M is any constant nonsingular matrix, then there exists a fundamental matrix $\Psi_1(x)$ of the iso-differential system (6) such that (11) holds.

Definition 11.0.294. *Let $A(x)$ satisfies the conditions (3), $\Psi(x)$ be a fundamental matrix of the iso-differential system (6). Then the nonsingular constant matrix C such that*

$$\Psi(x+\omega) = \Psi(x)C$$

will be called iso-monodromy matrix of the iso-differential system (6).

The eigenvalues of the matrix C are called iso-multipliers of (6).

The eigenvalues of the matrix R are called iso-exponents of (6).

Let

$$\sigma_1, \sigma_2, \ldots, \sigma_n$$

and

$$\lambda_1, \lambda_2, \ldots, \lambda_n,$$

respectively, be the iso-multipliers and iso-exponents of the iso-differential system (6), then we have the relation

$$\sigma_i = e^{\lambda_i \omega}, \qquad 1 \leq i \leq n.$$

Also, since the matrix C is nonsingular, none of the iso-multipliers

$$\sigma_1, \sigma_2, \ldots, \sigma_n$$

of the iso-differential system (6) is zero.

From the relation

$$\Psi(x+\omega) = \Psi(x) e^{R\omega}$$

we get

$$\Psi(\omega) = \Psi(0) e^{R\omega}$$

and hence we conclude that

$$\sigma_1, \sigma_2, \ldots, \sigma_n$$

are the eigenvalues of the matrix

$$\Psi^{-1}(0)\Psi(\omega)$$

or the matrix $\Phi(\omega, 0)$ if

$$\Psi(x) = \Phi(x, 0),$$

i.e., $\Psi(x)$ is the principal fundamental matrix of the iso-differential system (6).

We have, using the iso-Abel's formula,

$$\det\Phi(\omega, 0) = \prod_{i=1}^{n} \sigma_i$$

$$= \det\Phi(0,0) e^{\int_0^{\omega} \mathrm{Tr}\left(A(t) \frac{\hat{T}(t) - t\hat{T}'(t)}{\hat{T}(t)} + \frac{\hat{T}'(t)}{\hat{T}(t)}\right) dt}$$

$$= e^{\sum_{i=1}^{n} \lambda_i \omega}.$$

Theorem 11.0.295. *Let* $A(x)$ *satisfies the conditions* (3). *Then a complex number* λ *is an exponent of the iso-differential system* (6) *if and only if there exists a nontrivial solution of* (6) *of the form* $e^{\lambda x} p(x)$, *where* $p(x+\omega) = p(x)$. *In particular, there exists a periodic solution of* (6) *of period* $\omega(2\omega)$ *if and only if there is a multiplier* $1(-1)$ *of* (6).

Proof. Let

$$u(x) = e^{\lambda x} p(x), \qquad p(x+\omega) = p(x),$$

is a nontrivial solution of the iso-differential system (6) for which

$$u(0) = u^0.$$

Then

$$u(x) = \Phi(x,0)u^0 = e^{\lambda x}p(x),$$

where $\Phi(x,0)$ is the principal fundamental matrix of (6).

Also, we have

$$u(x) = \Phi(x,0)u^0 = P(x)e^{Rx}u^0,$$

where $P(x)$ is a periodic matrix of period ω. Therefore

$$\begin{aligned} e^{\lambda(x+\omega)}p(x) &= e^{\lambda\omega}e^{\lambda x}p(x) \\ &= e^{\lambda\omega}P(x)e^{Rx}u^0 \\ &= P(x)e^{R(x+\omega)}u^0, \end{aligned}$$

which is the same as

$$e^{\lambda x}e^{\lambda\omega}p(x) = P(x)e^{R(x+\omega)}u^0$$

or

$$e^{\lambda\omega}P(x)e^{Rx}u^0 = P(x)e^{R(x+\omega)}u^0$$

or

$$P(x)e^{Rx}\left(e^{\lambda\omega}I - e^{R\omega}\right)u^0 = 0$$

and hence

$$\det\left(e^{\lambda\omega}I - e^{R\omega}\right) = 0,$$

i.e., λ is an exponent of the iso-differential system (6). Then we have

$$e^{Rx}u^0 = e^{\lambda x}u^0$$

for all x, and hence

$$P(x)e^{Rx}u^0 = P(x)u^0e^{\lambda x}.$$

We note that

$$u(x) = P(x)e^{Rx}u^0$$

is the solution of (6).

The multiplier of (6) is 1 provided $\lambda = 0$, and therefore the solution $e^{\lambda x}p(x)$ reduces to $p(x)$, which is a periodic of period ω.

The multiplier of (6) is -1 provided $\lambda = \frac{\pi i}{\omega}$, therefore the solution $e^{\lambda x}p(x)$ reduces to

$$e^{\frac{\pi i x}{\omega}}p(x),$$

which is periodic of period 2ω. □

Chapter 12

Asymptotic Behaviour of Solutions of Linear Systems

In this chapter we will suppose that $x_0 \in \mathbb{R}$,

$$\hat{T} \in C^1([x_0, +\infty)), \qquad \hat{T}(x) > 0 \qquad \text{for} \qquad \forall x \in [x_0, +\infty).$$

We consider the iso-differential system

$$v'(x) = \Big(A(x)\frac{\hat{T}(x) - x\hat{T}'(x)}{\hat{T}(x)} + B(x)\frac{\hat{T}(x) - x\hat{T}'(x)}{\hat{T}(x)} + \frac{\hat{T}'(x)}{\hat{T}(x)}\Big)v(x), \tag{1}$$

where $A(x)$ and $B(x)$ are $n \times n$ matrices with continuous elements $a_{ij}(x)$, $b_{ij}(x)$, $1 \leq i, j \leq n$, respectively, in the interval $[x_0, +\infty)$.

Theorem 12.0.296. *Let all solutions of the iso-differential system*

$$v'(x) = \Big(A(x)\frac{\hat{T}(x) - x\hat{T}'(x)}{\hat{T}(x)} + \frac{\hat{T}'(x)}{\hat{T}(x)}\Big)v(x) \tag{2}$$

be bounded in the interval $[x_0, +\infty)$ *and*

$$\int_{x_0}^{\infty} \Big|\Big|B(t)\frac{\hat{T}(t) - t\hat{T}'(t)}{\hat{T}(t)}\Big|\Big| dt < \infty. \tag{3}$$

Then all solutions of the iso-differential system (1) *are bounded in the interval* $[x_0, +\infty)$ *provided*

$$\liminf_{x \longrightarrow \infty} \int_{x_0}^{x} \mathrm{Tr}\Big(A(t)\frac{\hat{T}(t) - t\hat{T}'(t)}{\hat{T}(t)} + \frac{\hat{T}'(t)}{\hat{T}(t)}\Big) dt > -\infty \tag{4}$$

or

$$\mathrm{Tr}\Big(A(x)\frac{\hat{T}(x) - x\hat{T}'(x)}{\hat{T}(x)} + \frac{\hat{T}'(x)}{\hat{T}(x)}\Big) = 0 \tag{5}$$

for every $x \in [x_0, +\infty)$.

Proof. Let $\Psi(x)$ be a fundamental matrix of the iso-differential system (2). Since all solutions of the iso differential system (2) are bounded, we have

$$||\Psi(x)|| < \infty.$$

From the iso-Abel's formula we have

$$\det\Psi(x) = \det\Psi(x_0)e^{\int_{x_0}^{x} \mathrm{Tr}\left(A(t)\frac{\hat{T}(t)-t\hat{T}'(t)}{\hat{T}(t)}+\frac{\hat{T}'(t)}{\hat{T}(t)}\right)dt}$$

and hence

$$\Psi^{-1}(x) = \frac{\mathrm{adj}\Psi(x)}{\det\Psi(x)}$$

$$= \frac{\mathrm{adj}\Psi(x)}{\det\Psi(x_0)e^{\int_{x_0}^{x} \mathrm{Tr}\left(A(t)\frac{\hat{T}(t)-t\hat{T}'(t)}{\hat{T}(t)}+\frac{\hat{T}'(x)}{\hat{T}(x)}\right)dt}}.$$

From here and (4), (5) we get

$$||\Psi^{-1}(x)|| < \infty.$$

For the solution $v(x)$ of the iso-differential system (1) we have that it satisfies the integral equation

$$v(x) = \Psi(x)\Psi^{-1}(x_0)v^0 + \int_{x_0}^{x} \Psi(x)\Psi^{-1}(t)B(t)\frac{\hat{T}(t)-t\hat{T}'(t)}{\hat{T}(t)}v(t)dt \tag{6}$$

provided $v(x_0) = v^0$.

Let

$$c = \max\left\{\sup_{x\geq x_0}||\Psi(x)||, \quad \sup_{x\geq x_0}||\Psi^{-1}(x)||\right\},$$

$$c_0 = c||\Psi^{-1}(x_0)v^0||.$$

Then, using (6), we have

$$||v(x)|| \leq ||\Psi(x)||||\Psi^{-1}(x_0)v^0||$$

$$+\int_{x_0}^{x} ||\Psi(x)||||\Psi^{-1}(t)||\left|\left|B(t)\frac{\hat{T}(t)-t\hat{T}'(t)}{\hat{T}(t)}\right|\right|||v(t)||dt$$

$$\leq c||\Psi^{-1}(x_0)v^0|| + c^2\int_{x_0}^{x}\left|\left|B(t)\frac{\hat{T}(t)-t\hat{T}'(t)}{\hat{T}(t)}\right|\right|||v(t)||dt.$$

$$= c_0 + c^2\int_{x_0}^{x}\left|\left|B(t)\frac{\hat{T}(t)-t\hat{T}'(t)}{\hat{T}(t)}\right|\right|||v(t)||dt.$$

The last inequality immediately implies that

$$||v(x)|| \leq c_0 e^{c^2\int_{x_0}^{x}\left|\left|B(t)\frac{\hat{T}(t)-t\hat{T}'(t)}{\hat{T}(t)}\right|\right|dt}.$$

From here and (3) it follows that all solutions of the iso-differential system (1) are bounded in the interval $[x_0, +\infty)$. □

Theorem 12.0.297. *Let all solutions of the iso-differential system*

$$v'(x) = \Big(A(x)\frac{\hat{T}(x) - x\hat{T}'(x)}{\hat{T}(x)}\Big)v(x) \tag{7}$$

be bounded in the interval $[x_0, +\infty)$ *and*

$$\int_{x_0}^{\infty} \Big|\Big|B(t)\frac{\hat{T}(t) - t\hat{T}'(t)}{\hat{T}(t)} + \frac{\hat{T}'(t)}{\hat{T}(t)}\Big|\Big| dt < \infty. \tag{8}$$

Then all solutions of the iso-differential system (1) *are bounded in the interval* $[x_0, +\infty)$ *provided*

$$\liminf_{x \longrightarrow \infty} \int_{x_0}^{x} \mathrm{Tr}\Big(A(t)\frac{\hat{T}(t) - t\hat{T}'(t)}{\hat{T}(t)}\Big) dt > -\infty \tag{9}$$

or

$$\mathrm{Tr}\Big(A(x)\frac{\hat{T}(x) - x\hat{T}'(x)}{\hat{T}(x)}\Big) = 0 \tag{10}$$

for every $x \in [x_0, +\infty)$.

Proof. Let $\Psi_1(x)$ be a fundamental matrix of the iso-differential system (7). Since all solutions of the iso-differential system (7) are bounded, we have

$$||\Psi_1(x)|| < \infty.$$

From the iso-Abel's formula we have

$$\det\Psi_1(x) = \det\Psi_1(x_0)e^{\int_{x_0}^{x} \mathrm{Tr}\left(A(t)\frac{\hat{T}(t) - t\hat{T}'(t)}{\hat{T}(t)}\right) dt}$$

and hence

$$\Psi_1^{-1}(x) = \frac{\mathrm{adj}\Psi_1(x)}{\det\Psi_1(x)}$$

$$= \frac{\mathrm{adj}\Psi_1(x)}{\det\Psi_1(x_0)e^{\int_{x_0}^{x} \mathrm{Tr}\left(A(t)\frac{\hat{T}(t) - t\hat{T}'(t)}{\hat{T}(t)}\right) dt}}.$$

From here and (9), (10) we get

$$||\Psi_1^{-1}(x)|| < \infty.$$

For the solution $v(x)$ of the iso-differential system (1) we have that it satisfies the integral equation

$$v(x) = \Psi_1(x)\Psi_1^{-1}(x_0)v^0 + \int_{x_0}^{x} \Psi_1(x)\Psi_1^{-1}(t)\Big(B(t)\frac{\hat{T}(t) - t\hat{T}'(t)}{\hat{T}(t)} + \frac{\hat{T}'(t)}{\hat{T}(t)}\Big)v(t)dt \tag{11}$$

provided $v(x_0) = v^0$.

Let

$$c_1 = \max\Big\{\sup_{x \geq x_0} ||\Psi_1(x)||, \quad \sup_{x \geq x_0} ||\Psi_1^{-1}(x)||\Big\},$$

$$c_2 = c_1||\Psi_1^{-1}(x_0)v^0||.$$

Then, using (11), we have

$$||v(x)|| \le ||\Psi_1(x)||||\Psi_1^{-1}(x_0)v^0||$$

$$+\int_{x_0}^x ||\Psi_1(x)||||\Psi_1^{-1}(t)||\left|\left|B(t)\frac{\hat{T}(t)-t\hat{T}'(t)}{\hat{T}(t)}+\frac{\hat{T}'(t)}{\hat{T}(t)}\right|\right|||v(t)||dt$$

$$\le c_1||\Psi_1^{-1}(x_0)v^0||+c_1^2\int_{x_0}^x\left|\left|B(t)\frac{\hat{T}(t)-t\hat{T}'(t)}{\hat{T}(t)}+\frac{\hat{T}'(t)}{\hat{T}(t)}\right|\right|||v(t)||dt$$

$$=c_2+c_1^2\int_{x_0}^x\left|\left|B(t)\frac{\hat{T}(t)-t\hat{T}'(t)}{\hat{T}(t)}+\frac{\hat{T}'(t)}{\hat{T}(t)}\right|\right|||v(t)||dt.$$

The last inequality immediately implies that

$$||v(x)|| \le c_2 e^{c_1^2\int_{x_0}^x\left|\left|B(t)\frac{\hat{T}(t)-t\hat{T}'(t)}{\hat{T}(t)}+\frac{\hat{T}'(t)}{\hat{T}(t)}\right|\right|dt}.$$

From here and (8) it follows that all solutions of the iso-differential system (1) are bounded in the interval $[x_0,+\infty)$. □

Theorem 12.0.298. *Let the fundamental matrix $\Psi(x)$ of the iso-differential system (2) be such that*

$$||\Psi(x)\Psi^{-1}(x)|| \le c$$

for all $x_0 \le t \le x < \infty$, where c is a positive constant. Let also, the condition (3) holds. Then all solutions of the iso-differential system (1) are bounded in the interval $[x_0,+\infty)$. Moreover, if all solutions of the system (2) tend to zero as $x \longrightarrow \infty$, then all solutions of the iso-differential system (1) tend to zero as $x \longrightarrow \infty$.

Proof. We have that the solutions $v(x)$ of the iso-differential system (1) for which $v(x_0) = v^0$ satisfy the integral equation (6). Therefore

$$||v(x)|| \le c||v^0||+c\int_{x_0}^x\left|\left|B(t)\frac{\hat{T}(t)-t\hat{T}'(t)}{\hat{T}(t)}\right|\right|||v(t)||dt$$

and hence

$$||v(x)|| \le c||v^0||e^{c\int_{x_0}^\infty\left|\left|B(t)\frac{\hat{T}(t)-t\hat{T}'(t)}{\hat{T}(t)}\right|\right|dt} := M < \infty.$$

Thus each solution of the iso-differential system (1) is bounded in the interval $[x_0,\infty)$.

Now, from the integral equation (6) we obtain

$$v(x) = \Psi(x)\Psi^{-1}(x_0)v^0+\int_{x_0}^{x_1}\Psi(x)\Psi^{-1}(t)B(t)\frac{\hat{T}(t)-t\hat{T}'(t)}{\hat{T}(t)}v(t)dt$$

$$+\int_{x_1}^x\Psi(x)\Psi^{-1}(t)B(t)\frac{\hat{T}(t)-t\hat{T}'(t)}{\hat{T}(t)}v(t)dt,$$

therefore it follows that

$$||v(x)|| \le ||\Psi(x)||||\Psi^{-1}(x_0)||||v^0||$$

$$+||\Psi(x)||\int_{x_0}^{x_1}||\Psi^{-1}(t)||\left|\left|B(t)\frac{\hat{T}(t)-t\hat{T}'(t)}{\hat{T}(t)}\right|\right|||v(t)||dt$$

$$+cM\int_{x_1}^\infty\left|\left|B(t)\frac{\hat{T}(t)-t\hat{T}'(t)}{\hat{T}(t)}\right|\right|dt.$$

Let $\varepsilon > 0$ be a given number.

From the condition (3) it follows that the last term in the above inequality can be made less than $\frac{\varepsilon}{2}$ by choosing x_1 sufficiently large.

Also, since all solutions of the iso-differential system (2) tend to zero when $x \longrightarrow \infty$, it is necessary that

$$||\Psi(x)|| \longrightarrow 0$$

as $x \longrightarrow \infty$.

Thus, the sum of the first two terms on the right side can be made arbitrarily small by choosing x large enough, say, less than $\frac{\varepsilon}{2}$. Hence,

$$||v(x)|| < \varepsilon$$

for large x.

This immediately implies that

$$||v(x)|| \longrightarrow 0$$

as $x \longrightarrow \infty$. □

Theorem 12.0.299. *Let the fundamental matrix $\Psi_1(x)$ of the iso-differential system (7) be such that*

$$||\Psi_1(x)\Psi_1^{-1}(x)|| \leq c$$

for all $x_0 \leq t \leq x < \infty$, where c is a positive constant. Let also, the condition (8) holds. Then all solutions of the iso-differential system (1) are bounded in the interval $[x_0, +\infty)$. Moreover, if all solutions of the system (7) tend to zero as $x \longrightarrow \infty$, then all solutions of the iso-differential system (1) tend to zero as $x \longrightarrow \infty$.

Proof. We have that the solutions $v(x)$ of the iso-differential system (1) for which $v(x_0) = v^0$ satisfy the integral equation (11). Therefore

$$||v(x)|| \leq c||v^0|| + c\int_{x_0}^{x} \left|\left|B(t)\frac{\hat{T}(t) - t\hat{T}'(t)}{\hat{T}(t)} + \frac{\hat{T}'(t)}{\hat{T}(t)}\right|\right| ||v(t)||dt$$

and hence

$$||v(x)|| \leq c||v^0||e^{c\int_{x_0}^{\infty}\left|\left|B(t)\frac{\hat{T}(t)-t\hat{T}'(t)}{\hat{T}(t)} + \frac{\hat{T}'(t)}{\hat{T}(t)}\right|\right|dt} := M_1 < \infty.$$

Thus each solution of the iso-differential system (1) is bounded in the interval $[x_0, \infty)$.

Now, from the integral equation (11) we obtain

$$v(x) = \Psi_1(x)\Psi_1^{-1}(x_0)v^0 + \int_{x_0}^{x_1} \Psi_1(x)\Psi_1^{-1}(t)\left(B(t)\frac{\hat{T}(t)-t\hat{T}'(t)}{\hat{T}(t)} + \frac{\hat{T}'(t)}{\hat{T}(t)}\right)v(t)dt$$

$$+\int_{x_1}^{x} \Psi_1(x)\Psi_1^{-1}(t)\left(B(t)\frac{\hat{T}(t)-t\hat{T}'(t)}{\hat{T}(t)} + \frac{\hat{T}'(t)}{\hat{T}(t)}\right)v(t)dt,$$

therefore it follows that

$$||v(x)|| \leq ||\Psi_1(x)||||\Psi_1^{-1}(x_0)||||v^0||$$

$$+||\Psi_1(x)||\int_{x_0}^{x_1}||\Psi_1^{-1}(t)||\left|\left|B(t)\frac{\hat{T}(t)-t\hat{T}'(t)}{\hat{T}(t)}+\frac{\hat{T}'(t)}{\hat{T}(t)}\right|\right|||v(t)||dt$$

$$+cM_1\int_{x_1}^{\infty}\left|\left|B(t)\frac{\hat{T}(t)-t\hat{T}'(t)}{\hat{T}(t)}+\frac{\hat{T}'(t)}{\hat{T}(t)}\right|\right|dt.$$

Let $\varepsilon > 0$ be a given number.

From the condition (8) it follows that the last term in the above inequality can be made less than $\frac{\varepsilon}{2}$ by choosing x_1 sufficiently large.

Also, since all solutions of the iso-differential system (7) tend to zero when $x \longrightarrow \infty$, it is necessary that

$$||\Psi_1(x)|| \longrightarrow 0$$

as $x \longrightarrow \infty$.

Thus, the sum of the first two terms on the right side can be made arbitrarily small by choosing x large enough, say, less than $\frac{\varepsilon}{2}$. Hence,

$$||v(x)|| < \varepsilon$$

for large x.

This immediately implies that

$$||v(x)|| \longrightarrow 0$$

as $x \longrightarrow \infty$. □

Theorem 12.0.300. *Let $\hat{T}(x)$ and $A(x)\frac{\hat{T}(x)-x\hat{T}'(x)}{\hat{T}(x)}$ be periodic of period ω in the interval $[x_0,+\infty)$. Let also, the condition (3) holds. Then the following hold.*

(i) *All solutions of the iso-differential system (1) are bounded in the interval $[x_0,+\infty)$ provided all solutions of (2) are bounded in $[x_0,+\infty)$.*

(ii) *All solutions of the iso-differential system (1) tend to zero as $x \longrightarrow \infty$ provided all solutions of (2) tend to zero as $x \longrightarrow \infty$.*

Proof. Let $\Psi(x)$ is a fundamental matrix of the iso-differential system (2). From the iso-Floquet's theorem it follows that

$$\Psi(x) = P(x)e^{Rx},$$

where $P(x)$ is a nonsingular periodic matrix of period ω and R is a constant matrix. Then from (6) it follows that

$$v(x) = P(x)e^{R(x-x_0)}P^{-1}(x_0)v^0 + \int_{x_0}^{x} P(x)e^{Rx}e^{-Rt}P^{-1}(t)B(t)\frac{\hat{T}(t)-t\hat{T}'(t)}{\hat{T}(t)}v(t)dt.$$

Hence, it follows

$$||v(x)|| \leq ||P(x)|| ||e^{Rx}|| ||e^{-Rx_0}P^{-1}(x_0)v^0|| \\ + \int_{x_0}^{x} ||P(x)|| ||e^{R(x-t)}|| ||P^{-1}(t)|| \left|\left| B(t)\frac{\hat{T}(t)-t\hat{T}'(t)}{\hat{T}(t)} \right|\right| ||v(t)||dt \tag{12}$$

Because $P(x)$ is nonsingular and periodic, $\det P(x)$ is periodic and is not vanish, i.e., it is bounded away from zero in the interval $[x_0, +\infty)$. Hence, $P(x)$ and its inverse

$$P^{-1}(x) = \frac{\operatorname{adj}P(x)}{\det P(x)}$$

are bounded in the interval $[x_0, +\infty)$.

Let

$$c_3 = \max\{\sup_{x\geq x_0} ||P(x)||, \sup_{x\geq x_0} ||P^{-1}(x)||\},$$

$$c_4 = c_3 ||e^{-Rx_0}P^{-1}(x_0)v^0||.$$

From (12) we get

$$||v(x)|| \leq c_4 ||e^{Rx}|| + c_3^2 \int_{x_0}^{x} ||e^{R(x-t)}|| \left|\left| B(t)\frac{\hat{T}(t)-t\hat{T}'(t)}{\hat{T}(t)} \right|\right| ||v(t)||dt. \tag{13}$$

Since all solutions of the iso-differential system (2) are bounded, then it is necessary that

$$||e^{Rx}|| \leq c_5$$

for all $x \geq x_0$, and from (13) it follows that

$$||v(x)|| \leq c_4c_5 + c_3^2c_5 \int_{x_0}^{x} \left|\left| B(t)\frac{\hat{T}(t)-t\hat{T}'(t)}{\hat{T}(t)} \right|\right| dt,$$

which immediately gives that

$$||v(x)|| \leq c_4c_5 e^{c_3^2 c_5 \int_{x_0}^{x} \left|\left| B(t)\frac{\hat{T}(t)-t\hat{T}'(t)}{\hat{T}(t)} \right|\right| dt}.$$

On the other hand, if all solutions of the iso-differential system (2) tend to zero as $x \longrightarrow \infty$, then there exist constants c_6 and α such that

$$||e^{Rx}|| \leq c_6 e^{-\alpha(x-x_0)}$$

for all $x \geq x_0$.

From here and (13) it follows

$$||v(x)|| \leq c_4c_6e^{-\alpha x} + c_3^2c_6 \int_{x_0}^{x} e^{-\alpha(x-t)} \left|\left| B(t)\frac{\hat{T}(t)-t\hat{T}'(t)}{\hat{T}(t)} \right|\right| ||v(t)||dt,$$

which easily gives that

$$||v(x)|| \leq c_4c_6 e^{c_3^2 c_6 \int_{x_0}^{x} \left|\left| B(t)\frac{\hat{T}(t)-t\hat{T}'(t)}{\hat{T}(t)} \right|\right| dt - \alpha x}.$$

From the last inequality and (3) it follows that

$$v(x) \longrightarrow 0$$

as $x \longrightarrow \infty$. □

Theorem 12.0.301. *Let $\hat{T}(x)$ and $A(x)\frac{\hat{T}(x)-x\hat{T}'(x)}{\hat{T}(x)}$ be periodic of period ω in the interval $[x_0, +\infty)$. Let also, the condition (8) holds. Then the following hold.*

(i) *All solutions of the iso-differential system (1) are bounded in the interval $[x_0, +\infty)$ provided all solutions of (7) are bounded in $[x_0, +\infty)$.*

(ii) *All solutions of the iso-differential system (1) tend to zero as $x \longrightarrow \infty$ provided all solutions of (7) tend to zero as $x \longrightarrow \infty$.*

Proof. Let $\Psi_1(x)$ is a fundamental matrix of the iso-differential system (7). From the iso-Floquet's theorem it follows that

$$\Psi_1(x) = P_1(x)e^{R_1 x},$$

where $P_1(x)$ is a nonsingular periodic matrix of period ω and R_1 is a constant matrix. Then from (11) it follows that

$$v(x) = P_1(x)e^{R_1(x-x_0)}P_1^{-1}(x_0)v^0 + \int_{x_0}^x P_1(x)e^{R_1 x}e^{-R_1 t}P_1^{-1}(t)\left(B(t)\frac{\hat{T}(t)-t\hat{T}'(t)}{\hat{T}(t)} + \frac{\hat{T}'(t)}{\hat{T}(t)}\right)v(t)dt.$$

Hence, it follows

$$\begin{aligned} ||v(x)|| &\leq ||P_1(x)||||e^{R_1 x}||||e^{-R_1 x_0}P_1^{-1}(x_0)v^0|| \\ &+ \int_{x_0}^x ||P_1(x)||||e^{R_1(x-t)}||||P_1^{-1}(t)||\left|\left|B(t)\frac{\hat{T}(t)-t\hat{T}'(t)}{\hat{T}(t)} + \frac{\hat{T}'(t)}{\hat{T}(t)}\right|\right|||v(t)||dt \end{aligned} \tag{14}$$

Because $P_1(x)$ is nonsingular and periodic, $\det P_1(x)$ is periodic and does not vanish, i.e., it is bounded away from zero in the interval $[x_0, +\infty)$. Hence, $P_1(x)$ and its inverse

$$P_1^{-1}(x) = \frac{\text{adj}P_1(x)}{\det P_1(x)}$$

are bounded in the interval $[x_0, +\infty)$.

Let

$$c_7 = \max\{\sup_{x \geq x_0}||P_1(x)||, \sup_{x \geq x_0}||P_1^{-1}(x)||\},$$

$$c_8 = c_7||e^{-R_1 x_0}P_1^{-1}(x_0)v^0||.$$

From (14) we get

$$||v(x)|| \leq c_8||e^{R_1 x}|| + c_7^2\int_{x_0}^x ||e^{R_1(x-t)}||\left|\left|B(t)\frac{\hat{T}(t)-t\hat{T}'(t)}{\hat{T}(t)} + \frac{\hat{T}'(t)}{\hat{T}(t)}\right|\right|||v(t)||dt. \tag{15}$$

Since all solutions of the iso-differential system (7) are bounded, then it is necessary that

$$||e^{R_1 x}|| \leq c_9$$

for all $x \geq x_0$, and from (15) it follows that

$$||v(x)|| \leq c_8c_9 + c_7^2c_9 \int_{x_0}^{x} \left|\left|B(t)\frac{\hat{T}(t) - t\hat{T}'(t)}{\hat{T}(t)} + \frac{\hat{T}'(t)}{\hat{T}(t)}\right|\right| dt,$$

which immediately gives that

$$||v(x)|| \leq c_8c_9 e^{c_7^2c_9 \int_{x_0}^{x} \left|\left|B(t)\frac{\hat{T}(t) - t\hat{T}'(t)}{\hat{T}(t)} + \frac{\hat{T}'(t)}{\hat{T}(t)}\right|\right| dt}.$$

On the other hand, if all solutions of the iso-differential system (7) tend to zero as $x \longrightarrow \infty$, then there exist constants c_{10} and α_1 such that

$$||e^{R_1x}|| \leq c_{10}e^{-\alpha_1(x-x_0)}$$

for all $x \geq x_0$.

From here and (15) it follows

$$||v(x)|| \leq c_8c_{10}e^{-\alpha_1x} + c_7^2c_{10} \int_{x_0}^{x} e^{-\alpha_1(x-t)} \left|\left|B(t)\frac{\hat{T}(t) - t\hat{T}'(t)}{\hat{T}(t)} + \frac{\hat{T}'(t)}{\hat{T}(t)}\right|\right| ||v(t)|| dt,$$

which easily gives that

$$||v(x)|| \leq c_8c_{10} e^{c_7^2c_{10} \int_{x_0}^{x} \left|\left|B(t)\frac{\hat{T}(t) - t\hat{T}'(t)}{\hat{T}(t)} + \frac{\hat{T}'(t)}{\hat{T}(t)}\right|\right| dt - \alpha_1 x}.$$

From the last inequality and (8) it follows that

$$v(x) \longrightarrow 0$$

as $x \longrightarrow \infty$. □

Now we consider the iso-differential system

$$v'(x) = \left(A(x)\frac{\hat{T}(x) - x\hat{T}'(x)}{\hat{T}(x)} + \frac{\hat{T}'(x)}{\hat{T}(x)}\right) + b(x)\frac{\hat{T}(x) - x\hat{T}'(x)}{\hat{T}(x)}, \tag{16}$$

where $b(x)$ is a $n \times 1$ vector whose components are continuous in the interval $[x_0, +\infty)$.

Theorem 12.0.302. *Suppose every solution of the iso-differential system (2) is bounded in the interval $[x_0, +\infty)$. Then every solution of the iso-differential system (16) is bounded provided at least one of its solutions is bounded.*

Proof. Let $u^1(x)$ and $u^2(x)$ be two solutions of the iso-differential system (16). Then

$$\phi(x) = u^1(x) - u^2(x)$$

is a solution of the iso-differential system (2) in the interval $[x_0, \infty)$.

Hence,

$$u^1(x) = \phi(x) + u^2(x), \qquad x \in [x_0, +\infty).$$

Now since $\phi(x)$ is bounded in the interval $[x_0, \infty)$, if $u^2(x)$ is a bounded solution of the iso-differential system (16), it immediately follows that $u^1(x)$ is also a bounded solution of (16). □

Theorem 12.0.303. *Suppose every solution of the iso-differential system* (2) *is bounded in the interval* $[x_0,\infty)$*, and the condition* (4) *or* (5) *holds. Then every solution of* (16) *is bounded provided*

$$\int_{x_0}^{\infty} ||b(t)||dt < \infty. \tag{17}$$

Proof. Let $\Psi(x)$ be a fundamental matrix of the iso-differential system (2). Since every solution of (2) is bounded in the interval $[x_0,\infty)$, we have that

$$||\Psi(x)|| \qquad \text{and} \qquad ||\Psi^{-1}(x)||$$

are bounded in the interval $[x_0,\infty)$. Thus, there exists a finite constant $c>0$ such that

$$c = \max\{\sup_{x\geq x_0} ||\Psi(x)||, \sup_{x\geq x_0} ||\Psi^{-1}(x)||\}.$$

Hence, for any solution $u(x)$ of the iso-differential system (16) such that $u(x_0)=u^0$, using its integral representation, we have that

$$||u(x)|| \leq c||\Psi^{-1}(x_0)u^0|| + c^2 \int_{x_0}^{x} \left|\left| b(t)\frac{\hat{T}(t)-t\hat{T}'(t)}{\hat{T}(t)}\right|\right| dt.$$

From here and (17) it follows that every solution of the iso-differential system (16) is bounded in the interval $[x_0,+\infty)$. □

Chapter 13

Stability of Solutions

Here we will suppose that $x_0 \in \mathbb{R}$, $u^0 \in \mathbb{R}^n$,

$$\hat{T} \in C^1([x_0,\infty)), \qquad \hat{T}(x) > 0 \qquad \text{for} \qquad \forall x \in [x_0,\infty).$$

We will investigate the following initial value problem for iso-differential systems

$$\left(\hat{u}^\wedge(\hat{x})\right)^{\circledast} = \hat{g}^\wedge(\hat{x},\hat{u}^\wedge(\hat{x})), \tag{1'}$$

$$u(x_0) = u^0, \tag{2}$$

where

$$u(x) = (u_1(x), u_2(x), \dots, u_n(x)),$$

$$u^0 = (u_1^0, u_2^0, \dots, u_n^0),$$

$$g(x,u) = (g_1(x,u), g_2(x,u), \dots, g_n(x,u)).$$

The iso-differential system $(1')$ we can rewrite in the following form.

$$u'(x) = \frac{\hat{T}'(x)}{\hat{T}(x)}u(x) + g(x,u(x))\frac{\hat{T}(x) - x\hat{T}'(x)}{\hat{T}(x)}. \tag{1}$$

Below we will use the following notation

$$\Delta u^0 = (\Delta u_1^0, \Delta u_2^0, \dots, \Delta u_n^0),$$

where $\Delta u_i^0 \in \mathbb{R}$, $1 \le i \le n$.

The following definitions were introduced by A. M. Lyapunov in 1892.

Definition 13.0.304. *A solution $u(x) = u(x,x_0,u^0)$ of the initial value problem (1), (2), existing in the interval $[x_0,\infty)$, is said to be stable or Lyapunov stable, or stable in the sense of Lyapunov, if for each $\varepsilon > 0$ there is a $\delta = \delta(\varepsilon, x_0) > 0$ such that the inequality*

$$||\Delta u^0|| < \delta$$

implies

$$||u(x,x_0,u^0+\Delta u^0) - u(x,x_0,u^0)|| < \varepsilon$$

for every $x \in [x_0,\infty)$.

Definition 13.0.305. *A solution* $u(x) = u(x, x_0, u^0)$ *of the initial value problem* (1), (2) *is said to be unstable if it is not stable.*

Definition 13.0.306. *A solution* $u(x) = u(x, x_0, u^0)$ *of the initial value problem* (1), (2) *is said to be asymptotically stable if it is stable and there exists a* $\delta_0 > 0$ *such that the inequality*

$$||\Delta u^0|| < \delta_0$$

implies that

$$||u(x, x_0, u^0 + \Delta u^0) - u(x, x_0, u^0)|| \longrightarrow 0$$

as $x \longrightarrow \infty$.

We will note that the concepts of stability and boundedness are independent. We will see this in the following example.

Example 13.0.307. *Let* $n = 1$, $\hat{T}(x) = e^x$, $x_0 = 2$, $u^0 \in \mathbb{R}$ *be arbitrary chosen,*

$$g(x, u) = \frac{u - x}{x - 1}.$$

Then

$$\frac{\hat{T}'(x)}{\hat{T}(x)} = 1,$$

$$\frac{\hat{T}(x) - x\hat{T}'(x)}{\hat{T}(x)} = \frac{e^x - xe^x}{e^x}$$

$$= 1 - x.$$

Then the equation (1) *takes the form*

$$u'(x) = u + \frac{u-x}{x-1}(1 - x)$$

$$= u + x - u$$

$$= x.$$

Consequently we consider the initial value problem

$$u'(x) = x \qquad x > 2,$$

$$u(2) = u^0.$$

Its solution is

$$u(x) = u^0 - 2 + \frac{x^2}{2},$$

which is defined and unbounded in $[2, \infty)$.

Let $\varepsilon > 0$ *be arbitrary chosen. Then for every* $\delta \in (0, \varepsilon]$ *the inequality*

$$||\Delta u^0|| < \delta$$

implies

$$||u(x,x_0,u^0+\Delta u^0)-u(x,x_0,u^0)|| = ||\Delta u^0||$$

$$< \delta$$

$$\leq \varepsilon.$$

Therefore this $u(x)$ is stable in the sense of Lyapunov.

However, in the case of homogeneous linear iso-differential system

$$u'(x) = \left(A(x)\frac{\hat{T}(x)-x\hat{T}'(x)}{\hat{T}(x)} + \frac{\hat{T}'(x)}{\hat{T}(x)}\right)u(x) \tag{3}$$

the concepts of stability and boundedness are equivalent. We will see this in the following theorem.

Theorem 13.0.308. *All solutions of the iso-differential system* (3) *are stable if and only if they are bounded.*

Proof. Let $\Psi(x)$ be a fundamental matrix of the iso-differential system (3). If all solutions of (3) are bounded, then there exists a positive constant c such that

$$||\Psi(x)|| \leq c$$

for all $x \geq x_0$.

Let now $\varepsilon > 0$ be arbitrary chosen. We take

$$\delta(\varepsilon) = \frac{\varepsilon}{c||\Psi^{-1}(x_0)||} > 0.$$

Then the inequality

$$||\Delta u^0|| < \delta(\varepsilon)$$

implies

$$||u(x,x_0,u^0+\Delta u^0)-u(x,x_0,u^0)|| = ||\Psi(x)\Psi^{-1}(x_0)(u^0+\Delta u^0)-\Psi(x)\Psi^{-1}(x_0)u^0||$$

$$= ||\Psi(x)\Psi^{-1}(x_0)\Delta u^0||$$

$$\leq ||\Psi(x)||||\Psi^{-1}(x_0)||||\Delta u^0||$$

$$\leq c||\Psi^{-1}(x_0)||||\Delta u^0||$$

$$\leq c\delta(\varepsilon)||\Psi^{-1}(x_0)||$$

$$= \frac{\varepsilon}{c||\Psi^{-1}(x_0)||}c||\Psi^{-1}(x_0)||$$

$$= \varepsilon.$$

Let now all solutions of (3) are stable. Then, in particular, the trivial solution is stable, i.e., $u(x,x_0,0)=0$ is stable. From here, for arbitrary chosen $\varepsilon>0$, there exists a $\delta=\delta(\varepsilon)>0$ such that the inequality

$$||\Delta u^0||<\delta$$

implies that

$$||u(x,x_0,\Delta u^0)||<\varepsilon$$

for all $x\geq x_0$.

Since

$$u(x,x_0,\Delta u^0)=\Psi(x)\Psi^{-1}(x_0)\Delta u^0$$

we find that

$$||\Psi(x)\Psi^{-1}(x_0)\Delta u^0||<\varepsilon.$$

We choose

$$\Delta u^0=\frac{\delta}{2}e^j,$$

where

$$e^j=(0,0,\ldots,0,1,0,\ldots,0).$$

Let $\psi^j(x)$ is the jth column of $\Psi(x)\Psi^{-1}(x_0)$. Then

$$\begin{aligned}||\Psi(x)\Psi^{-1}(x_0)\Delta u^0|| &= ||\psi^j(x)||\tfrac{\delta}{2}\\ &< \varepsilon.\end{aligned}$$

Therefore

$$\begin{aligned}||\Psi(x)\Psi^{-1}(x_0)|| &= \max_{1\leq j\leq n}||\psi^j(x)||\\ &\leq 2\tfrac{\varepsilon}{\delta}.\end{aligned}$$

From here, for any solution $u(x,x_0,u^0)$ of the iso-differential system (3) we have

$$\begin{aligned}||u(x,x_0,u^0)|| &= ||\Psi(x)\Psi^{-1}(x_0)u^0||\\ &\leq ||\Psi(x)\Psi^{-1}(x_0)||||u^0||\\ &\leq 2\tfrac{\varepsilon}{\delta}||u^0||,\end{aligned}$$

i.e., all solutions of (3) are bounded. □

Theorem 13.0.309. *Let $\Psi(x)$ be a fundamental matrix of the iso-differential system* (3). *Then all solutions of* (3) *are asymptotically stable if and only if*

$$||\Psi(x)||\longrightarrow 0 \tag{4}$$

as $x\longrightarrow\infty$.

Proof. We will note that every solution $u(x, x_0, u^0)$ of the iso-differential system can be represented in the form

$$u(x, x_0, u^0) = \Psi(x)\Psi^{-1}(x_0)u^0.$$

Since $\Psi(x)$ is continuous, then, using (4), it follows that there exists a constant $c > 0$ such that

$$||\Psi(x)|| \leq c$$

for all $x] \geq x_0$. From here,

$$||u(x, x_0, u^0)|| = ||\Psi(x)\Psi^{-1}(x_0)u^0||$$

$$\leq ||\Psi(x)|| ||\Psi^{-1}(x_0)u^0||$$

$$\leq c||\Psi^{-1}(x_0)u^0||.$$

From the last inequality we conclude that all solutions of the iso-differential system are bounded and therefore all solutions of (3) are stable.

Moreover, because

$$||u(x, x_0, \Delta u^0 + u^0) - u(x, x_0, u^0)|| = ||\Psi(x)\Psi^{-1}(x_0)\Delta u^0||$$

$$\leq ||\Psi(x)|| ||\Psi^{-1}(x_0)\Delta u^0|| \longrightarrow 0$$

as $x \longrightarrow \infty$, it follows that every solution of (3) is asymptotically stable.

Conversely, let all solutions of (3) are asymptotically stable. Then the trivial solution $u(x, x_0, 0) = 0$ is asymptotically stable. Therefore

$$||u(x, x_0, \Delta u^0)|| \longrightarrow 0$$

as $x \longrightarrow \infty$, from where

$$||\Psi(x)|| \longrightarrow 0$$

as $x \longrightarrow \infty$. □

Definition 13.0.310. *A solution $u(x) = u(x, x_0, u^0)$ of the initial value problem (1), (2) is said to be uniformly stable, if for each $\delta = \delta(\varepsilon) > 0$ such that for any solution $u^1(x) = u(x, x_0, u^1)$ of the initial value problem*

$$u'(x) = \frac{\hat{T}'(x)}{\hat{T}(x)}u(x) + g(x, u(x))\frac{\hat{T}(x) - x\hat{T}'(x)}{\hat{T}(x)}, \qquad x > x_1,$$

$$u(x_1) = u^1,$$

the inequalities

$$x_1 \geq x_0$$

and

$$||u^1(x_1) - u(x_1)|| < \delta$$

imply that

$$||u^1(x) - u(x)|| < \varepsilon$$

for all $x \geq x_1$.

Theorem 13.0.311. *Let $\Psi(x)$ be a fundamental matrix of the iso-differential system (3). Then all solutions of (3) are uniformly stable if and only if*

$$||\Psi(x)\Psi^{-1}(t)|| \leq c, \qquad x_0 \leq t \leq x < \infty, \tag{5}$$

where c is a positive constant.

Proof. Let $u(x) = u(x, x_0, u^0)$ be a solution of the iso-differential system (3). Then for any $x_1 \geq x_0$ we have

$$u(x) = \Psi(x)\Psi^{-1}(x_1)u(x_1).$$

If

$$u^1(x) = \Psi(x)\Psi^{-1}(x_1)u^1(x_1)$$

is any other solution of the iso-differential system (3), using the condition (5), we get

$$\begin{aligned}
||u^1(x) - u(x)|| &= ||\Psi(x)\Psi^{-1}(x_1)u^1(x_1) - \Psi(x)\Psi^{-1}(x_1)u(x_1)|| \\
&= ||\Psi(x)\Psi^{-1}(x_1)(u^1(x_1) - u(x_1))|| \\
&\leq ||\Psi(x)\Psi^{-1}(x_1)||||u^1(x_1) - u(x_1)|| \\
&\leq c||u^1(x_1) - u(x_1)||
\end{aligned} \tag{6}$$

for all $x_0 \leq x_1 \leq x < \infty$.

Let $\varepsilon > 0$ be arbitrarily chosen and $x_1 \geq x_0$,

$$||u^1(x_1) - u(x_1)|| < \frac{\varepsilon}{c}.$$

Consequently, using (6), we have

$$||u^1(x) - u(x)|| < \varepsilon,$$

and hence the solution $u(x)$ is uniformly stable.

Conversely, if all solutions of the iso-differential system (3) are uniformly stable, then the trivial solution $u(x, x_0, 0) = 0$ is uniformly stable.

Let now $\varepsilon > 0$ be arbitrary chosen. Then there exists a $\delta = \delta(\varepsilon) > 0$ such that inequalities $x_1 \geq x_0$ and

$$||u^1(x_1)|| < \delta$$

imply the inequality

$$||u^1(x)|| < \varepsilon$$

for all $x \geq x_1$.

In this way we obtain that

$$||\Psi(x)\Psi^{-1}(x_1)u^1(x_1)|| < \varepsilon \tag{7}$$

for all $x \geq x_1$.

Let

$$u^1(x_1) = \frac{\delta}{2} e^j,$$

where

$$e^j = (0, 0, \ldots, 0, 1, 0, \ldots, 0), \qquad j = 1, 2, \ldots, n,$$

and $\psi^j(x)$ to be the jth column of $\Psi(x)\Psi^{-1}(x_1)$. Then, from (7),

$$||\Psi(x)\Psi^{-1}(x_1)u^1(x_1)|| = ||\psi^j(x)||\frac{\delta}{2} < \varepsilon,$$

therefore

$$||\Psi(x)\Psi^{-1}(x_1)|| \leq \max_{1\leq j\leq n} ||\psi^j(x)||$$

$$\leq 2\tfrac{\varepsilon}{\delta},$$

and since $x_1 \geq x_0$ was arbitrarily chosen, then (5) holds. □

Example 13.0.312. *Let* $n = 1$, $\hat{T}(x) = e^x$, $x_0 \geq 2$,

$$g(x, u(x)) = \frac{(p(x) - 1)u(x)}{1 - x}, \qquad x \geq 2,$$

where $p(x)$ *is a continuous function in the interval* $[x_0, \infty)$. *We will find conditions for the function* $p(x)$ *so that the trivial solution of the initial value problem of* (1), (2), $x \geq x_0$ *to be uniformly stable.*

We have

$$\tfrac{\hat{T}'(x)}{\hat{T}(x)} = \tfrac{e^x}{e^x}$$

$$= 1,$$

$$\tfrac{\hat{T}(x) - x\hat{T}'(x)}{\hat{T}(x)} = \tfrac{e^x - xe^x}{e^x}$$

$$= 1 - x,$$

$$\tfrac{\hat{T}'(x)}{\hat{T}(x)} u(x) + \tfrac{\hat{T}(x) - x\hat{T}'(x)}{\hat{T}(x)} g(x, u(x)) = u(x) + (1 - x)\tfrac{(p(x) - 1)u(x)}{1 - x}$$

$$= u(x) + (p(x) - 1)u(x)$$

$$= u(x) + p(x)u(x) - u(x)$$

$$= p(x)u(x).$$

Consequently we obtain the initial value problem

$$u'(x) = p(x)u(x), \qquad x > x_0,$$

$$u(x_0) = u_0,$$

its solution is

$$u(x) = u_0 e^{\int_{x_0}^x p(t)dt}.$$

Let

$$u^1(x) = u^1(x_1) e^{\int_{x_1}^x p(t)dt}, \qquad x_1 \geq x_0.$$

Let $\varepsilon > 0$ *be arbitrarily chosen. We will search* $\delta = \delta(\varepsilon) > 0$ *such that the inequalities* $x_1 \geq x_0$ *and*

$$||u^1(x_1)|| < \delta$$

imply the inequality

$$\left|\left|u^1(x_1) e^{\int_{x_1}^x p(t)dt}\right|\right| < \varepsilon,$$

which is possible if and only if

$$\int_{x_1}^x p(t)dt \tag{8}$$

to be bounded above for all $x \geq x_1 \geq x_0$.

If (8) *is bounded above by the real constant* c_1, *then we can take*

$$\delta = \delta(\varepsilon) = \frac{\varepsilon}{e^{c_1}}.$$

Exercise 13.0.313. *Test the stability, asymptotic stability or instability for the trivial solution of the iso-differential system* (3) *in the cases*

1) $A(x) = \begin{pmatrix} \frac{1}{x-1} & 0 \\ \frac{2}{x-1} & \frac{1}{x-1} \end{pmatrix}$, 2) $A(x) = \begin{pmatrix} \frac{2}{x-1} & \frac{1-e^{2x}}{x-1} \\ \frac{1}{x-1} & \frac{2}{x-1} \end{pmatrix}$,

3) $A(x) = \begin{pmatrix} \frac{1}{x-1} & 0 & \frac{1}{x-1} \\ \frac{1}{x-1} & \frac{1}{x-1} & 0 \\ \frac{2}{x-1} & \frac{7}{x-1} & \frac{6}{x-1} \end{pmatrix}$, 4) $A(x) = \begin{pmatrix} 0 & \frac{1}{1-x} & \frac{1}{x-1} \\ \frac{1}{x-1} & 0 & 0 \\ 0 & \frac{2}{1-x} & 0 \end{pmatrix}$,

5) $A(x) = \begin{pmatrix} 0 & \frac{2}{x-1} & \frac{2}{x-1} \\ 0 & 0 & \frac{4}{x-1} \\ 0 & \frac{6}{x-1} & \frac{4}{x-1} \end{pmatrix}$.

Answer. 1) stable, 2) unstable, 3) asymptotically stable, 4) unstable, 5) unstable.

Now we will consider the iso-differential system

$$u'(x) = \Big(A(x)\frac{\hat{T}(x) - x\hat{T}'(x)}{\hat{T}(x)} + \frac{\hat{T}'(x)}{\hat{T}(x)}\Big)u(x) + g(x,u(x))\frac{\hat{T}(x) - x\hat{T}'(x)}{\hat{T}(x)}, \qquad x \geq x_0, \tag{9}$$

in the case when $A(x)$ is $n \times n$ matrix with continuous elements in the interval $[x_0, \infty)$.

We will suppose that

$$||g(x,u(x))|| \leq \lambda(x)||u(x)||, \tag{10}$$

where $\lambda(x)$ is a nonnegative function such that

$$\int_{x_0}^x \lambda(t)\frac{|\hat{T}(t) - t\hat{T}'(t)|}{\hat{T}(t)}dt < \infty.$$

We will note that the condition (10) implies that $v(x) \equiv 0$ is a solution of (9).

Theorem 13.0.314. *Suppose that the solutions of the iso-differential system* (3) *are uniformly (uniformly and asymptotically) stable, and the function* $g(x,u(x))$ *satisfies the condition* (10). *Then the trivial solution of the iso-differential system* (9) *is uniformly(uniformly and asymptotically) stable.*

Proof. Since all solutions of the iso-differential system (3) are uniformly stable, then, if $\Psi(x)$ is a fundamental matrix of (3), there exists a constant $c>0$ such that

$$||\Psi(x)\Psi^{-1}(t)||\leq c$$

for all $x_0\leq t\leq x<\infty$.

If $v(x)$ is a solution of (9) for which $v(x_1)=v^1$, $x_1\geq x_0$, then it satisfies the integral equation

$$v(x)=\Psi(x)\Psi^{-1}(x_1)v^1+\int_{x_1}^{x}\Psi(x)\Psi^{-1}(t)g(t,v(t))\frac{\hat{T}(t)-t\hat{T}'(t)}{\hat{T}(t)}dt.$$

Thus it follows that

$$||v(x)||=\left|\left|\Psi(x)\Psi^{-1}(x_1)v^1+\int_{x_1}^{x}\Psi(x)\Psi^{-1}(t)g(t,v(t))\frac{\hat{T}(t)-t\hat{T}'(t)}{\hat{T}(t)}dt\right|\right|$$

$$\leq||\Psi(x)\Psi^{-1}(x_1)v^1||+\int_{x_1}^{x}||\Psi(x)\Psi^{-1}(t)||||g(t,v(t))||\frac{|\hat{T}(t)-t\hat{T}'(t)|}{\hat{T}(t)}dt$$

$$\leq||\Psi(x)\Psi^{-1}(x_1)||||v^1||+c\int_{x_1}^{x}\lambda(t)\frac{|\hat{T}(t)-t\hat{T}'(t)|}{\hat{T}(t)}||v(t)||dt$$

$$\leq c||v^1||+c\int_{x_1}^{x}\lambda(t)\frac{|\hat{T}(t)-t\hat{T}'(t)|}{\hat{T}(t)}||v(t)||dt,$$

from where

$$||v(t)||\leq c||v^1||e^{c\int_{x_1}^{x}\frac{|\hat{T}(t)-t\hat{T}'(t)|}{\hat{T}(t)}dt}\leq K||v^1||, \tag{11}$$

where

$$K=ce^{c\int_{x_1}^{x}\frac{|\hat{T}(t)-t\hat{T}'(t)|}{\hat{T}(t)}dt}.$$

Let now $\varepsilon>0$ be arbitrarily chosen and let also

$$||v^1||\leq K^{-1}\varepsilon.$$

Then, from (11), we get

$$||v(x)||<\varepsilon,$$

for all $x\geq x_1$, i.e., the trivial solution of the iso-differential system (9) is uniformly stable.

Let now the solutions of the iso-differential system (9) are, in addition, asymptotically stable. Then

$$||\Psi(x)||\longrightarrow 0$$

as $x\longrightarrow\infty$.

Let $\varepsilon>0$ be arbitrarily chosen.

Then we can choose x_2 large enough so that

$$||\Psi(x)\Psi^{-1}(x_0)v^0|| \leq \varepsilon$$

for all $x \geq x_2$.

From here, for the solution $v(x) = v(x, x_0, v^0)$, we have

$$||v(x)|| \leq ||\Psi(x)\Psi^{-1}(x_0)v^0|| + \int_{x_0}^{x} ||\Psi(x)\Psi^{-1}(t)|| ||g(t, v(t))|| \frac{|\hat{T}(t) - t\hat{T}'(t)|}{\hat{T}(t)} dt$$

$$\leq \varepsilon + c \int_{x_0}^{x} \int_{x_0}^{x} \lambda(t) ||v(t)|| \frac{|\hat{T}(t) - t\hat{T}'(t)|}{\hat{T}(t)} dt, \qquad x \geq x_2.$$

From this, we get

$$||v(x)|| \leq c e^{c \int_{x_0}^{x} \lambda(t) \frac{|\hat{T}(t) - t\hat{T}'(t)|}{\hat{T}(t)} dt}$$

$$\leq L\varepsilon, \qquad x \geq x_2,$$

where

$$L = e^{c \int_{x_0}^{x} \lambda(t) \frac{|\hat{T}(t) - t\hat{T}'(t)|}{\hat{T}(t)} dt}.$$

Since ε is arbitrary and L does not depend on ε and x_2 we conclude that

$$||v(x)|| \longrightarrow 0$$

as $x \longrightarrow \infty$, i.e., the trivial solution of the iso-differential system (9) is, in addition, asymptotically stable. □

Chapter 14

Lyapunov's Direct Method for Iso-Differential Systems

Let $x_0 \in \mathbb{R}$, $\hat{T} \in \mathcal{C}^1([x_0,\infty))$, $\hat{T}(x) > 0$ in $[x_0,\infty)$. For $\rho > 0$ we define the set

$$S_\rho = \{u \in \mathbb{R}^n : ||u|| < \rho\}.$$

For $M \subset \mathbb{R}^n$, $N \subset \mathbb{R}^m$ with $\mathcal{C}^1(M,N)$ we will denote the set of all continuous-differentiable functions $f : M \longrightarrow N$. With $\mathcal{C}(M,N)$ we will denote the set of all continuous functions $f : M \longrightarrow N$.

Definition 14.0.315. *A function $\phi(r)$ is said to belong to the class $\mathcal{K}$ if and only if $\phi \in \mathcal{C}([0,\rho],\mathbb{R}^+)$, $\phi(0) = 0$, and $\phi(r)$ is strictly monotonically increasing in r.*

Example 14.0.316. *The function $a(r) = 3r^2 \in \mathcal{K}$.*

Definition 14.0.317. *A real-valued function $V(x,u)$ defined in $[x_0,\infty) \times S_\rho$ is said to be positive definite if and only if $V(x,0) \equiv 0$, $x \geq x_0$, and there exists a function $a(r) \in \mathcal{K}$ such that $a(r) \leq V(x,u)$, $||u|| = r$, $(x,u) \in [x_0,\infty) \times S_\rho$. It is negative definite if $V(x,u) \leq -a(r)$.*

Example 14.0.318. *The function*

$$V(x,u_1,u_2) = (2+\sin^2 x)u_1^2 + (3+2\cos^2 x)u_2^2$$

is positive definite in $[0,\infty) \times \mathbb{R}^2$.

Indeed, we have

$$V(x,0,0) = (2+\sin^2 x)0^2 + (3+2\cos^2 x)0^2$$

$$= 0.$$

Also, if $||u|| = r$ and $a(r) = r^2$, we have

$$r^2 \leq 2u_1^2 + 3u_2^2$$

$$\leq (2+\sin^2 x)u_1^2 + (3+2\cos^2 x)u_2^2$$

$$= V(x,u_1,u_2).$$

Therefore the function $V(x,u_1,u_2)$ is positive definite function in $[x_0,\infty) \times \mathbb{R}^2$.

Definition 14.0.319. *A real-valued function $V(x,u)$ defined in $[x_0,\infty)\times S_\rho$ is said to be decrescent if and only if $V(x,0)=0$, $x\geq x_0$, and there exists an h, $0<h\leq\rho$ and a function $b(r)\in\mathcal{K}$ such that*

$$V(x,u)\leq b(||u||)$$

for $||u||<h$ and $x\geq x_0$.

Example 14.0.320. *We consider the function*

$$V(x,u_1,u_2)=(4+3\sin^2(2x))u_1^2+(5+\cos^2(3x))u_2^2$$

in $[0,\infty)\times\mathbb{R}^2$. Let $b(r)=15r^2$. Then

$$\begin{aligned}V(x,u_1,u_2)&\leq 7u_1^2+6u_2^2\\ &\leq 7r^2+6r^2\\ &\leq 15r^2\\ &=b(r).\end{aligned}$$

Here we will investigate the initial value problem

$$u'(x)=\frac{\hat{T}'(x)}{\hat{T}(x)}u(x)+g(x,u(x))\frac{\hat{T}(x)-x\hat{T}'(x)}{\hat{T}(x)},\qquad x>x_0,\tag{1}$$

$$u(x_0)=u^0,\tag{2}$$

where $u^0\in\mathbb{R}^n$,

$$u(x)=(u_1(x),u_2(x),\ldots,u_n(x)),$$

$$g(x,u(x))=(g_1(x,u(x)),g_2(x,u(x)),\ldots,g_n(x,u(x)),$$

$g_i(x,u(x))\in C([x_0,\infty)\times S_\rho,\mathbb{R})$, $i=1,2,\ldots,n$.

We will assume that $V(x,u)\in C^1([x_0,\infty)\times S_\rho,\mathbb{R})$. Using the chain rule we get

$$\frac{dV}{dx}(x,u)=\frac{\partial V}{\partial x}(x,u)+\sum_{i=1}^{N}\frac{\partial V}{\partial u_i}(x,u)\frac{du_i(x)}{dx}.\tag{3}$$

Our interest is in the derivative of $V(x,u)$ along a solution $u(x)=u(x,x_0,u^0)$ of the initial value problem (1), (2), in this case the equality (3) we can rewrite in the following form.

$$\frac{dV}{dx}(x,u(x))=\frac{\partial V}{\partial x}(x,u(x))+\sum_{i=1}^{n}\frac{\partial V}{\partial u_i}(x,u(x))\left(g_i(x,u(x))\frac{\hat{T}(x)-x\hat{T}'(x)}{\hat{T}(x)}+\frac{\hat{T}'(x)}{\hat{T}(x)}u_i(x)\right)$$

$$=\frac{\partial V}{\partial x}(x,u(x))+\frac{\hat{T}(x)-x\hat{T}'(x)}{\hat{T}(x)}\mathrm{grad}V(x,u(x))\cdot g(x,u(x))$$

$$+\frac{\hat{T}'(x)}{\hat{T}(x)}\mathrm{grad}V(x,u(x))\cdot u(x):=V^*(x,u(x)).$$

Obviously, the function $V^*(x,u(x))$ we will connect with so called Lyapunov' s function, which plays important role in the investigation for stability of the trivial solution of the iso-differential system (1).

Example 14.0.321. *Let* $n=1$, $\hat{T}(x)=e^x$, $g(x,u)=\frac{u}{1-x}$, $x_0=2$, $u_0=1$, $V(x,u)=x^2+u^2$. *Then*

$$\frac{\hat{T}'(x)}{\hat{T}(x)}=\frac{e^x}{e^x}$$

$$=1,$$

$$\frac{\hat{T}(x)-x\hat{T}'(x)}{\hat{T}(x)}=\frac{e^x-xe^x}{e^x}$$

$$=1-x,$$

$$\frac{\partial V}{\partial x}(x,u)=2x,$$

$$\frac{\partial V}{\partial u}(x,u)=2u,$$

$$u'(x)=u(x)+\frac{u(x)}{1-x}(1-x)$$

$$=2u(x).$$

We have the following initial value problem

$$u'(x)=2u(x), \qquad x>0,$$

$$u(0)=1.$$

Its solution is

$$u(x)=e^{2x}.$$

Consequently

$$V^*(x,u(x))=2x+(1-x)2e^{2x}\frac{e^{2x}}{1-x}+e^{2x}(2e^{2x})$$

$$=2x+4e^{4x}.$$

Theorem 14.0.322. *If there exists a positive definite scalar function* $V(x,u)\in C^1([x_0,\infty)\times S_\rho,\mathbb{R}^+)$, *called a Lyapunov function, such that* $V^*(x,u)\leq 0$ *in* $[x_0,\infty)\times S_\rho$, *then the trivial solution of the iso-differential system* (1) *is stable.*

Proof. Since $V(x,u)$ is positive definite then there exists a function $a\in\mathcal{K}$ such that

$$a(||u||)\leq V(x,u)$$

for all $(x,u)\in[x_0,\infty)\times S_\rho$.

Let $0<\varepsilon<\rho$ be given.

Since $V(x,u)$ is continuous in $[x_0,\infty)\times S_\rho$ and $V(x,0)=0$, then we can find a $\delta-\delta(\varepsilon)>0$ such that the inequality

$$||u^0||<\delta$$

implies that

$$V(x,u^0)<a(\varepsilon)$$

for every $x\in[x_0,\infty)$.

Let us assume that the trivial solution of the iso-differential system (1) is unstable. Then there exists a solution $u(x)=u(x,x_0,u^0)$ of the iso-differential system (1) such that

$$||u^0||<\delta$$

and

$$||u(x_1)||=\varepsilon$$

for some $x_1>x_0$.

However, since

$$V^*(x,u)\leq 0$$

for every $(x,u)\in[x_0,\infty)\times S_\rho$, we have

$$V(x_1,u(x_1))\leq V(x_0,u^0),$$

and hence,

$$\begin{aligned}a(\varepsilon)&=a(||u(x_1)||)\\&\leq V(x_1,u(x_1))\\&\leq V(x_0,u^0)\\&<a(\varepsilon),\end{aligned}$$

which is impossible.

Thus, if $||u^0||<\delta$ then

$$||u(x)||<\varepsilon$$

for all $x\geq x_0$.

Therefore the trivial solution of (1) is stable. □

Theorem 14.0.323. *If there exists a positive definite and decrescent scalar function $V(x,u)\in C^1([x_0,\infty)\times S_\rho,\mathbb{R}^+)$ such that $V^*(x,u)$ is negative definite in $[x_0,\infty)\times S_\rho$, then the trivial solution of the iso-differential system (1) is asymptotically stable.*

Proof. Since all the conditions of the previous theorem are satisfied, the trivial solution of (1) is stable. Consequently, for given $0<\varepsilon<\rho$, we suppose that there exist $\delta>0$ and $\lambda>0$, and a solution $u(x)=u(x,x_0,u^0)$ of the iso-differential system (1), such that

$$\lambda\leq||u(x)||<\varepsilon,\qquad x\geq x_0,\qquad ||u^0||<\delta. \tag{4}$$

Because $V^*(x,u)$ is negative definite, then there exists a function $a \in \mathcal{K}$ such that

$$V^*(x,u(x)) \leq -a(||u(x)||).$$

From

$$||u(x)|| \geq \lambda > 0$$

for $x \geq x_0$, it follows that there exists a constant $d > 0$ such that

$$a(||u(x)||) \geq d$$

for $x \geq x_0$. Hence, we have

$$V^*(x,u(x)) \leq -d < 0, \qquad x \geq x_0.$$

This implies that

$$V(x,u(x)) = V(x_0,u^0) + \int_{x_0}^{x} V^*(t,u(t))dt$$

$$\leq V(x_0,u^0) - xd,$$

from the last inequality we conclude that for sufficiently large x the right side of the last inequality will become negative, which contradicts with the fact that $V(x,u)$ being positive definite. Consequently, there is no such λ for which (4) holds.

Moreover, since $V(x,u(x))$ is positive definite and a decreasing function of x, and decrescent, it follows that

$$\lim_{x \longrightarrow \infty} V(x,u(x)) = 0.$$

Therefore,

$$\lim_{x \longrightarrow \infty} ||u(x)|| = 0,$$

which implies that the trivial solution of the iso-differential system (1) is asymptotically stable. □

Example 14.0.324. *Let* $n = 2$, $\hat{T}(x) = e^x$, $x_0 = 2$,

$$g_1(x,u_1,u_2) = \frac{u_2 - u_1}{1-x},$$

$$g_2(x,u_1,u_2) = \frac{(2+e^{-x})u_1 + u_2}{x-1}.$$

Then

$$\frac{\hat{T}'(x)}{\hat{T}(x)} = \frac{e^x}{e^x}$$

$$= 1,$$

$$\frac{\hat{T}(x)-x\hat{T}'(x)}{\hat{T}(x)} = \frac{e^x-xe^x}{e^x}$$

$$= 1-x,$$

$$\frac{\hat{T}'(x)}{\hat{T}(x)}u_1(x)+g_1(x,u_1(x),u_2(x))\frac{\hat{T}(x)-x\hat{T}'(x)}{\hat{T}(x)} = u_1(x)+\frac{u_2(x)-u_1(x)}{1-x}(1-x)$$

$$= u_1(x)+u_2(x)-u_1(x)$$

$$= u_2(x),$$

$$\frac{\hat{T}'(x)}{\hat{T}(x)}u_2(x)+g_2(x,u_1(x),u_2(x))\frac{\hat{T}(x)-x\hat{T}'(x)}{\hat{T}(x)} = u_2(x)+\frac{(2+e^{-x})u_1(x)+u_2(x)}{x-1}(1-x)$$

$$= u_2(x)-u_1(x)(2+e^{-x})-u_2(x)$$

$$= -u_1(x)(2+e^{-x}).$$

In this way we obtain the iso-differential system

$$\begin{aligned} u_1'(x) &= u_2(x) \\ u_2'(x) &= -(2+e^{-x})u_1(x). \end{aligned} \tag{5}$$

Let us consider the function

$$V(x,u_1,u_2) = (2+e^{-x})u_1^2+u_2^2.$$

We have

$$V(x,u_1(x),u_2(x)) \in \mathcal{C}^1([2,\infty)\times S_\rho,\mathbb{R}^+),$$

$$V(x,0,0) = (2+e^{-x})0^2+0^2$$

$$= 0,$$

$$V(x,u_1(x),u_2(x)) \geq ||u||^2.$$

Therefore the function $V(x,u_1(x),u_2(x))$ *is positive definite in* $[2,\infty)\times S_\rho$.

Also,

$$\frac{\partial V}{\partial x}(x,u_1(x),u_2(x)) = -e^{-x}u_1^2(x),$$

$$\frac{\partial V}{\partial u_1}(x,u_1(x),u_2(x)) = 2(2+e^{-x})u_1(x),$$

$$\frac{\partial V}{\partial u_2}(x,u_1(x),u_2(x)) = 2u_2(x).$$

Then

$$V^*(x,u_1(x),u_2(x)) = -e^{-x}u_1^2(x)+2(2+e^{-x})u_1(x)\left(\frac{u_2(x)-u_1(x)}{1-x}(1-x)+u_1(x)\right)$$

$$+2u_2(x)\left(\frac{(2+e^{-x})u_1(x)+u_2(x)}{x-1}(1-x)+u_2(x)\right)$$

$$= -e^{-x}u_1^2(x)+2(2+e^{-x})u_1(x)(u_2(x)-u_1(x)+u_1(x))$$

$$+2u_2(x)\left(-(2+e^{-x})u_1(x)-u_2(x)+u_2(x)\right)$$

$$= -e^{-x}u_1^2(x)+2(2+e^{-x})u_1(x)u_2(x)-2(2+e^{-x})u_1(x)u_2(x)$$

$$= -e^{-x}u_1^2(x)$$

$$\leq 0$$

in $[2,\infty)\times\mathbb{R}^2$.

Therefore the trivial solution of the iso-differential system (5) *is stable.*

Example 14.0.325. *Let* $n=2$, $\hat{T}(x)=1+x^2$, $x_0=2$,

$$g_1(x,u_1,u_2) = \frac{-(2u_1+u_2)(1+x^2)-2xu_1}{x^2-1},$$

$$g_2(x,u_1,u_2) = \frac{(u_1-2u_2)(1+x^2)-2xu_2}{1-x^2}.$$

Then

$$\frac{\hat{T}'(x)}{\hat{T}(x)} = \frac{2x}{1+x^2},$$

$$\frac{\hat{T}(x)-x\hat{T}'(x)}{\hat{T}(x)} = \frac{1+x^2-2x^2}{1+x^2}$$

$$= \frac{1-x^2}{1+x^2},$$

$$\frac{\hat{T}'(x)}{\hat{T}(x)}u_1(x)+g_1(x,u_1(x),u_2(x))\frac{\hat{T}(x)-x\hat{T}'(x)}{\hat{T}(x)} = \frac{2x}{1+x^2}iu_1(x)+\frac{(2u_1(x)+u_2(x))(1+x^2)-2xu_1(x)}{x^2-1}\frac{1-x^2}{1+x^2}$$

$$= \frac{2x}{1+x^2}u_1(x)-\frac{(2u_1(x)+u_2(x))(1+x^2)+2xu_1(x)}{1+x^2}$$

$$= -2u_1(x)-u_2(x),$$

$$\frac{\hat{T}'(x)}{\hat{T}(x)}u_2(x)+g_2(x,u_1(x),u_2(x))\frac{\hat{T}(x)-x\hat{T}'(x)}{\hat{T}(x)} = \frac{2x}{1+x^2}u_2(x)+\frac{(u_1(x)-2u_2(x))(1+x^2)-2xu_2(x)}{1-x^2}\frac{1-x^2}{1+x^2}$$

$$= \frac{2x}{1+x^2}u_2(x)+\frac{(u_1(x)-2u_2(x))(1+x^2)-2xu_2(x)}{1+x^2}$$

$$= u_1(x)-2u_2(x).$$

In this way we obtain the following iso-differential system

$$u_1'(x) = -2u_1(x) - u_2(x)$$
$$u_2'(x) = u_1(x) - 2u_2(x). \tag{6}$$

Let us consider the function

$$V(x,u_1,u_2) = u_1^2 + u_2^2.$$

We have

$$V \in C^1([2,\infty) \times S_\rho, \mathbb{R}^+),$$

$$V(x,0,0) = 0,$$

$$V(x,u_1,u_2) \geq ||u||^2,$$

therefore the function $V(x,u_1,u_2)$ *is positive definite in* $[2,\infty) \times S_\rho$.

Also,

$$V^*(x,u_1(x),u_2(x)) = 2u_1(x)\Big(\tfrac{-(2u_1(x)+u_2(x))(1+x^2)-2xu_1(x)}{1-x^2}\tfrac{1-x^2}{1+x^2} + \tfrac{2x}{1+x^2}u_1(x)\Big)$$

$$+2u_2(x)\Big(\tfrac{(u_1(x)-2u_2(x))(1+x^2)-2xu_2(x)}{1-x^2}\tfrac{1-x^2}{1+x^2} + \tfrac{2x}{1+x^2}u_2(x)\Big)$$

$$= -2u_1(x)(2u_1(x)+u_2(x)) + 2u_2(x)(u_1(x)-2u_2(x))$$

$$= 4(u_1^2(x)+u_2^2(x))$$

in $[2,\infty) \times \mathbb{R}^2$.

Consequently the trivial solution of the iso-differential system (6) *is asymptotically stable.*

Theorem 14.0.326. *If there exists a scalar function* $V(x,u) \in C^1([x_0,\infty) \times S_\rho, \mathbb{R})$ *such that*

(i) $|V(x,u)| \leq b(||u||)$ *for all* $(x,u) \in [x_0,\infty) \times S_\rho$, *where* $b \in \mathcal{K}$,

(ii) *for every* $\delta > 0$ *there exists and* u^0 *with* $||u^0|| < \delta$ *such that* $V(x_0,u^0) < 0$,

(iii) $V^*(x,u) \leq -a(||u||)$ *for* $(x,u) \in [x_0,\infty) \times S_\rho$, *where* $a \in \mathcal{K}$,

then the trivial solution of the iso-differential system (1) *is unstable.*

Proof. Let the trivial solution of (1) is stable. Then for every $\varepsilon > 0$ such that $\varepsilon < \rho$, there exists a $\delta = \delta(\varepsilon) > 0$ such that $||u^0|| < \delta$ implies that

$$||u(x)|| = ||u(x,x_0,u^0)|| < \varepsilon$$

for all $x \geq x_0$.

Let u^0 be such that

$$||u^0|| < \delta \tag{7}$$

and

$$V(x_0, u^0) < 0.$$

Since (7) we have

$$||u(x)|| < \varepsilon.$$

Hence, the condition (i) gives

$$|V(x, u(x))| \leq b(||u(x)||) < b(\varepsilon) \tag{8}$$

for all $x \geq x_0$.

Now from (iii), it follows that $V(x, u(x))$ is a decreasing function, and therefore

$$V(x, u(x)) \leq V)(x_0, u^0) < 0$$

for every $x \geq x_0$.

Consequently

$$|V(x, u(x))| \geq |V(x_0, u^0)|.$$

From the last inequality and (i) we obtain

$$||u(x)|| \geq b^{-1}(|V(x_0, u^0)|). \tag{9}$$

Now we apply the condition (iii) and we get

$$V^*(x, u(x)) \leq -a(||u(x)||),$$

from where, for $x \geq x_0$,

$$\int_{x_0}^{x} V^*(t, u(t))dt \leq -\int_{x_0}^{x} a(||u(t)||)dt$$

or

$$V(x, u(x)) \leq V(x_0, u^0) - \int_{x_0}^{x} a(||u(t)||)dt. \tag{10}$$

From (9) we get

$$a(||u(x)||) \geq a\Big(b^{-1}(|V(x_0, u^0)|)\Big).$$

Thus, we obtain, using (9) and (10),

$$V(x, u(x)) \leq V(x_0, u^0) - \textstyle\int_{x_0}^{x} a\Big(b^{-1}(|V(x_0, u^0)|)\Big)dt$$

$$= V(x_0, u^0) - (x - x_0)a\Big(b^{-1}(|V(x_0, u^0)|)\Big),$$

from where

$$\lim_{x \longrightarrow \infty} V(x, u(x)) = -\infty,$$

which contradicts with (8).

Consequently the trivial solution of (1) is unstable. □

Theorem 14.0.327. *If there exists a positive definite and decrescent scalar function $V(x,u) \in C^1([x_0,\infty) \times S_\rho, \mathbb{R}^+)$ such that $V^*(x,u) \leq 0$ in $[x_0,\infty) \times S_\rho$, then the trivial solution of the iso-differential system* (1) *is uniformly stable.*

Proof. Because $V(x,u)$ is positive definite and decrscent, there exist functions $a, b \in \mathcal{K}$ such that

$$a(||u||) \leq V(x,u) \leq b(||u||)$$

for all $(x,u) \in [x_0,\infty) \times S_\rho$.

For each ε, $0 < \varepsilon < \rho$, let $\delta = \delta(\varepsilon) > 0$ be chosen so that

$$b(\delta) < a(\varepsilon).$$

Suppose that the trivial solution of the iso-differential system (1) does not uniformly stable. Then there exists some $x_2 > x_1$ such that the inequalities $x_1 \geq x_0$ and $||u(x_1)|| < \delta$ imply that

$$||u(x_2)|| = \varepsilon.$$

Integrating the inequality

$$V^*(x,u(x)) \leq 0$$

from x_1 to x, $x \geq x_1$, we get

$$V(x,u(x)) \leq V(x_1,u(x_1))$$

and from here, for $x = x_2$, we have

$$\begin{aligned} a(\varepsilon) &= a(||u(x_2)||) \\ &\leq V(x_2,u(x_2)) \\ &\leq V(x_1,u(x_1)) \\ &\leq b(||u(x_1)||) \\ &\leq b(\delta) \\ &< a(\varepsilon), \end{aligned}$$

which is a contradiction. Therefore the trivial solution of (1) is uniformly stable. □

Advanced Practical Exercises

Problem 14.0.328. *Show that the function*

$$V(x,u_1,u_2) = (u_1^2 + u_2^2)\cos^2 x$$

is decrescent in $[0,\infty) \times \mathbb{R}^2$.

Problem 14.0.329. *Show that the function*

$$V(x,u_1,u_2) = u_1^2 + e^{-x}u_2^2$$

is decrescent in $[0,\infty)\times\mathbb{R}^2$.

Problem 14.0.330. *Show that the function*

$$V(x,u_1,u_2) = u_1^2 + e^{x}u_2^2$$

is positive definite but not decrescent in $[0,\infty)\times\mathbb{R}^2$.

Problem 14.0.331. *Show that the function*

$$V(x,u_1,u_2) = \left(1+\cos^2 x + e^{-2x}\right)(u_1^4+u_2^4)$$

is positive definite and decrescent in $[0,\infty)\times\mathbb{R}^2$.

Problem 14.0.332. *Let* $n=1$, $\hat{T}(x)=e^x$,

$$g(x,u) = \frac{\left(\sin(\log x)+\cos(\log x) - a - 1\right)u}{1-x},$$

$x_0 = 2$.

Consider the function

$$V(x,u) = u^2 e^{2\left(a-\sin(\log x)\right)x}.$$

Show the following.

(i) *The function* $V(x,y)$ *is positive definite but not decrescent in* $[0,\infty)\times\mathbb{R}$.

(ii) *The trivial solution of* (1) *is stable.*

Problem 14.0.333. *Let* $n=2$, $\hat{T}(x)=e^x$, $x_0=2$,

$$g(x,u) = \frac{x\sin x - 2x - u}{1-x}.$$

Consider the function

$$V(x,u) = u^2 e^{\int_2^x (2t-\sin t)dt}.$$

Show the following.

(i) *The function* $V(x,u)$ *is positive definite but not decrescent in* $[2,\infty)\times\mathbb{R}$.

(ii) $V^*(x,y) \le -\lambda V(x,y)$ *for all* $x \ge \lambda > 0$.

(iii) *The trivial solution of* (1) *is asymptotically stable.*

Problem 14.0.334. *Let* $n = 1$, $\hat{T}(x) = e^x$, $x_0 = 2$,

$$g_1(x, u_1, u_2) = \frac{-u_1 + xu_2 - x^2 u_1(u_1^2 + u_2^2)}{1 - x},$$

$$g_2(x, u_1, u_2) = \frac{-u_2 - xu_1 - x^2 u_2(u_1^2 + u_2^2)}{1 - x}.$$

Show that the trivial solution of (1) *is stable.*

Problem 14.0.335. *Let* $n = 1$, $\hat{T}(x) = e^x$, $x_0 = 2$,

$$g_1(x, u_1, u_2) = \frac{-u_1 + xu_2 - (1 + x^2) u_1(u_1^2 + u_2^2)}{1 - x},$$

$$g_2(x, u_1, u_2) = \frac{-u_2 - xu_1 - (1 + x^2) u_2(u_1^2 + u_2^2)}{1 - x}.$$

Show that the trivial solution of (1) *is asymptotically stable.*

Problem 14.0.336. *Let* $n = 1$, $\hat{T}(x) = e^x$, $x_0 = 2$,

$$g_1(x, u_1, u_2) = \frac{-u_1 + xu_2 + (1 + x^2) u_1(u_1^2 + u_2^2)}{1 - x},$$

$$g_2(x, u_1, u_2) = \frac{-u_2 - xu_1 + (1 + x^2) u_2(u_1^2 + u_2^2)}{1 - x}.$$

Show that the trivial solution of (1) *is unstable.*

Chapter 15

Second Order Linear Iso-Differential Equations

We suppose that J is an interval in $\mathbb{R}$, $\hat{T}(x) \in C^2(J)$, $\hat{T}(x) > 0$ for every $x \in J$. Let also $p_1(x), p_2(x) \in C(J)$.

Here we will consider the following second-order linear iso-differential equation

$$\left(\left(\hat{y}^{\wedge}(\hat{x})\right)^{\circledast}\right)^{\circledast} + \hat{p}_1^{\wedge}(\hat{x})\hat{\times}\left(\hat{y}^{\wedge}(\hat{x})\right)^{\circledast} + \hat{p}_2^{\wedge}(\hat{x})\hat{\times}\hat{y}^{\wedge}(\hat{x}) = 0. \tag{1}$$

Definition 15.0.337. *The equation* (1) *will be called second-order linear iso-differential equation.*

The equation (1) we can rewrite in the following form

$$y''(x) + y'(x)\left(p_1(x)\frac{\hat{T}(x)-x\hat{T}'(x)}{\hat{T}(x)} - \frac{\hat{T}(x)\hat{T}'(x)-x\hat{T}'^2(x)-x\hat{T}(x)\hat{T}'(x)}{\hat{T}(x)(\hat{T}(x)-x\hat{T}'(x))}\right)$$

$$+y(x)\left(\frac{-\hat{T}^2(x)\hat{T}''(x)+\hat{T}(x)\hat{T}'^2(x)-x\hat{T}'^3(x)}{\hat{T}^2(x)(\hat{T}(x)-x\hat{T}'(x))}\right.$$

$$\left.-p_1(x)\frac{\hat{T}'(x)(\hat{T}(x)-x\hat{T}'(x))}{\hat{T}^2(x)} + p_2(x)\frac{(\hat{T}(x)-x\hat{T}'(x))^2}{\hat{T}^2(x)}\right) = 0.$$

Let

$$p(x) = p_1(x)\frac{\hat{T}(x)-x\hat{T}'(x)}{\hat{T}(x)} - \frac{\hat{T}(x)\hat{T}'(x)-x\hat{T}'^2(x)-x\hat{T}(x)\hat{T}'(x)}{\hat{T}(x)(\hat{T}(x)-x\hat{T}'(x))},$$

$$q(x) = \frac{-\hat{T}^2(x)\hat{T}''(x)+\hat{T}(x)\hat{T}'^2(x)-x\hat{T}'^3(x)}{\hat{T}^2(x)(\hat{T}(x)-x\hat{T}'(x))}$$

$$-p_1(x)\frac{\hat{T}'(x)(\hat{T}(x)-x\hat{T}'(x))}{\hat{T}^2(x)} + p_2(x)\frac{(\hat{T}(x)-x\hat{T}'(x))^2}{\hat{T}^2(x)}.$$

Consequently the equation (1) we can rewrite in the form

$$y''(x) + p(x)y'(x) + q(x)y(x) = 0.$$

The equation (1) is exact if

$$y''(x) + p(x)y'(x) + q(x)y(x) = \left(y'(x) + f(x)y(x)\right)', \tag{2}$$

where $f(x)$ is a differentiable function in J.

Expanding (2) we get

$$y''(x)+p(x)y'(x)+q(x)y(x) = y''(x)+f(x)y'(x)+f'(x)y(x),$$

from where we obtain the system

$$f(x) = p(x)$$

$$f'(x) = q(x).$$

Therefore

$$p'(x) = q(x),$$

which is equivalent of

$$\left(\hat{T}(x)(\hat{T}(x)-x\hat{T}'(x))\left(\frac{1}{\hat{T}(x)(\hat{T}(x)-x\hat{T}'(x))}\right)'\right)'$$

$$+p_1'(x)\frac{\hat{T}(x)-x\hat{T}'(x)}{\hat{T}(x)}+p_1(x)\left(\frac{\hat{T}(x)-x\hat{T}'(x)}{\hat{T}(x)}\right)'$$

$$= p_2(x)\frac{(\hat{T}(x)-x\hat{T}'(x))^2}{\hat{T}^2(x)} - p_1(x)\frac{\hat{T}'(x)(\hat{T}(x)-x\hat{T}'(x))}{\hat{T}^2(x)}$$

$$-\frac{\hat{T}''(x)\hat{T}^2(x)-\hat{T}(x)\hat{T}'^2(x)+x\hat{T}'^3(x)}{\hat{T}^2(x)}$$

or

$$\left(\hat{T}(x)(\hat{T}(x)-x\hat{T}'(x))\left(\frac{1}{\hat{T}(x)(\hat{T}(x)-x\hat{T}'(x))}\right)'\right)'$$

$$+p_1'(x)\frac{\hat{T}(x)-x\hat{T}'(x)}{\hat{T}(x)}-xp_1(x)\frac{\hat{T}''(x)}{\hat{T}(x)}$$

$$= p_2(x)\frac{(\hat{T}(x)-x\hat{T}'(x))^2}{\hat{T}^2(x)} - \frac{\hat{T}''(x)\hat{T}^2(x)-\hat{T}(x)\hat{T}'^2(x)+x\hat{T}'^3(x)}{\hat{T}^2(x)}.$$

Now we consider the equation

$$\left(\left(\hat{y}^\wedge(\hat{x})\right)^{\circledast}\right)^{\circledast}+\hat{p}_1^\wedge(\hat{x})\hat{\times}\left(\hat{y}^\wedge(\hat{x})\right)^{\circledast}+\hat{p}_2^\wedge(\hat{x})\hat{\times}\hat{y}^\wedge(\hat{x}) = h(x), \tag{3}$$

where $h(x)$ is a differentiable function in J.

The equation (3) we can rewrite in the form

$$y''(x)+p(x)y'(x)+q(x)y(x) = h(x)\frac{(\hat{T}(x)-x\hat{T}'(x))^2}{\hat{T}(x)}.$$

If the equation (3) is an exact equation then there exists a differentiable function $f(x)$ in the interval J such that

$$\left(y'(x)+f(x)y(x)\right)' = h(x)\frac{(\hat{T}(x)-x\hat{T}'(x))^2}{\hat{T}(x)},$$

whereupon

$$y'(x)+f(x)y(x)=\int h(x)\frac{(\hat{T}(x)-x\hat{T}'(x))^2}{\hat{T}(x)}dx+C,$$

here C is a real constant.

Example 15.0.338. *Let* $\hat{T}(x)=e^x$,

$$p_1(x)=\tfrac{1-2x^2}{x(1-x)^2},\qquad p_2(x)=\tfrac{-1+2x-x^3}{x^2(1-x)^3},$$

$$h(x)=\tfrac{x^2e^{-x}}{(1-x)^2}.$$

Then

$$p(x)=\tfrac{e^{2x}+xe^{2x}+xe^{2x}}{e^{2x}(1-x)}+\tfrac{1-2x^2}{x(1-x)^2}\tfrac{e^x-xe^x}{e^x}$$

$$=\tfrac{-1+2x}{1-x}+\tfrac{1-2x^2}{x(1-x)}$$

$$=\tfrac{-x+2x^2+1-2x^2}{x(1-x)}$$

$$=\tfrac{1}{x},$$

$$q(x)=\tfrac{-1+2x-x^3}{x^2(1-x)^3}\tfrac{(e^x-xe^x)^2}{e^{2x}}-\tfrac{1-2x^2}{x(1-x)^2}\tfrac{e^x(e^x-xe^x)}{e^{2x}}$$

$$-\tfrac{e^xe^{2x}-e^xe^{2x}+xe^{3x}}{e^{2x}(e^x-xe^x)}$$

$$=\tfrac{-1+2x-x^3}{x^2(1-x)}-\tfrac{1-2x^2}{x(1-x)}-\tfrac{x}{1-x}$$

$$=\tfrac{-1+2x-x^3-x+2x^3-x^3}{x^2(1-x)}$$

$$=-\tfrac{1}{x^2},$$

$$h(x)\tfrac{(\hat{T}(x)-x\hat{T}'(x))^2}{\hat{T}(x)}=\tfrac{x^2e^{-x}}{(1-x)^2}\tfrac{e^{2x}(1-x)^2}{e^x}$$

$$=x^2.$$

In this way the equation (3) *takes the form*

$$y'(x)+\frac{1}{x}y'(x)-\frac{1}{x^2}y(x)=x^2$$

or

$$y''(x)+\frac{xy'(x)-y(x)}{x^2}=x^2,$$

or

$$\left(y'(x)+\frac{y(x)}{x}\right)'=x^2,$$

whereupon

$$y'(x) + \frac{1}{x}y(x) = \frac{1}{3}x^3 + C,$$

or

$$y'(x) = -\frac{1}{x}y(x) + \frac{1}{3}x^3 + C.$$

Therefore

$$\begin{aligned} y(x) &= \tfrac{1}{x}\left(C_1 + \int \left(\tfrac{1}{3}x^3 + C\right) x dx\right) \\ &= \tfrac{1}{x}\left(C_1 + \tfrac{1}{15}x^5 + \tfrac{C}{2}x^2\right) \\ &= C_1 \tfrac{1}{x} + C_2 x + \tfrac{1}{15}x^4, \end{aligned}$$

where $C_2 = \frac{1}{2}C$ and C, and C_1 are real constants.

If the equation (1) is not exact, then we will seek an integrating factor $z(x)$ that will make it exact. In other words, we multiply (1) with twice continuous-differentiable function $z(x)$ in the interval J and we get

$$z(x)y''(x) + z(x)p(x)y'(x) + z(x)q(x)y(x) = 0. \tag{4}$$

If (4) is an exact equation, then there exist differentiable functions $f(x)$ and $g(x)$ in the interval Jsuch that

$$z(x)y''(x) + z(x)p(x)y'(x) + z(x)q(x)y(x) = \left(f(x)y'(x) + g(x)y(x)\right)'$$

or

$$z(x)y''(x) + z(x)p(x)y'(x) + z(x)q(x)y(x) = f(x)y''(x) + (f'(x) + g(x))y'(x) + g'(x)y(x).$$

Therefore we obtain the system

$$\begin{aligned} z(x) &= f(x) \\ z(x)p(x) &= f'(x) + g(x) \\ z(x)q(x) &= g'(x). \end{aligned}$$

From the last system we obtain

$$\begin{aligned} &z'(x) = f'(x) = z(x)p(x) - g(x) \implies \\ &z'(x) - z(x)p(x) + g(x) = 0. \end{aligned}$$

We differentiate the last equality with respect to the variable x and we get

$$z''(x) - (z(x)p(x))' + g'(x) = 0 \implies$$

$$z''(x) - (z(x)p(x))' + z(x)q(x) = 0. \tag{5}$$

The equation (5) is a second-order iso-differential equation in $z(x)$ and it can be written

$$z''(x) + q_1(x)z'(x) + q_2(x)z(x) = 0, \tag{6}$$

where

$$\begin{aligned} q_1(x) &= -p(x) \\ q_2(x) &= -p'(x) + q(x). \end{aligned} \tag{7}$$

The equation (5) is the adjoint equation of the equation (1). To see this, we observe that the equation (1) can be written in the form.

$$\begin{pmatrix} u_1'(x) \\ u_2'(x) \end{pmatrix} = \begin{pmatrix} 0 & 1 \\ -q(x) & -p(x) \end{pmatrix} \begin{pmatrix} u_1(x) \\ u_2(x) \end{pmatrix}$$

and its adjoint is

$$\begin{pmatrix} v_1'(x) \\ v_2'(x) \end{pmatrix} = \begin{pmatrix} 0 & q(x) \\ -1 & p(x) \end{pmatrix} \begin{pmatrix} v_1(x) \\ v_2(x) \end{pmatrix},$$

or

$$v_1'(x) = q(x)v_2(x)$$

$$v_2'(x) = -v_1(x) + p(x)v_2(x),$$

from where

$$v_1(x) = -v_2'(x) + p(x)v_2(x)$$

and

$$\left(-v_2'(x) + p(x)v_2(x)\right)' = q(x)v_2(x) \qquad \Longrightarrow$$

$$-v_2''(x) + p'(x)v_2(x) + p(x)v_2'(x) = q(x)v_2(x) \qquad \Longrightarrow$$

$$v_2''(x) + q_1(x)v_2'(x) + q_2(x)v_2(x) = 0,$$

which is the same as the iso-differential equation (6).

Definition 15.0.339. *When an iso-differential equation and its adjoint are the same, it is said to be self-adjoint.*

Thus the equation (1) is self-adjoint if

$$q_1(x) = p(x)$$

$$q_2(x) = q(x).$$

In such a situation, the relations (7) give

$$p(x) = 0$$

$$q_2(x) = q(x).$$

Thus the self-adjoint equation takes the form

$$y''(x) + q(x)y(x) = 0. \tag{8}$$

Moreover, any self-adjoint equation can be written in the form (8).

Exercise 15.0.340. *Let $\hat{T}(x) = e^x$,*

$$p_1(x) = \frac{-x^2 - x + 1}{(1-x)^2}, \qquad p_2(x) = \frac{x+2}{(1-x)^2}.$$

Verify that the iso-differential equation (1) *is an exact equation and find its general solution.*

Solution. We have

$$\begin{aligned} p(x) &= \tfrac{-e^{2x}+xe^{2x}+xe^{2x}}{e^x(e^x-xe^x)} + \tfrac{-x^2-x+1}{(1-x)^2}\tfrac{e^x-xe^x}{e^x} \\ &= \tfrac{2x-1}{1-x} + \tfrac{-x^2-x+1}{1-x} \\ &= \tfrac{-x^2+x}{1-x} \\ &= x, \\ q(x) &= \tfrac{x+2}{(1-x)^2}\tfrac{(e^x-xe^x)^2}{e^{2x}} - \tfrac{-x^2-x+1}{(1-x)^2}\tfrac{e^x(e^x-xe^x)}{e^{2x}} \\ &\quad - \tfrac{e^{3x}-e^{3x}+xe^{3x}}{e^{2x}(e^x-xe^x)} \\ &= x+2+\tfrac{x^2+x-1}{1-x} - \tfrac{x}{1-x} \\ &= x+2-x-1 \\ &= 1. \end{aligned}$$

Thus, the iso-differential equation (1) takes the form

$$y''(x) + xy'(x) + y(x) = 0$$

or

$$y''(x) + (xy(x))' = 0,$$

or

$$\left(y'(x) + xy(x)\right)' = 0,$$

whereupon

$$y'(x) = -xy(x) + C,$$

its general solution is

$$y(x) = e^{-\frac{x^2}{2}}\left(C_1 + C\int e^{\frac{x^2}{2}}dx\right),$$

where C and C_1 are real constants.

Remark 15.0.341. *If the equation* (1) *is self-adjoint, then*

$$p(x) = 0 \qquad \Longrightarrow$$

$$p_1(x)\frac{\hat{T}(x)-x\hat{T}'(x)}{\hat{T}(x)} + \frac{-\hat{T}(x)\hat{T}'(x)+x\hat{T}'^2(x)+x\hat{T}(x)\hat{T}''(x)}{\hat{T}(x)(\hat{T}(x)-x\hat{T}'(x))} = 0$$

or

$$p_1(x) = \frac{\hat{T}(x)\hat{T}'(x) - x\hat{T}'^2(x) - x\hat{T}(x)\hat{T}''(x)}{(\hat{T}(x) - x\hat{T}'(x))^2},$$

from where

$$q(x) = p_2(x)\frac{(\hat{T}(x)-x\hat{T}'(x))^2}{\hat{T}^2(x)}$$

$$-\frac{\hat{T}(x)\hat{T}'(x)-x\hat{T}'^2(x)-x\hat{T}(x)\hat{T}''(x)}{(\hat{T}(x)-x\hat{T}'(x))^2}\frac{(\hat{T}(x)-x\hat{T}'(x))\hat{T}'(x)}{\hat{T}^2(x)}$$

$$-\frac{\hat{T}''(x)\hat{T}^2(x)-\hat{T}(x)\hat{T}'^2(x)+x\hat{T}'^3(x)}{\hat{T}^2(x)(\hat{T}(x)-x\hat{T}'(x))}$$

$$= p_2(x)\frac{(\hat{T}(x)-x\hat{T}'(x))^2}{\hat{T}^2(x)} - \frac{\hat{T}(x)\hat{T}'^2(x)-x\hat{T}'^3(x)-x\hat{T}(x)\hat{T}'(x)\hat{T}''(x)}{\hat{T}^2(x)(\hat{T}(x)-x\hat{T}'(x))}$$

$$-\frac{\hat{T}''(x)\hat{T}^2(x)-\hat{T}(x)\hat{T}'^2(x)+x\hat{T}'^3(x)}{\hat{T}^2(x)(\hat{T}(x)-x\hat{T}'(x))}$$

$$= p_2(x)\frac{(\hat{T}(x)-x\hat{T}'(x))^2}{\hat{T}^2(x)} - \frac{\hat{T}''(x)\hat{T}(x)-x\hat{T}'(x)\hat{T}''(x)}{\hat{T}(x)(\hat{T}(x)-x\hat{T}'(x))}$$

$$= p_2(x)\frac{(\hat{T}(x)-x\hat{T}'(x))^2}{\hat{T}^2(x)} - \frac{\hat{T}''(x)}{\hat{T}(x)}.$$

Consequently the equation (1) *is self-adjoint if*

$$\begin{aligned} p_1(x) &= \frac{\hat{T}(x)\hat{T}'(x)-x\hat{T}'^2(x)-x\hat{T}(x)\hat{T}''(x)}{(\hat{T}(x)-x\hat{T}'(x))^2}, \\ q(x) &= p_2(x)\frac{(\hat{T}(x)-x\hat{T}'(x))^2}{\hat{T}^2(x)} - \frac{\hat{T}''(x)}{\hat{T}(x)}. \end{aligned} \tag{9}$$

Below we will investigate the self-adjoint equation (8), i.e., we will suppose (9).

Theorem 15.0.342. *(iso-Sturm's comparison theorem) If* $\alpha,\ \beta \in J$ *are consecutive zeros of the nontrivial solution* $y(x)$ *of the self-adjoint iso-differential equation* (8), *and if* $\overline{q}(x)$ *is a continuous function in* J *and* $\overline{q}(x) \geq q(x)$, *then every nontrivial solution* $z(x)$ *of the iso-differential equation*

$$z''(x) + \overline{q}(x)z(x) = 0 \tag{10}$$

has a zero in (α, β).

Proof. We multiply the equation (8) by $z(x)$ and we get

$$y''(x)z(x) + q(x)y(x)z(x) = 0. \tag{11}$$

Now we multiply the equation (10) by $y(x)$ and we obtain

$$z''(x)y(x) + \overline{q}(x)y(x)z(x) = 0. \tag{12}$$

We subtract from (11) the equation (12) and we go to

$$y''(x)z(x) - z''(x)y(x) + (q(x) - \overline{q}(x))y(x)z(x) = 0,$$

which is the same as

$$\left(y'(x)z(x) - z'(x)y(x)\right)' + (q(x) - \overline{q}(x))y(x)z(x) = 0.$$

Since $y(\alpha) = y(\beta) = 0$ we have

$$y'(\beta)z(\beta) - y'(\alpha)z(\alpha) + \int_{\alpha}^{\beta} (q(x) - \overline{q}(x))y(x)z(x)dx = 0.$$

Without loss of generality we can suppose that $y(x) > 0$ in (α, β). From here, since $y(\alpha) = 0$ and $y(\beta) = 0$ we have that

$$y'(\alpha) > 0, \qquad y'(\beta) < 0.$$

If we assume that $z(x) > 0$ in the interval (α, β), then

$$y'(\beta)z(\beta) < 0,$$

$$y'(\alpha)z(\alpha) > 0,$$

and from here

$$y'(\beta)z(\beta) - y'(\alpha)z(\alpha) < 0. \tag{13}$$

Also,

$$y(x)z(x) > 0 \quad \text{in} \quad (\alpha, \beta),$$

$$q(x) - \overline{q}(x) \le 0 \quad \text{in} \quad (\alpha, \beta),$$

therefore

$$\int_{\alpha}^{\beta} (q(x) - \overline{q}(x))y(x)z(x)dx \le 0.$$

From the last inequality and (13) it follows that

$$y'(\beta)z(\beta) - y'(\alpha)z(\alpha) + \int_{\alpha}^{\beta} (q(x) - \overline{q}(x))y(x)z(x)dx < 0,$$

which is a contradiction.

Let us assume that $z(x) < 0$ in (α, β).

Then

$$y'(\beta)z(\beta) > 0,$$

$$y'(\alpha)z(\alpha) < 0,$$

$$y'(\beta)z(\beta) - y'(\alpha)z(\alpha) > 0, \tag{14}$$

also,

$$y(x)z(x) \le 0 \quad \text{in} \quad (\alpha, \beta)$$

and

$$\int_{\alpha}^{\beta} (q(x) - \overline{q}(x))y(x)z(x)dx \ge 0,$$

from the last inequality and the inequality (14) it follows that

$$y'(\beta)z(\beta)-y'(\alpha)z(\alpha)+\int_{\alpha}^{\beta}(q(x)-\overline{q}(x))y(x)z(x)dx>0,$$

which is a contradiction.

Consequently $z(x)$ cannot be of fixed sign in (α,β), i.e., there exists $x_1\in(\alpha,\beta)$ such that $z(x_1)=0$. □

Lemma 15.0.343. *(Picone's identity) Let the functions*

$$y,\qquad z,\qquad py',\qquad p_1z'$$

be differentiable in the interval J and $z(x)\neq 0$ in J. Then the following identity holds.

$$\left(\frac{y}{z}(zpy'-yp_1z')\right)'=y(py')'-\frac{y^2}{z}(p_1z')'+(p-p_1)y'^2+p_1\left(y'-\frac{y}{z}z'\right)^2.$$

Proof. We have

$$\begin{aligned}\left(\tfrac{y}{z}(zpy'-yp_1z')\right)'&=\left(\tfrac{y}{z}\right)'(zpy'-yp_1z')+\tfrac{y}{z}(zpy'-yp_1z')'\\
&=\tfrac{y'z-yz'}{z^2}(zpy'-yp_1z')+\tfrac{y}{z}\left(z'py'+z(py')'-y'(p_1z')-y(p_1z')'\right)\\
&=\left(\tfrac{y'}{z}-y\tfrac{z'}{z^2}\right)(zpy'-yp_1z')+y\tfrac{z'}{z}py'+y(py')'-\tfrac{yy'}{z}(p_1z')-\tfrac{y^2}{z}(p_1z')'\\
&=p(y')^2-\tfrac{y'}{z}yp_1z'-yp\tfrac{z'}{z}y'+y^2\tfrac{p_1}{z^2}(z')^2\\
&\quad+\tfrac{y}{z}z'py'+y(py')'-\tfrac{yy'}{z}(p_1z')-\tfrac{y^2}{z}(p_1z')'\\
&=p(y')^2-2\tfrac{yy'}{z}(p_1z')+y^2\tfrac{p_1}{z^2}(z')^2+y(py')'-\tfrac{y^2}{z}(p_1z')'\\
&=p(y')^2+p_1\left(y^2\tfrac{(z')^2}{z^2}-2y\tfrac{y'}{z}z'+y'^2-y'^2\right)+y(py')'-\tfrac{y^2}{z}(p_1z')'\\
&=(p-p_1)(y')^2+p_1\left(y\tfrac{z'}{z}-y'\right)^2+y(py')'-\tfrac{y^2}{z}(p_1z')'.\end{aligned}$$

□

Theorem 15.0.344. *(iso-Sturm-Picone's theorem) If $\alpha,\ \beta\in J$ are the consecutive zeros of a nontrivial solution $y(x)$ of (8), and if $\overline{q}(x)$ is continuous in J and $\overline{q}(x)\geq q(x)$ in $[\alpha,\beta]$, then every nontrivial solution $z(x)$ of the equation*

$$z''(x)+\overline{q}(x)z(x)=0$$

has a zero in $[\alpha,\beta]$.

Proof. Let $z(x) \neq 0$ in $[\alpha, \beta]$. Then from the Picone's identity we have

$$\left(\frac{y}{z}(zy' - yz')\right)' = (\overline{q} - q)y^2 + \left(y' - \frac{y}{z}z'\right)^2.$$

Integrating the above identity from α to β and using that $y(\alpha) = y(\beta) = 0$ we get

$$\int_\alpha^\beta \left((\overline{q} - q)y^2 + \left(y' - \frac{y}{z}z'\right)^2\right)dx = 0,$$

which is a contradiction unless

$$\overline{q}(x) \equiv q(x) \qquad \text{in} \qquad [\alpha, \beta],$$

$$y'(x) - \frac{y(x)}{z(x)}z'(x) = 0 \qquad \text{in} \qquad [\alpha, \beta].$$

The last identity is the same as

$$\left(\frac{y(x)}{z(x)}\right)' = 0$$

and hence

$$\frac{y(x)}{z(x)} \equiv \text{const.}$$

However, since $y(\alpha) = 0$, this constant must be zero, therefore $y(x) \equiv 0$ in $[\alpha, \beta]$. This contradiction implies that there exists a $x_1 \in [\alpha, \beta]$ such that $z(x_1) = 0$. □

Corollary 15.0.345. *(iso-Sturm-Picone's theorem) If $y_1(x)$ and $y_2(x)$ are two linearly independent solutions of the equation (8) in the interval J, then their zeros are interlaced, i.e. between two consecutive zeros of one there is exactly one zero of the other.*

Proof. Since $y_1(x)$ and $y_2(x)$ cannot be common zeros, then the iso-Sturm-Picone's theorem implies that $y_2(x)$ has at least one zero between two consecutive zeros of $y_1(x)$. Interchanging $y_1(x)$ and $y_2(x)$ we conclude that $y_2(x)$ has at most one zero between two consecutive zeros of $y_1(x)$. □

Theorem 15.0.346. *The only solution of the equation (8) which vanishes infinitely often in $J_1 = [\alpha, \beta]$ is the trivial solution.*

Proof. We suppose that the solution $y(x)$ of (8) has infinite number of zeros in the interval J_1. The set of the zeros of $y(x)$ then will have a limit point $x^* \in J_1$. Then there exists a sequence $\{x_m\}_{m=1}^\infty$ of zeros of y^* which converges to x^*, $x_m \neq x^*$.

From the continuity of the solution $y(x)$ implies that

$$y(x^*) = \lim_{m \longrightarrow \infty} y(x_m) = 0.$$

Also, from the differentiability of the solution $y(x)$, we have

$$y'(x^*) = \lim_{m \longrightarrow \infty} \frac{y(x_m) - y(x^*)}{x_m - x^*} = 0.$$

Consequently

$$y(x^*) = y'(x^*) = 0.$$

Then from the uniqueness of the solutions it follows that

$$y(x) \equiv 0 \qquad \text{in} \qquad J_1.$$

□

Now we will consider the equation

$$y''(x) + q(x)y(x) = r(x) \tag{15}$$

in the interval $J_1 = [\alpha, \beta]$, where $q(x)$ and $r(x)$ are continuous functions in J_1.

Together with the equation (15) we shall consider the boundary conditions.

$$\begin{aligned} l_1(y) &= a_0 y(\alpha) + a_1 y'(\alpha) + b_0 y(\beta) + b_1 y'(\beta) = A, \\ l_2(y) &= c_0 y(\alpha) + c_1 y'(\alpha) + d_0 y(\beta) + d_1 y'(\beta) = B, \end{aligned} \tag{16}$$

where a_i, b_i, c_i d_i, $i = 0, 1$, and A, B are given constants.

Throughout, we will suppose that there does not exist a constant c such that

$$(a_0, a_1, b_0, b_1) = c(c_0, c_1, d_0, d_1).$$

Definition 15.0.347. *The boundary value problem* (15), (16) *is called nonhomogeneous two point linear boundary value problem.*

Definition 15.0.348. *The homogeneous equation* (8) *together with the homogeneous boundary conditions*

$$l_1(y) = 0, \qquad l_2(y) = 0 \tag{17}$$

will be called a homogeneous boundary value problem.

Boundary conditions (16) are quite general and, in particular, include the

(i) first boundary conditions (Dirichlet's boundary conditions)

$$y(\alpha) = A, \qquad y(\beta) = B, \tag{18}$$

(ii) second boundary conditions (mixed boundary conditions)

$$y(\alpha) = A, \qquad y'(\beta) = B, \tag{19}$$

or

$$y'(\alpha) = A, \qquad y(\beta) = B, m \tag{20}$$

(iii) separated boundary conditions (third boundary conditions)

$$\begin{aligned} a_0 y(\alpha) + a_1 y'(\alpha) &= A \\ d_0 y(\beta) + d_1 y'(\beta) &= B, \end{aligned} \tag{21}$$

where $a_0^2 + a_1^2 \neq 0$ and $d_0^2 + d_1^2 \neq 0$,

(iv) periodic boundary conditions

$$y(\alpha) = y(\beta), \qquad y'(\alpha) = y'(\beta). \tag{22}$$

Theorem 15.0.349. *Let $y_1(x)$ and $y_2(x)$ be any two linearly independent solutions of the equation* (8). *Then the homogeneous boundary value problem* (8), (17) *has only the trivial solution if and only if*

$$\Delta = \begin{vmatrix} l_1(y_1(x)) & l_1(y_2(x)) \\ l_2(y_1(x)) & l_2(y_2(x)) \end{vmatrix} \neq 0.$$

Proof. Any solution of the equation (8) can be written as

$$y(x) = c_1 y_1(x) + c_2 y_2(x),$$

where c_1 and c_2 are real constants.

This is a solution of the homogeneous boundary value problem (8), (17) if and only if

$$\begin{aligned} l_1(y(x)) &= l_1(c_1 y_1(x) + c_2 y_2(x)) \\ &= c_1 l_1(y_1(x)) + c_2 l_1(y_2(x)) \\ &= 0, \\ l_2(y(x)) &= l_2(c_1 y_1(x) + c_2 y_2(x)) \\ &= c_1 l_2(y_1(x)) + c_2 l_2(y_2(x)) \\ &= 0, \end{aligned}$$

i.e., if and only if the system

$$\begin{aligned} c_1 l_1(y_1(x)) + c_2 l_1(y_2(x)) &= 0 \\ c_1 l_2(y_1(x)) + c_2 l_2(y_2(x)) &= 0 \end{aligned}$$

has only the trivial solutions.

Consequently the problem (8), (17) has only the trivial solution if and only if $\Delta \neq 0$. □

Corollary 15.0.350. *The homogeneous boundary value problem* (8), (17) *has an infinite number of nontrivial solutions if and only if* $\Delta = 0$.

Theorem 15.0.351. *The nonhomogeneous boundary value problem* (15), (16) *has a unique solution if and only if the homogeneous boundary problem* (8), (17) *has only the trivial solution.*

Proof. Let $y_1(x)$ and $y_2(x)$ be any two linearly independent solutions of the equation (8) and $z(x)$ be a particular solution of (15). Then the general solution of (15) is given by

$$y(x) = c_1 y_1(x) + c_2 y_2(x) + z(x),$$

where c_1 and c_2 are real constants.

This $y(x)$ is a solution of the problem (15), (16) if and only if

$$l_1(y(x)) = c_1 l_1(y_1(x)) + c_2 l_1(y_2(x)) + l_1(z(x))$$

$$= A,$$

$$l_2(y(x)) = c_1 l_1(y_1(x)) + c_2 l_1(y_2(x)) + l_2(z(x))$$

$$= B.$$

The last system has a unique solution if and only if $\Delta \neq 0$, i.e., if and only if the homogeneous boundary problem (8), (17) has only the trivial solution. □

Chapter 16

Green's Functions

We suppose that $J = [\alpha, \beta]$, α, $\beta \in \mathbb{R}$, $\alpha < \beta$, $\hat{T} \in \mathcal{C}^1(J)$, $\hat{T}(x) > 0$ for every $x \in J$.

We will investigate the equations

$$\left(\left(\hat{y}^\wedge(\hat{x})\right)^{\circledast}\right)^{\circledast} + \hat{p}_1^\wedge(\hat{x}) \hat{\times} \left(\hat{y}^\wedge(\hat{x})\right)^{\circledast} + \hat{p}_2^\wedge(\hat{x}) \hat{\times} \hat{y}^\wedge(\hat{x}) = 0 \tag{1'}$$

and

$$\left(\left(\hat{y}^\wedge(\hat{x})\right)^{\circledast}\right)^{\circledast} + \hat{p}_1^\wedge(\hat{x}) \hat{\times} \left(\hat{y}^\wedge(\hat{x})\right)^{\circledast} + \hat{p}_2^\wedge(\hat{x}) \hat{\times} \hat{y}^\wedge(\hat{x}) = \hat{h}^\wedge(\hat{x}), \tag{2'}$$

where $p_1(x)$, $p_2(x)$, $h(x) \in \mathcal{C}(J)$.

The equations $(1')$ and $(2')$ we can rewrite in the form

$$y''(x) + p(x)y'(x) + q(x)y(x) = 0 \tag{1}$$

and

$$y''(x) + p(x)y'(x) + q(x)y(x) = r(x), \tag{2}$$

where

$$p(x) = p_1(x)\frac{\hat{T}(x) - x\hat{T}'(x)}{\hat{T}(x)} - \frac{\hat{T}(x)\hat{T}'(x) - x\hat{T}'^2(x) - x\hat{T}(x)\hat{T}'(x)}{\hat{T}(x)(\hat{T}(x) - x\hat{T}'(x))},$$

$$q(x) = \frac{-\hat{T}^2(x)\hat{T}''(x) + \hat{T}(x)\hat{T}'^2(x) - x\hat{T}'^3(x)}{\hat{T}^2(x)(\hat{T}(x) - x\hat{T}'(x))}$$

$$-p_1(x)\frac{\hat{T}'(x)(\hat{T}(x) - x\hat{T}'(x))}{\hat{T}^2(x)} + p_2(x)\frac{(\hat{T}(x) - x\hat{T}'(x))^2}{\hat{T}^2(x)},$$

$$r(x) = h(x)\frac{(\hat{T}(x) - x\hat{T}'(x))^2}{\hat{T}(x)}.$$

For $y \in \mathcal{C}^1(J)$ we define

$$l_1(y) = a_0 y(\alpha) + a_1 y'(\alpha) + b_0 y(\beta) + b_1 y'(\beta),$$

$$l_2(y) = c_0 y(\alpha) + c_1 y'(\alpha) + d_0 y(\beta) + d_1 y'(\beta),$$

where a_i, b_i, c_i, d_i, $i = 0, 1$, are given real constants, and there does not exist a constant c such that

$$(a_0, a_1, b_0, b_1) = c(c_0, c_1, d_0, d_1).$$

We will consider the equation (1) under the boundary conditions

$$l_1(y) = 0, \qquad l_2(y) = 0. \tag{3}$$

Below we will suppose that the problem (1), (3) has only the trivial solution.

Definition 16.0.352. *(Green's function) Green's function $G(x,t)$ for the boundary problem* (1), (3) *is defined in $[\alpha,\beta]\times[\alpha,\beta]$ and satisfies the following properties.*

(i) *$G(x,t)$ is continuous in $[\alpha,\beta]\times[\alpha,\beta]$.*

(ii) *$\frac{\partial G}{\partial x}(x,t)$ is continuous in the triangles $\alpha \le x \le t \le \beta$ and $\alpha \le t \le x \le \beta$, moreover,*

$$\frac{\partial G}{\partial x}(t^+,t) - \frac{\partial G}{\partial x}(t^-,t) = 1,$$

where

$$\frac{\partial G}{\partial x}(t^+,t) = \lim_{x\longrightarrow t, x>t}\frac{\partial G}{\partial x}(x,t), \qquad \frac{\partial G}{\partial x}(t^-,t) = \lim_{x\longrightarrow t, x<t}\frac{\partial G}{\partial x}(x,t).$$

(iii) *For every $t\in[\alpha,\beta]$, the function $z(x) = G(x,t)$ is a solution of the equation* (1) *in the intervals $[\alpha,t)$ and $(t,\beta]$.*

(iv) *For every $t\in[\alpha,\beta]$, the function $z(x) = G(x,t)$ satisfies the boundary conditions* (3).

Theorem 16.0.353. *There exists a unique Green's function $G(x,t)$ for the problem* (1), (3).

Proof. Let $y_1(x)$ and $y_2(x)$ be two linearly independent solutions of the equation (1). Then, using the property *(iii)* of the Green's function $G(x,t)$, we conclude that there exist functions $\lambda_1(t)$, $\lambda_2(t)$, $\mu_1(t)$ and $\mu_2(t)$ such that

$$G(x,t) = \begin{cases} y_1(x)\lambda_1(t) + y_2(x)\lambda_2(t), & \alpha \le x \le t, \\ y_2(x)\mu_1(t) + y_2(x)\mu_2(t), & t \le x \le \beta. \end{cases}$$

Since $G(x,t)$ is a continuous function in $[\alpha,\beta]\times[\alpha,\beta]$, we get

$$y_1(x)\lambda_1(t) + y_2(x)\lambda_2(t) = y_1(x)\mu_1(t) + y_2(x)\mu_2(t), \tag{4}$$

Also,

$$\frac{\partial G}{\partial x}(x,t) = \begin{cases} y_1'(x)\lambda_1(t) + y_2'(x)\lambda_2(t), & \alpha \le x \le t, \\ y_1'(x)\mu_1(t) + y_2'(x)\mu_2(t), & t \le x \le \beta. \end{cases}$$

From here,

$$\tfrac{\partial G}{\partial x}(t^-,t) = y_1'(t)\lambda_1(t) + y_2'(t)\lambda_2(t),$$

$$\tfrac{\partial G}{\partial x}(t^+,t) = y_1'(t)\mu_1(t) + y_2'(t)\mu_2(t).$$

From the last expressions and the second property of the Green's function $G(x,t)$ we get

$$y_1'(t)(\mu_1(t) - \lambda_1(t)) + y_2'(t)(\mu_2(t) - \lambda_2(t)) = 1. \tag{5}$$

Let

$$\nu_1(t) = \mu_1(t) - \lambda_1(t),$$
$$\nu_2(t) = \mu_2(t) - \lambda_2(t). \tag{6}$$

Then, using (4) and (5), we obtain the system

$$y_1(t)\nu_1(t) + y_2(t)\nu_2(t) = 0$$
$$y_1'(t)\nu_1(t) + y_2'(t)\nu_2(t) = 1.$$

Since $y_1(x)$ and $y_2(x)$ are linearly independent in $[\alpha, \beta]$, then the Wronskian

$$W(y_1(t), y_2(t)) = \begin{vmatrix} y_1(t) & y_2(t) \\ y_1'(t) & y_2'(t) \end{vmatrix} \neq 0$$

for every $t \in [\alpha, \beta]$. Thus the system (6) uniquely determines the functions $\nu_1(t)$ and $\nu_2(t)$ for every $t \in [\alpha, \beta]$.

From (6) it follows

$$\mu_1(t) = \nu_1(t) + \lambda_1(t)$$
$$\mu_2(t) = \nu_2(t) + \lambda_2(t).$$

In this way we can rewrite the Green's function $G(x,t)$ in the following way

$$G(x,t) = \begin{cases} y_1(x)\lambda_1(t) + y_2(x)\lambda_2(t), & \alpha \leq x \leq 1, \\ y_1(x)\lambda_1(t) + y_2(x)\lambda_2(t) & \\ +y_1(x)\nu_1(t) + y_2(x)\nu_2(t), & t \leq x \leq \beta. \end{cases}$$

From the property (iv) of the Green's function $G(x,t)$ we find that

$$l_1(G(x,t)) = 0, \qquad l_2(G(x,t)) = 0.$$

Let us consider

$$l_1(G(x,t)) = 0.$$

We have

$$
\begin{aligned}
0 &= l_1(G(x,t)) \\
&= a_0 G(\alpha,t) + a_1 \tfrac{\partial G}{\partial x}(\alpha,t) + b_0 G(\beta,t) + b_1 \tfrac{\partial G}{\partial x}(\beta,t) \\
&= a_0\Big(y_1(\alpha)\lambda_1(t) + y_2(\alpha)\lambda_2(t)\Big) \\
&+ a_1\Big(y_1'(\alpha)\lambda_1(t) + y_2'(\alpha)\lambda_2(t)\Big) \\
&+ b_0\Big(y_1(\beta)\lambda_1(t) + y_2(\beta)\lambda_2(t) + y_1(\beta)\nu_1(t) + y_2(\beta)\nu_2(t)\Big) \\
&+ b_1\Big(y_1'(\beta)\lambda_1(t) + y_2'(\beta)\lambda_2(t) + y_1'(\beta)\nu_1(t) + y_2'(\beta)\nu_2(t)\Big) \\
&= \Big(a_0 y_1(\alpha) + a_1 y_1'(\alpha) + b_0 y_1(\beta) + b_1 y_1'(\beta)\Big)\lambda_1(t) \\
&+ \Big(a_0 y_2(\alpha) + a_1 y_2'(\alpha) + b_0 y_2(\beta) + b_1 y_2'(\beta)\Big)\lambda_2(t) \\
&+ b_0 y_1(\beta)\nu_1(t) + b_0 y_2(\beta)\nu_2(t) \\
&+ b_1 y_1'(\beta)\nu_1(t) + b_1 y_2'(\beta)\nu_2(t) \\
&= l_1(y_1(x))\lambda_1(t) + l_1(y_2(x))\lambda_2(t) \\
&+ b_0(y_1(\beta)\nu_1(t) + y_2(\beta)\nu_2(t)) \\
&+ b_1(y_1'(\beta)\nu_1(t) + y_2'(\beta)\nu_2(t)),
\end{aligned}
$$

from where

$$
\begin{aligned}
& l_1(y_1(x))\lambda_1(t) + l_1(y_2(x))\lambda_2(t) \\
&= -b_0(y_1(\beta)\nu_1(t) - y_2(\beta)\nu_2(t)) - b_1(y_1'(\beta)\nu_1(t) - y_2'(\beta)\nu_2(t)).
\end{aligned}
$$

As in above, using that

$$l_2(G(x,t)) = 0,$$

we get

$$
\begin{aligned}
& l_2(y_1(x))\lambda_1(t) + l_2(y_2(x))\lambda_2(t) \\
&= -d_0(y_1(\beta)\nu_1(t) - y_2(\beta)\nu_2(t)) - d_1(y_1'(\beta)\nu_1(t) - y_2'(\beta)\nu_2(t)).
\end{aligned}
$$

Thus we obtain the system

$$\begin{aligned} & l_1(y_1(x))\lambda_1(t)+l_1(y_2(x))\lambda_2(t) \\ & =-b_0(y_1(\beta)\nu_1(t)+y_2(\beta)\nu_2(t))-b_1(y_1'(\beta)\nu_1(t)+y_2'(\beta)\nu_2(t)) \\ & l_2(y_1(x))\lambda_1(t)+l_2(y_2(x))\lambda_2(t) \\ & =-d_0(y_1(\beta)\nu_1(t)+y_2(\beta)\nu_2(t))-d_1(y_1'(\beta)\nu_1(t)+y_2'(\beta)\nu_2(t)). \end{aligned} \tag{7}$$

Since the problem (1), (3) has only the trivial solution, it follows that the system (7) uniquely determines $\lambda_1(t)$ and $\lambda_2(t)$. □

Theorem 16.0.354. *The unique solution of the problem* (2), (3) *can be represented by*

$$\begin{aligned} y(x) &= \int_\alpha^\beta G(x,t)r(t)dt \\ &= \int_\alpha^x G(x,t)r(t)dt+\int_x^\beta G(x,t)r(t)dt. \end{aligned}$$

Proof. Since $G(x,t)$ is differentiable with respect to x in each of the intervals, we find

$$\begin{aligned} & y'(x)=G(x,x)r(x)+\int_\alpha^x \tfrac{\partial G}{\partial x}(x,t)r(t)dt \\ & -G(x,x)r(x)+\int_x^\beta \tfrac{\partial G}{\partial x}(x,t)r(t)dt \\ & =\int_\alpha^x \tfrac{\partial G}{\partial x}(x,t)r(t)dt+\int_x^\beta \tfrac{\partial G}{\partial x}(x,t)r(t)dt \\ & =\int_\alpha^\beta \tfrac{\partial G}{\partial x}(x,t)r(t)dt. \end{aligned} \tag{8}$$

Also, because $\frac{\partial G}{\partial x}(x,t)$ is a continuous function of (x,t) in the triangles $\alpha \le t \le x \le \beta$, $\alpha \le x \le t \le \beta$, then for any point (s,s) on the diagonal of the square, i.e., $t=x$ it is necessary to have

$$\frac{\partial G}{\partial x}(s,s^-)=\frac{\partial G}{\partial x}(s^+,s)$$

and

$$\frac{\partial G}{\partial x}(s,s^+)=\frac{\partial G}{\partial x}(s^-,s).$$

Now we differentiate the equality (8) with respect to x and we get

$$\begin{aligned} & y''(x)=\tfrac{\partial G}{\partial x}(x,x^-)r(x)+\int_\alpha^x \tfrac{\partial^2 G}{\partial x^2}(x,t)r(t)dt \\ & -\tfrac{\partial G}{\partial x}(x,x^+)r(x)+\int_x^\beta \tfrac{\partial^2 G}{\partial x^2}(x,t)r(t)dt, \end{aligned}$$

or

$$\begin{aligned} & y''(x)=\left(\tfrac{\partial G}{\partial x}(x,x^-)-\tfrac{\partial G}{\partial x}(x,x^+)\right)dx \\ & +\int_\alpha^\beta \tfrac{\partial^2 G}{\partial x^2}(x,t)r(t)dt. \end{aligned}$$

Using the property (ii) of the Green's function $G(x,t)$, from the last equality, we get

$$y''(x) = r(x) + \int_{\alpha}^{\beta} \frac{\partial^2 G}{\partial x^2}(x,t)r(t)dt.$$

Therefore

$$y''(x) + p(x)y'(x) + q(x)y(x)$$

$$= r(x) + \int_{\alpha}^{\beta} \left(\frac{\partial^2 G}{\partial x^2}(x,t) + p(x)\frac{\partial G}{\partial x}(x,t) + q(x)G(x,t) \right) r(t)dt$$

$$= r(x).$$

Consequently $y(x)$ is a solution to the equation (2).

Also, using the property (iv) of the Green's function $G(x,t)$,

$$l_1(y(x) = l_1\left(\int_{\alpha}^{\beta} G(x,t)r(t)dt \right)$$

$$= \int_{\alpha}^{\beta} L_1(G(x,t))r(t)dt$$

$$= 0,$$

$$l_2(y(x) = l_2\left(\int_{\alpha}^{\beta} G(x,t)r(t)dt \right)$$

$$= \int_{\alpha}^{\beta} l_2(G(x,t))r(t)dt$$

$$= 0.$$

□

Example 16.0.355. *Let*

$$\hat{T}(x) = e^x, \qquad p_1(x) = -\frac{2x-1}{(1-x)^2}, \qquad p_2(x) = \frac{2}{(1-x)^2}.$$

We will consider the periodic boundary conditions

$$y(0) = y(\pi),$$

$$y'(0) = y'(\pi).$$

We have

$$p(x) = \frac{-e^{2x}+xe^{2x}+xe^{2x}}{e^x(e^x-xe^x)} - \frac{2x-1}{(1-x)^2}\frac{e^x-xe^x}{e^x}$$

$$= \frac{2x-1}{1-x} - \frac{2x-1}{1-x}$$

$$= 0,$$

$$q(x) = \frac{2}{(1-x)^2}\frac{(e^x-xe^x)^2}{e^{2x}} + \frac{2x-1}{(1-x)^2}\frac{e^x(e^x-xe^x)}{e^x}$$

$$-\frac{e^{3x}-e^{3x}+xe^{3x}}{e^{2x}(e^x-xe^x)}$$

$$= \frac{2}{(1-x)^2}(1-x)^2 + \frac{2x-1}{(1-x)^2}(1-x) - \frac{x}{1-x}$$

$$= 2 + \frac{2x-1}{1-x} - \frac{x}{1-x}$$

$$= 2 + \frac{x-1}{1-x}$$

$$= 2-1$$

$$= 1.$$

Thus the equation (1) *takes the form*

$$y''(x) + y(x) = 0.$$

Two linearly independent solutions are

$$y_1(x) = \cos x, \qquad y_2(x) = \sin x.$$

We will search Green's function in the form

$$G(x,t) = \begin{cases} \lambda_1(t)\cos x + \lambda_2(t)\sin x, & 0 \le x \le t, \\ \mu_1(t)\cos x + \mu_2(t)\sin x, & t \le x \le \pi. \end{cases}$$

From the properties (i) *and* (ii) *of the Green's function we get the system*

$$\lambda_1(t)\cos t + \lambda_2(t)\sin t = \mu_1(t)\cos t + \mu_2(t)\sin t$$

$$-\mu_1(t)\sin t + \mu_2(t)\cos t\lambda_1(t)\cos t - \lambda_2(t)\cos t = 1.$$

Let

$$\nu_1(t) = \mu_1(t) - \lambda_1(t)$$

$$\nu_2(t) = \mu_2(t) - \lambda_2(t).$$

Thus we obtain the system

$$\nu_1(t)\cos t + \nu_2(t)\sin t = 0$$

$$-\nu_1(t)\sin t + \nu_2(t)\cos t = 1,$$

whereupon

$$\nu_1(t) = -\sin t,$$

$$\nu_2(t) = \cos t.$$

Therefore

$$\mu_1(t) = \lambda_1(t) - \sin t,$$

$$\mu_2(t) = \lambda_2(t) + \cos t.$$

Consequently

$$G(x,t) = \begin{cases} \lambda_1(t)\cos x + \lambda_2(t)\sin x, & 0 \le x \le t, \\ \lambda_1(t)\cos x + \lambda_2(t)\sin x - \cos x \sin t + \sin x \cos t. \end{cases}$$

From the boundary conditions we find

$$\lambda_1(t) = -\lambda_1(t) + \sin t$$

or

$$\lambda_1(t) = \frac{1}{2}\sin t$$

and

$$\lambda_2(t) = -\lambda_2(t) - \cos t$$

or

$$\lambda_2(t) = -\frac{1}{2}\cos t.$$

Consequently

$$G(x,t) = \begin{cases} \frac{1}{2}\sin(t-x), & 0 \le x \le t, \\ -\frac{1}{2}\sin(t-x), & t \le x \le \pi. \end{cases}$$

References

[1] Abraham, R. and J. E. Marsden (1978). *Foundations of Mechanics*, 2nd ed. Reading: The Benjamin/Cummings Pub. Co.

[2] Abraham, R., J. E. Marsden, and T. Ratiu (1988. *Manifolds, Tensor Analysis, and Applications*, 2nd ed. New York: Springer-Verlag.

[3] Abraham, R. and J. Robbin (1967). *Transversal Mappings and Flows*. New York: W. A. Benjamin, Inc.

[4] Andersen, C. and J. Geer (1982). Power series expansions for frequency and period of the limit cycle of the van der Pol equation. *SIAM J. Appl. Math.* 42, 678-693.

[5] Andronov, A. A., E. A. Leontovich, I. I. Gordon, and A. G. Maier (1973). *Qualitative Theory of Second-Order Dynamic Systems*. New York: John Wiley & Sons.

[6] Arnold, V. I. (1973). *Ordinary Differential Equations*. Cambridge: M. I. T. Press.

[7] Arnold, V. I. (1982). *Geometric Methods in the Theory of Ordinary Differential Equations*. New York: Springer-Verlag.

[8] Arnold, V. I. (1978). *Mathematical Methods of Celestial Mechanics*. New York: Springer-Verlag.

[9] Arnold, V. I., ed. (1988). *Dynamical Systems I*. New York: SpringerVerlag.

[10] Arnold, V. I., ed. (1988). *Dynamical Systems III*. New York: SpringerVerlag.

[11] Arnold, V. I., ed. (1990). *Huygens and Barrow*, Newton and Hooke, E. J. F. Primrose, trans. Basel: Birkhäuser-Verlag.

[12] Arnold, V. I. and B. A. Khesin (1998). *Topological Methods in Hydrodynamics*. New York: Springer-Verlag.

[13] Aronson D. G. and H. F. Weinberger (1978). Multidimensional nonlinear diffusion arising in population genetics. *Adv. Math.* 30, 33-76.

[14] Brieskorn, E. and H. Knörrer (1986). *Plane Algebraic Curves*. Basel: Birkhäuser-Verlag.

[15] Byrd, P. F. and M. D. Friedman (1971). *Handbook of Elliptic Integrals for Engineers and Scientists*, 2nd ed. Berlin: Springer-Verlag.

[16] Carathéodory, C. (1954). *Theory of Functions of a Complex Variable*, 1, pp. 229-236. New York: Chelsea.

[17] Chen, M., X. Y. Chen, and J. K. Hale (1992). Structural stability for time-periodic one-dimensional parabolic equations. *J. Diff. Eqs.* 96(2), 355-418.

[18] Chicone, C. (1986). Limit cycles of a class of polynomial vector fields in the plane. *J. Diff. Eqs.* 63(1), 68-87.

[19] Chicone, C. (1987). The monotonicity of the period function for planar Hamiltonian vector fields. *J. Diff. Eqs.* 69, 310-321.

[20] Chicone, C. (1993). review of Finiteness Theorems for Limit Cycles, by Yu. Ilyashenko. *Bull. Amer. Math. Soc.* 28(1), 123-130.

[21] Chicone, C. (1992). Bifurcation of nonlinear oscillations and frequency entrainment near resonance. *SIAM J. Math. Anal.* 23(6), 1577-1608.

[22] Chicone, C. (1994). Lyapunov Schmidt reduction and Melnikov integrals for bifurcation of periodic solutions in coupled oscillators. *J. Diff. Eqs.* 112, 407-447.

[23] Chicone, C. (1995). A geometric approach to regular perturbation theory with an application to hydrodynamics. *Trans. Amer. Math. Soc.* 374(12), 4559-4598.

[24] Chicone, C. (1997). Invariant tori for periodically perturbed oscillators. *Publ. Mat.* 41, 57-83.

[25] Dunford, N. and J. T. Schwartz (1958). *Linear Operators*. New York: Interscience Publications, Inc.

[26] Ewing, G. M. (1969). *Calculus of Variations with Applications*. New York: W.W. Norton&Company, Inc.

[27] Falkner, V. M. and S. W. Skan (1931). Solution of boundary layer equations. *Philos. Mag.* 7(12), 865-896.

[28] Farkas, M. (1994). *Periodic Motions*. New York: Springer-Verlag.

[29] Fenichel, N. (1971). Persistence and smoothness of invariant manifolds for flows. *Indiana U. Math. J.* 21, 193-226.

[30] Fermi, E., S. Ulam, and J. Pasta (1974). Studies of nonlinear problems I. In Nonlinear Wave Motion, *Lect. Appl. Math., Amer. Math. Soc.* 15, 143-155.

[31] Feynman, R. P., R. B. Leighton, and M. Sands (1964). *The Feynman Lectures on Physics* I-III. Reading: Addison-Wesley Pub. Co.

[32] Georgiev, S.(2014). *Foundations of Iso-Differential Calculus*, Vol. 1., Vol. 2, Nova Science Publishers, Inc.

[33] Gutzwiller, M. C. (1990). *Chaos in Classical and Quantum Mechanics*. New York: Springer-Verlag.

[34] Hale, J. K. (1980). *Ordinary Differential Equations*, 2nd ed. Malabar: R. E. Krieger Pub. Co.

[35] Hale, J. K., L.T. Magaláes, and W. M. Oliva (1984). *An Introduction to Infinite Dimensional Dynamical Systems-Geometric Theory.* New York: Springer-Verlag.

[36] Hale, J. K. (1988). *Asymptotic Behavior of Dissipative Systems.* Providence: American Mathematical Society.

[37] Hall, R. (1984). Resonance zones in two-parameter families of circle homeomorphisms. *SIAM J. Math. Analy.* 15, 1075-1081.

[38] Hartman, P. (1960). A lemma in the theory of structural stability of differential equations. *Proc. Amer. Math. Soc.* 11, 610-620.

[39] Hastings, S. P. and W. C. Troy (1988). Oscillating solutions of the Falknerkan equation for positive. *J. Diff. Eqs.* 71, 123-144.

[40] Hayashi, C. (1964). *Nonlinear Oscillations in Physical Systems.* New York: McGraw-Hill Book Co.

[41] Henry, D. (1981). Geometric Theory of Semilinear Parabolic Equations. New York: Springer-Verlag.

[42] Howe, R. (1983). Very basic Lie theory. *Amer. Math. Monthly.* 90, 600-623; Correction (1984). 91, 247.

[43] Hirsch, M. (1976). *Differential Topology*. New York: Springer-Verlag.

[44] Hirsch, M. and C. Pugh (1970). Stable manifolds and hyperbolic sets. InGlobal Analysis XIV, *Amer. Math. Soc.* 133-164.

[45] Hirsch, M., C. Pugh, and M. Shub (1977). Invariant Manifolds, *Lecture Notes in Math.*, 583. New York: Springer-Verlag.

[46] Hirsch, M. and S. Smale (1974). *Differential Equations, Dynamical Systems, and Linear Algebra.* New York: Academic Press.

[47] Holmes, P. and J. Marsden (1981). A partial differential equation with infinitely many periodic orbits and chaotic oscillations of a forced beam. *Arch. Rat. Mech. Analy.* 76, 135-165.

[48] Kolmogorov, A. N., I. G. Petrovskii, and N. S. Piskunov (1991). *A study of the diffusion equation with increase in the amount of substance, and its applications to a biological problem.* In Selected Works of A. N. Kolmogorov, V. M. Tikhomirov, ed. Dordrecht: Kluwer Academic Pub. Group.

[49] Landau, L. D. and E. M. Lifshitz (1951). *Classical Theory of Fields.* Reading: Addison-Wesley Press.

[50] Landau L. D. and E. M. Lifshitz (1987). *Fluid Mechanics*, 2nd ed., J. B. Sykes and W. H. Reid, trans. New York: Pergamon Press.

[51] Lang, S. (1962). *Introduction to Differentiable Manifolds*. New York: Interscience Pub.

[52] Lefschetz, S. (1977). *Differential Equaitions: Geometric Theory*, 2nd ed. New York: Dover Publications, Inc.

[53] Levi, M. (1988). Stability of the inverted pendulum - A topological explanation. *SIAM Rev.* 30, 639-644.

[54] Levi, M., C. Hoppensteadt, and M. Miranker (1978). Dynamics of the Josephson junction. *Quart. Appl. Math.* 36, 167-198.

[55] Lochak, P. and C. Meunier (1988). *Multiphase Averaging for Classical Systems*, H. S. Dumas, trans. New York: Springer-Verlag.

[56] Lorenz, E. N. (1963). Deterministic non-periodic flow. *J. Atmos. Sci.* 20, 131-141.

[57] Loud, W. S. (1964). Behavior of the periods of solutions of certain plane autonomous systems near centers. *Contributions Diff. Eqs.* 3, 21-36.

[58] Lichtenberg, A. J. and M. A. Lieberman (1982). *Regular and Stochastic Motion*. New York: Springer-Verlag.

[59] Liapounoff, M. A. (1947). *Problème gènèral de la stabilité du movement*. Princeton: Princeton University Press.

[60] Magnus, W. and S. Winkler (1979). *Hill's Equation*. New York: Dover Publications, Inc.

[61] Massey, W. S. (1967). *Algebraic Topology: An Introduction*. New York: Springer-Verlag.

[62] Marsden, J. E. and M. McCracken (1976). *The Hopf Bifurcation and Its Applications*. New York: Springer-Verlag.

[63] McCuskey, S. W. (1963). *Introduction to Celestial Mechanics*. Reading: Addison-Wesley Pub. Co.

[64] McLachlan, N. W. (1958). *Ordinary Non-linear Differential Equations in Engineering and Physical Sciences*, 2nd ed. New York: Oxford University Press.

[65] Melnikov, V. K. (1963). On the stability of the center for time periodic perturbations. *Trans. Moscow Math. Soc.* 12, 1-57.

[66] Milnor, J. (1963). *Morse Theory*. Princeton: Princeton University Press.

[67] Milnor, J. (1978). Analytic proofs of the 'hairy ball theorem' and the Brouwer fixed-point theorem. *Amer. Math. Monthly* 85(7), 521-524.

[68] Meyer, K. R. and G. R. Hall (1992). *Introduction to Hamiltonian Dynamical Systems and theN-Body Problem.* New York: SpringerVerlag.

[69] Meyer, K. R. and D. S. Schmidt (1977). Entrainment domains. *Funkcialaj Edvacioj.* 20, 171-192.

[70] Minorsky, N. (1962). *Nonlinear Oscillations.* Princeton: D. Van Nostrand Company, Inc.

[71] Moser, J. (1969). On a theorem of Anosov. *J. Diff. Eqns.* 5, 411-440.

[72] Moser, J. (1973).Stable and Random Motions in *Dynamical Systems, Annals of Math. Studies*, 77. Princeton: Princeton University Press.

[73] Moser, J. (1978/79). Is the solar system stable? *Math. Intelligencer* 1(2), 65-71.

[74] Munkres, J. R. (1975). *Topology: A First Course.* Englewood Cliffs: Prentice-Hall, Inc.

[75] Murdock, J. A. (1991). *Perturbations: Theory and Methods.* New York: Wiley-Interscience Pub.

[76] Murdock, J. A. and C. Robinson (1980). Qualitative dynamics from asymptotic expansions: Local theory. *J. Diff. Eqns.* 36, 425-441.

[77] Murray, J. D. (1980). *Mathematical Biology.* New York: Springer Verlag.

[78] Nayfeh, A. H. (1973). *Perturbation Methods.* New York: John Wiley &Sons.

[79] Nemytskii, V. V. and V. V. Stepanov (1989). *Qualitative Theory of Differential Equations.* New York: Dover Publications, Inc.

[80] Palais, R. P. (1997). The symmetries of solitons. *Bull. Amer. Math. Soc.* 34, 339-403.

[81] Palis, J. and W. deMelo (1982). *Geometric Theory of Dynamical Systems: An Introduction.* New York: Springer-Verlag.

[82] Pao, C. V. (1992). *Nonlinear Parabolic and Elliptic Equations.* New York: Plenum Press.

[83] Perko, L. (1996). *Differential Equations and Dynamical Systems*, 2nd ed. New York: Springer-Verlag.

[84] Petty, C. M. and J. V. Breakwell (1960). Satellite orbits about a planet with rotational symmetry. *J. Franklin Inst.* 270(4), 259-282.

[85] Poincaré, H. (1881). Mèmoire sur les courbes définies par une équation diff erentielle. *J. Math. Pures et Appl.* 7(3), 375-422; (1882). 8, 251-296; (1885). 1(4), 167-244; (1886). 2, 151-217; all reprinted (1928). Oeuvre, Tome I. Paris: Gauthier-Villar.

[86] Pugh, C. (1969). On a theorem of P. Hartman. *Amer. J. Math.* 91, 363-367.

[87] Reyn, J. W. (1994). *A Bibliography of the Qualitative Theory of Quadratic Systems of Differential Equations in the Plane*, 3rd ed., Tech. Report 94-02. Delft: TU Delft.

[88] Rayleigh, Lord (1883). On maintained vibrations. *Phil. Mag.* 15, 229.

[89] Rhouma, M. B. H. and C. Chicone (1999). On the continuation of periodic orbits. *Meth. Appl. Analy.* To appear.

[90] Robbin, J. W. (1968). On the existence theorem for differential equations. *Proc. Amer. Math. Soc.* 19, 1005-1006.

[91] Robbin, J. W. (1981). Algebraic Kupka-Smale theory. *Lecture Notes in Math.* 898, 286-301.

[92] Robbin, J. W. (1972). Topological conjugacy and structural stability for discrete dynamical systems. *Bull. Amer. Math. Soc.* 78(6), 923-952.

[93] Robinson, C. (1995). *Dynamical Systems: Stability, Symbolic Dynamics, and Chaos*. Boca Raton: CRC Press.

[94] Roberts, A. J. (1997). *Low-dimensional modelling of dynamical systems.* http://xxx.lanl.gov/abs/chao-dyn/9705010.

[95] Rothe, F. (1993). Remarks on periods of planar Hamiltonian systems. *SIAM J. Math. Anal.* 24(1), 129-154.

[96] Rousseau, C. and B. Toni (1997). Local bifurcations of critical periods in the reduced Kukles system. *Can. J. Math.* 49(2), 338-358.

[97] Rudin, W. (1966). *Real and Complex Analysis.* New York: McGraw-Hill Book Co.

[98] Sanders, J. A. and F. Verhulst (1985). *Averaging Methods in Nonlinear Dynamical Systems.* New York: Springer-Verlag.

[99] Schecter, S. (1990). Simultaneous equilibrium and heteroclinic bifurcation of planar vector fields via the Melnikov integral. *Nonlinearity*. 3, 79-99.

[100] Smale, S. (1965). Diffeomorphisms with many periodic points. In *Differential and Combinatorial Topology*, S. S. Cairns, ed. Princeton: Princeton Univ. Press 63-80.

[101] Smale, S. (1980). *The Mathematics of Time: Essays on Dynamical Systems, Economic Processes and Related Topics*. New York: Springer-Verlag.

[102] Smale, S. (1998). Finding a horseshoe on the beaches of Rio. *Math. Intelligencer.* 20, 39-62. 542 References

[103] Smoller, J. (1980). *Shock Waves and Reaction-Diffusion Equations.* New York: Springer-Verlag.

[104] Sotomayor, J. (1974). Generic one-parameter families of vector fields on two-dimensional manifolds. *Publ. Math. IHES*43, 5-46.

[105] Sotomayor, J. (1979). *Li cóes de equa cóes diferenciais ordinarias.* Rio de Janerio: IMPA.

[106] Sternberg, S. (1969). *Celestial Mechanics*, Parts I and II. New York: W. A. Benjamin, Inc.

[107] Stoker, J. J. (1950). *Nonlinear Vibrations.* New York: John Wiley & Sons.

[108] Strang, G. and G. J. Fix (1973). *An Analysis of the Finite Element Method.* Englewood Cliffs: Prentice-Hall, Inc.

[109] Strauss, W. A. (1992). *Partial Differential Equations: An Introduction.* New York: John Wiley & Sons.

[110] Strogatz, S. (1994). *Nonlinear Dynamics and Chaos.* Reading: AddisonWesley Pub. Co.

[111] Takens, F. (1974). Singularities of vector fields. *Publ. Math. IHES* 43, 47100.

[112] Temam, R. (1988). *Infinite-dimensional Dynamical Systems in Mechanics and Physics.* New York: Springer-Verlag.

[113] Trotter, H. F. (1959). On the product of semi-groups of operators. *Proc. Amer. Math. Soc.* 10, 545-551.

[114] van der Pol, B. (1926). On relaxation oscillations. *Phil. Mag.* 2, 978-992.

[115] van der Pol, B. and J. van der Mark (1927). Frequency demultiplication. *Nature.* 120, 363-364.

[116] van der Pol, B. and J. van der Mark (1929). The heartbeat considered as a relaxation oscillation and an electrical model of the heart. *Arch. Néerl. de Physiol. de L'homme et des Animaux.* 14, 418-443.

[117] Verhulst, F. (1989). *Nonlinear Differential Equations and Dynamical Systems*, New York: Springer-Verlag.

[118] Watson, G. N. (1966). *A Treatise on the Theory of Bessel Functions*, 2nd ed. Cambridge: Cambridge University Press.

[119] Weinberger, H. (1999). An example of blowup produced by equal diffusions. *J. Diff. Eqns.* To appear.

[120] Weissert, T. (1997). *The Genesis of Simulation in Dynamics: Pursuing the Fermi-Ulam-Pasta Problem.* New York: Springer-Verlag.

[121] Weyl, H. (1952). *Space Time Matter*, H. L. Brose, trans. New York: Dover Publications, Inc.

[122] Wiggins, S. W. (1988). *Global Bifurcations and Chaos.* New York: Springer-Verlag.

[123] Wiggins, S. W. (1990). *Introduction to Applied Nonlinear Dynamical Systems and Chaos*. New York: Springer-Verlag.

[124] Whittaker, E. T. and G. N. Watson (1927). *A Course of Modern Analysis*, 1996 reprint. Cambridge: Cambridge University Press.

[125] Winkel, R. (1998). *A transfer principle in the real plane from non-singular algebraic curves to polynomial vector fields*. http://www.iram.rwth-aachen.de/winkel/.

[126] Yagasaki, K. (1994). Chaos in a pendulum with feedback control. *Nonlinear Dyn.* 6, 125-142.

[127] Yakubovich, V. A. and V. M. Starzhinskii (1975). *Linear Differential Equations with Periodic Coefficients*, D. Louvish, trans. New York: John Wiley & Sons.

Index